Stem Cell Biology and Regenerative Medicine

Series editor

Kursad Turksen, Ph.D.
e-mail: kursadturksen@gmail.com

More information about this series at http://www.springer.com/series/7896

Anandwardhan A. Hardikar
Editor

Pancreatic Islet Biology

Springer

Editor
Anandwardhan A. Hardikar
Diabetes and Islet-biology Group,
NHMRC Clinical Trials Centre,
Faculty of Medicine
The University of Sydney
Sydney, NSW
Australia

ISSN 2196-8985 ISSN 2196-8993 (electronic)
Stem Cell Biology and Regenerative Medicine
ISBN 978-3-319-45305-7 ISBN 978-3-319-45307-1 (eBook)
DOI 10.1007/978-3-319-45307-1

Library of Congress Control Number: 2016948220

Printed on acid-free paper

This Springer imprint is published by Springer Nature
The registered company is Springer International Publishing AG
The registered company address is: Gewerbestrasse 11, 6330 Cham, Switzerland

This book is dedicated to my family. Words cannot express the support received from my wife—Mugdha Hardikar, parents—Asha and Awadhoot Hardikar, in-laws—Sheela and Vinay Joglekar and our two daughters—Aditi and Eesha Hardikar who bring much joy into my life and support me during all the difficult times.
I would also like to dedicate this to my extended family of mentors—Professors Ramesh Bhonde and Chittaranjan Yajnik (India); Professor Claude Remacle and late Professor Joseph Hoet (Belgium); Professors Doris Stoffers and Marvin Gershengorn (USA); and Professors Bernie Tuch, Alicia Jenkins and Anthony Keech (Australia), who trained me in diabetes research.
Finally—to caffeine and sugar—a compound that keeps me awake and a molecule that keeps me passionate to research this area of islet biology and diabetes.

—Anandwardhan A. Hardikar, Ph.D.

Preface

A collection of chapters on pancreatic islet cells can easily span many diverse areas in islet cell biology and diabetes, extending into multiple volumes of a bulky book or gigabytes of an e-book. This is a very wide-ranging and highly researched area of biology that impacts on the health of multiple organs in the body—including the eyes, kidneys, arteries, heart, brain, nerves, feet and liver, as well as the psychological, social and economic aspects of an individual and those surrounding this person with diabetes. All of this starts with the dysfunction of insulin-producing β-cells in the pancreatic islets of Langerhans—the sole player in the game and the body's only insulin-producing factory. The diverse effects of the dysfunction of this single (pro-)hormone on multiple organs and in different age groups; starting with monogenic forms of diabetes in babies, type 1 diabetes, which often starts in childhood, and type 2 diabetes, which becomes increasingly common with age, attract many curious minds to invest their lifetime careers into training and specializing in one of the many areas that islet cell research has to offer.

The idea of putting together a book on pancreatic islet biology came to my mind when I realized that most students, whilst being well read in their field of speciality may lack the general understanding of islet cell function in development, health and disease. I thought it would be useful to have a collection of chapters discussing different areas of pancreatic islet biology. Such a book can help a burgeoning researcher by providing them with current knowledge in the field through a "tasting platter" of highly relevant, yet diverse, articles in pancreatic islet biology. This volume starts with an overview chapter from Professor Alicia Jenkins explaining insulin production, secretion and signalling under physiological and pathological conditions. This is followed by a set of very interesting contributions from the groups of Dr. Manami Hara and Dr. Vipul Periwal describing the diversity in architecture of pancreatic islets of Langerhans and a mathematical model to understand this diversity. These studies specifically describe the differences in human and rodent models, which we need to remember whilst addressing questions in human diabetes. Another important aspect of insulin production is the epigenetic regulation of gene expression. Professor Sanjeev Galande's contribution specifically sheds light upon multiple epigenetic mechanisms involved in efficient insulin production. The next series of contributions are organized to educate the reader in understanding the regenerative potential of endocrine pancreas (Professors Bonner-Weir and Takasawa) and the potential of other cell types to differentiate into insulin-producing cells (Professors Ferber & Simpson and Dr. Joglekar). We then discuss some of the newer aspects in understanding the role of microRNAs and viruses in the development of type 1 diabetes (Dr. Akil) and then close off by learning about islet cell death and current therapy for the treatment of type 1 diabetes (Professors Thomas and Hawthorne).

Overall, this book is aimed to educate young researchers who are starting a career in islet biology, senior researchers in understanding fundamental aspects in different areas of islet biology, and cross-disciplinary scientists and experts interested in multiple aspects of pancreatic islet cell development, lineage commitment, their differentiation, regeneration and function, the pathobiology of diabetes and clinical replacement of islets in diabetes. The contributions in this book are very diverse, yet are unified resources to support the novice, expert or cross-discipline researcher to establish a strong basis in understanding the biology of the pancreatic islets of Langerhans. Each chapter can be read in sequence after the other or in isolation, to serve as a quick reference for any researcher in islet biology.

I am sure that the tremendous contribution of time and effort that I greatly appreciate from around four dozen top-ranked researchers in their fields will also be recognized and valued by every reader of this book. I greatly acknowledge the time and patience in working with me to put this volume together for you. During the time when I was preparing to put together this book, I was hit by a truck in a pedestrian crossing accident (whilst carrying my 11-month-old daughter), and the resultant neck injuries restricted my work capacity. I greatly appreciate the support from my wife—Mugdha, my parents, in-laws and my two daughters during those difficult times. I would also like to thank all of my past and current mentors who prepared me to take up and progress in diabetes research. The key people in my career who have trained me during the past two decades in multiple areas of pancreatic islet biology, obesity and diabetes include Professors Ramesh Bhonde and Ranjan Yajnik (India); Professor Claude Remacle and late Professor Joseph Hoet (Belgium); Professors Doris Stoffers and Marvin Gershengorn (USA); and Professors Bernie Tuch, Alicia Jenkins and Anthony Keech (Australia). This work could not have been possible without the scholarly contributions from all the authors and their willingness to participate in this endeavour. I greatly acknowledge the support from the Australian Research Council (ARC) and the Juvenile Diabetes Research Foundation (JDRF) Australia, Type 1 Diabetes Clinical Research Network (T1DCRN) for their support. Finally, I thank the staff from Springer Publishers, USA, who deserve a special mention for their patience during the time taken in assembling this book to its present stage. I hope that you will find this book to be a great resource for your own laboratory, as well as to pass on to your new students in their welcome pack. This reminds me of a famous quote by Michelangelo "Every block of stone has a beautiful statue inside it and it is the task of the sculptor to discover it by hammering out all of the unwanted part". I am sure that the clarity of thoughts and ideas presented by each of the contributing authors to this book will serve as a hammer and a chisel to carve out the ignorance of novice readers, educating them to establish a solid background in understanding pancreatic islet cell biology and putting them on a path to discover a diabetes-free world in the coming years.

Sydney, NSW, Australia Anandwardhan Awadhoot Hardikar Ph.D.

Contents

Contributors

Ammira Al-Shabeeb Akil Division of Translational Medicine, Sidra Medical and Research Centre, Education City North Campus, Qatar Foundation, Doha, QatarFaculty of Medicine, School of Women's and Children's Health, Health University of New South Wales, Kensington, NSW, Australia; Virology Division, South Eastern Area Laboratory Services SEALS, Prince of Wales Hospital, University of New South Wales, Randwick, NSW, Australia

Hila Barash Sheba Regenerative Medicine, Stem Cells and Tissue Engineering Center, Sheba Medical Center, Tel-Hashomer, Israel

Susan Bonner-Weir Section on Islet Cell and Regenerative Biology, Research Division, Joslin Diabetes Center, Harvard Medical School, Boston, MA, USA

Chang-Yi Chen Institute of Oral Biology, School of Dentistry, National Yang-Ming University, Taipei, Taiwan

Hung-Yu Chien Department of Endocrinology and Metabolism, Taipei City Hospital, Ren-Ai Branch, Taipei, Taiwan

Maria E. Craig Faculty of Medicine, School of Women's and Children's Health, Health University of New South Wales, Kensington, NSW, Australia; The Children's Hospital, Institute of Endocrinology and Diabetes, Westmead, NSW, Australia; Virology Division, South Eastern Area Laboratory Services SEALS, Prince of Wales Hospital, University of New South Wales, Randwick, NSW, Australia; Discipline of Pediatrics and Child Health, Health University of Sydney, Westmead, NSW, Australia

Sarah Ferber Sheba Regenerative Medicine, Stem Cells and Tissue Engineering Center, Sheba Medical Center, Tel-Hashomer, Israel; Department of Human Genetics and Molecular Medicine, Sackler School of Medicine, Tel-Aviv University, Tel Aviv, Israel

Carah A. Figueroa-Crisostomo Faculty of Science, School of Biotechnology and Biomolecular Sciences, University of New South Wales, Kensington, NSW, Australia

Takanori Fujimura Department of Biochemistry, Nara Medical University (NMU), Shijo-cho, Kashihara, Japan

Sanjeev Galande Indian Institute of Science Education and Research, Pune, India; Centre of Excellence in Epigenetics, Indian Institute of Science Education and Research, Pune, India

Dario Gerace School of Life Sciences and the Centre for Health Technologies, University of Technology Sydney, Broadway, NSW, Australia

Kate L. Graham St. Vincent's Institute of Medical Research, Fitzroy, VIC, Australia; Department of Medicine, St. Vincent's Hospital, The University of Melbourne, Fitzroy, VIC, Australia

Manami Hara Department of Medicine, The University of Chicago, Chicago, IL, USA

Anandwardhan A. Hardikar Diabetes and Islet Biology Group, NHMRC Clinical Trials Centre, Faculty of Medicine, The University of Sydney, Camperdown, NSW, Australia

Wayne John Hawthorne Department of Surgery, Westmead Clinical School, University of Sydney, Westmead Hospital, Westmead, NSW, Australia; National Pancreas and Islet Transplant Laboratories, The Westmead Institute for MedicalResearch, Westmead, NSW, Australia

Andy Ho Faculty of Medicine, School of Biotechnology and Biomolecular Sciences, University of New South Wales, Kensington, NSW, Australia

Asako Itaya-Hironaka Department of Biochemistry, Nara Medical University (NMU), Shijo-cho, Kashihara, Japan

Andrzej S. Januszewski NHMRC Clinical Trials Centre, University of Sydney, Camperdown, Sydney, Australia; Department of Endocrinology, St Vincent's Hospital, Melbourne, Australia

Alicia J. Jenkins NHMRC Clinical Trials Centre, University of Sydney, Camperdown, Sydney, Australia; Centre for Experimental Medicine, Queens University, Belfast, Northern Ireland, UK; Department of Endocrinology, St Vincent's Hospital, Melbourne, Australia

Junghyo Jo Asia Pacific Center for Theoretical Physics, Pohang, Korea

Mugdha V. Joglekar Diabetes and Islet Biology Group, NHMRC Clinical Trials Centre, Faculty of Medicine, The University of Sydney, Camperdown, NSW, Australia

Prachi Katre Diabetes Unit, KEM Hospital Research Centre, Pune, India

Thomas W.H. Kay St. Vincent's Institute of Medical Research, Fitzroy, VIC, Australia; Department of Medicine, St. Vincent's Hospital, The University of Melbourne, Fitzroy, VIC, Australia

Satyajeet Khare Indian Institute of Science Education and Research, Pune, India

Wan-Chun Li Institute of Oral Biology, School of Dentistry, National Yang-Ming University, Taipei, Taiwan

Thomas Loudovaris St. Vincent's Institute of Medical Research, Fitzroy, VIC, Australia

Rosetta Martiniello-Wilks School of Life Sciences and the Centre for Health Technologies, University of Technology Sydney, Broadway, NSW, Australia; Translational Cancer Research Group, University of Technology Sydney, Sydney, Australia

Irit Meivar-Levy Sheba Regenerative Medicine, Stem Cells and Tissue Engineering Center, Sheba Medical Center, Tel-Hashomer, Israel

Christopher J. Nolan Department of Diabetes and Endocrinology, The Canberra Hospital and the Australian National University Medical School, Canberra, Australia

Hiroyo Ota Department of Biochemistry, Nara Medical University (NMU), Shijo-cho, Kashihara, Japan

David N. O'Neal NHMRC Clinical Trials Centre, University of Sydney, Camperdown, Sydney, Australia; Department of Endocrinology, St Vincent's Hospital, Melbourne, Australia

Vipul Periwal Laboratory of Biological Modeling, NIDDK, NIH, Bethesda, MD, USA

Ananta Poudel Department of Medicine, The University of Chicago, Chicago, IL, USA

William D. Rawlinson Virology Division, South Eastern Area Laboratory Services SEALS, Prince of Wales Hospital, University of New South Wales, Randwick, NSW, Australia; Faculty of Science, School of Biotechnology and Biomolecular Sciences, University of New South Wales, Kensington, NSW, Australia

Sumiyo Sakuramoto-Tsuchida Department of Biochemistry, Nara Medical University (NMU), Shijo-cho, Kashihara, Japan

Omid Savari Departments Surgery, The University of Chicago, Chicago, IL, USA

Ann M. Simpson School of Life Sciences and the Centre for Health Technologies, University of Technology Sydney, Broadway, NSW, Australia

Deborah A. Striegel Laboratory of Biological Modeling, NIDDK, NIH, Bethesda, MD, USA

Shin Takasawa Department of Biochemistry, Nara Medical University (NMU), Shijo-cho, Kashihara, Japan

Maiko Takeda Department of Biochemistry, Nara Medical University (NMU), Shijo-cho, Kashihara, Japan

Zehra Tekin Departments Surgery, The University of Chicago, Chicago, IL, USA

Helen E. Thomas St. Vincent's Institute of Medical Research, Fitzroy, VIC, Australia; Department of Medicine, St. Vincent's Hospital, The University of Melbourne, Fitzroy, VIC, Australia

Hiroki Tsujinaka Department of Biochemistry, Nara Medical University (NMU), Shijo-cho, Kashihara, Japan

Wilson Wong Diabetes and Islet Biology Group, NHMRC Clinical Trials Centre, Faculty of Medicine, The University of Sydney, Camperdown, NSW, Australia

Akiyo Yamauchi Department of Biochemistry, Nara Medical University (NMU), Shijo-cho, Kashihara, Japan

Chapter 1
The Pathobiology of Diabetes Mellitus

Alicia J. Jenkins, David N. O'Neal, Christopher J. Nolan and Andrzej S. Januszewski

Abbreviations

ACE	Angiotensin-converting enzyme
ADA	American Diabetes Association
AGEs	Advanced glycation end-products
ACTH	Adrenocorticotropic hormone
ADP	Adenosine diphosphate
ATP	Adenosine triphosphate
CCK	Cholecystokinin
CGM	Continuous glucose monitoring
CKD	Chronic kidney disease

A.J. Jenkins (✉) · D.N. O'Neal · A.S. Januszewski
NHMRC Clinical Trials Centre, University of Sydney, 92-94 Parramatta Rd, Camperdown, Sydney 2050, Australia
e-mail: alicia.jenkins@ctc.usyd.edu.au

A.J. Jenkins
Centre for Experimental Medicine, Queens University, Belfast, Northern Ireland, UK

A.J. Jenkins · D.N. O'Neal · A.S. Januszewski
Department of Endocrinology, St Vincent's Hospital, Melbourne, Australia

C.J. Nolan
Department of Diabetes and Endocrinology, The Canberra Hospital and the Australian National University Medical School, Canberra, Australia

A.A. Hardikar (ed.), *Pancreatic Islet Biology*, Stem Cell Biology and Regenerative Medicine, DOI 10.1007/978-3-319-45307-1_1

CRF	Corticotropin-releasing factor (or hormone)
DCCT	Diabetes control and complications trial
DNA	Deoxyribonucleic acid
DIDMOAD	Diabetes insipidus, diabetes mellitus, optic atrophy, deafness
DKA	Diabetic ketoacidosis
DPP-4	Dipeptidyl peptidase-4
EGP	Endogenous glucose production
EDIC	Epidemiology of diabetes interventions and complications
GFR	Glomerular filtration rate
GH	Growth hormone
GHRF	Growth hormone-releasing factor
GHIH	Growth hormone inhibitory hormone
GIP-1	Glucose-dependent insulinotropic polypeptide-1
GLP-1	Glucagon-like peptide-1
GLUT	Glucose transporter
GSIS	Glucose-stimulated insulin secretion
GTP	Guanosine triphosphate
HIF	Hypoxia inducible factor
IAPP	Islet amyloid polypeptide
IV	Intravenous
LADA	Latent autoimmune diabetes of adulthood
MODY	Maturity-onset diabetes of the young
mRNA	Messenger ribonucleic acid
NADPH	Nicotinamide adenine dinucleotide phosphate
oGTT	Oral glucose tolerance test
PKC	Protein kinase C
PP	Pancreatic polypeptide
RER	Rough endoplasmic reticulum
RRP	Readily releasable pool
SD	Standard deviation
T1D	Type 1 diabetes mellitus
T2D	Type 2 diabetes mellitus
TNF	Tumor necrosis factor
TSH	Thyroid-stimulating hormone
UKPDS	United Kingdom Prospective Diabetes Study
VEGF	Vascular endothelial growth factor
VIP	Vasoactive intestinal peptide

This chapter is dedicated to Marie-Elise and Liam McCallum and to other people who live with type 1 diabetes, with the hope that results of ongoing and future medical research will lessen the impact of diabetes on their lives.

Introduction

Diabetes mellitus is an increasingly common chronic condition characterized by an absolute [as in type 1 diabetes (T1D)] or relative lack of insulin [as in type 2 diabetes (T2D)], hyperglycemia, dyslipidemia, and neurovascular damage that can affect every organ system in an individual. It is associated with both acute and chronic complications which can be life-threatening. Every 6 s today someone dies from diabetes (International Diabetes Federation 2013). Diabetes also impacts on the affected individual's family, friends, colleagues and the community, including the healthcare system, and the national and global economies (Jenkins 2015). In 2013, there were ≈382 million people with diabetes globally, most in low- to middle-income countries, predicted to rise to over 592 million in the next 25 years, the majority of whom have T2D (International Diabetes Federation 2013). Almost half of the many people with T2D are undiagnosed, particularly in less affluent regions.

The increasing incidence and prevalence of T2D are contributed to by non-modifiable factors such as population growth, increased longevity and better screening and therefore case ascertainment. Importantly though, the T2D epidemic parallels increasing rates of overweight and obesity associated with changing lifestyles (urbanization and increased sedentary behavior) that potentially are modifiable (Eckel et al. 2011). Gestational diabetes (GDM), which is glucose intolerance diagnosed in pregnancy, and T2D in pregnancy are also increasing in incidence for the same reasons (Nolan et al. 2011). The incidence of the autoimmune condition T1D, which very often starts in childhood, has been increasing at about 3 % per annum in Australia (Insulin-treated diabetes in Australia 2000–2007; Shaw and Tanamas 2012) and similarly overseas over the past 30–40 years (Gale 2002). This increase is most likely due to environmental rather than changing genotypes. The 'hygiene hypothesis' postulates that it relates to changing immune function as a consequence of reduced early life immune stimuli due to our more hygienic environment (Versini et al. 2015). Not mutually exclusive is the 'accelerator hypothesis' which postulates that insulin resistance, as a consequence of the obesogenic environment, accelerates development of T1D in people at risk (Fourlanos et al. 2008).

In spite of the availability of modern therapies for glucose, blood pressure and lipid control, which are often not available or affordable to all who may benefit (Jenkins 2015), optimal treatment targets are often not met, hence increasing the risk of both acute and chronic complications of diabetes. The predominant acute complications of diabetes include both hypoglycemic and hyperglycemic crises which relate to a mismatch between blood glucose and insulin (and other related glucose-modulating hormones and neurogenic stimuli). Chronic complications are a consequence of hyperglycemic and mixed nutrient-induced damage to tissues of the body, particularly via the vascular supply involving both the larger arteries (macrovascular), causing myocardial infarctions, strokes and peripheral vascular disease, and the small vessel (microvascular) networks, causing diabetic eye disease (retinopathy), renal disease (nephropathy) and neuropathy. Peripheral neuropathy and/or peripheral vascular disease increase the risk of lower limb amputation. Acute and chronic complications of diabetes, many of which are life-threatening, cause even greater physical, emotional and socioeconomic demands on the person with diabetes. Furthermore, compared to persons with well-controlled complication-free diabetes, those with complications exponentially increase healthcare costs (Shaw and Tanamas 2012). Even 'pre-diabetes,' in which blood glucose levels are elevated above normal, but not to the level of those diagnostic of diabetes, is associated with accelerated atherosclerosis (Faerch et al. 2014), but not with renal, retinal or nerve damage. Better means to predict, detect, stage and prevent the various forms of diabetes and to predict, prevent and treat the short- and long-term complications of diabetes are highly desirable.

Clinical and basic science research has taught us much about diabetes, its etiologies, complications and treatments. Modern medicine has led to a wide range of oral and injectable glucose control agents, including insulin, first available for clinical use in 1922, but followed by improved engineered insulin analogues in recent years. Major advances have also been achieved in insulin delivery and blood glucose monitoring devices, including increasingly 'smarter' insulin pumps and subcutaneous continuous glucose monitors. In addition, today there are many other drugs and therapies to treat diabetes, risk factors and its complications, including islet cell, pancreas and kidney transplantation (which usually require immunosuppression). The challenges of living with diabetes have lessened, and its prognosis has improved substantially, at least in regions with access to modern diabetes care (Gregg et al. 2014). Nevertheless, much further work, including clinical and basic science research, biomedical engineering, population health, healthcare systems, policy and health economics research, is needed to lessen the major personal and economic burden of diabetes.

In this chapter, we describe the normal pancreas, islets of Langerhans, normal glucose homeostasis, with an emphasis on insulin, the types of diabetes and its complications, the underlying pathobiology of diabetes per se and its vascular and neurological complications, and current treatment modalities. We hope this chapter will complement other excellent chapters in this volume, which will inform and update the reader regarding many aspects of the pancreas, diabetes and its treatment.

The Pancreas

Located in the retroperitoneal space in the abdominal cavity, at the level of the first and second lumbar vertebrae, the pancreas is a J-shaped soft lobulated yellowish colored organ usually measuring about 15–20 cm long (in adults), 5 cm wide and with an average weight of about 90 g. The pancreas is usually described as consisting of a head, body and tail, the majority of which is located on the left side of

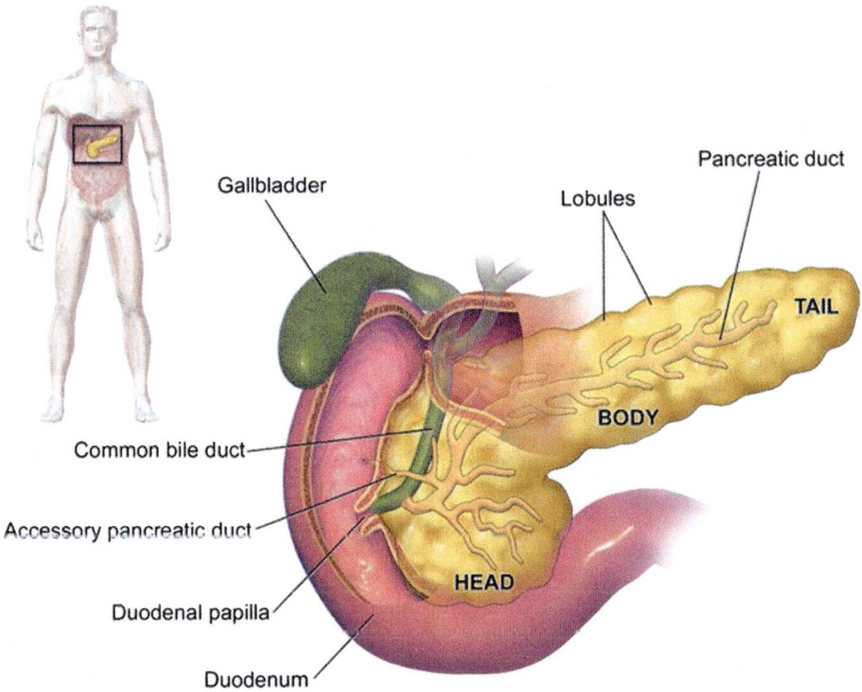

Fig. 1.1 Anatomy of the pancreas. Blausen.com staff. "Blausen gallery 2014". Wikiversity Journal of Medicine. doi:10.15347/wjm/2014.010. ISSN 20018762.—Own work. Licensed under CC BY 3.0 via Commons—https://commons.wikimedia.org/wiki/File:Blausen_0699_PancreasAnatomy2.png#/media/File:Blausen_0699_PancreasAnatomy2.png

the abdomen. The head of the pancreas is closely surrounded by the duodenum, and its tail abuts the spleen and left colic flexure (Fig. 1.1). In front of the pancreas are the stomach and loops of small intestine, and behind it is the left kidney and adrenal gland, inferior vena cava and aorta (Guyton and Hall 1996).

The *blood supply* to the pancreas is via the superior mesenteric and the common hepatic and splenic artery branches of the celiac trunk. Its venous drainage is via the superior mesenteric, portal and splenic veins. *Innervation* of the pancreas is by the vagal and spinal nerves. *Lymphatic drainage* is via the splenic, celiac and superior mesenteric lymph nodes.

In the pancreas parenchyma, stemming from many minor ducts in the pancreatic exocrine tissue, there is a major *pancreatic duct* (also called the duct of Wirsung) which drains the pancreatic exocrine secretions (bicarbonate and digestive enzymes) into the duodenum. A smaller, shorter and more anatomically variable accessory pancreatic duct often joins the main pancreatic duct in the head of the pancreas. The pancreas is a *dual-function gland*, having features of both exocrine and endocrine glands (Guyton and Hall 1996; Greenstein and Wood 2011).

The Exocrine Pancreas

The majority of the mass of the pancreas is dedicated to its exocrine function of producing (protein, fat and carbohydrate) digestive enzymes and bicarbonate, which are produced by cellular clusters (called acini) and secreted via the acinar lumen to intralobular ducts to the main pancreatic duct(s) and into the duodenum. The main digestive enzymes are secreted as inactive proenzymes or zymogens (to protect against pancreatic autodigestion) and are activated by enzymes and bile acids once in the small intestine.

The *digestive enzymes* are:

(i) trypsin, chymotrypsin, carboxypeptidases and elastase, which digest proteins;
(ii) lipase, phospholipase, lysophospholipase and cholesterol esterase, which digest lipids;
(iii) α-amylase, which (in conjunction with polysaccharidases produced by the intestinal mucosa) digests carbohydrates; and
(iv) ribonuclease and deoxyribonuclease, which digest nucleic acids.

These pancreatic secretions (including that of bicarbonate, which neutralizes acidic contents from the stomach) are stimulated via food in and distension of the stomach and duodenum, modulated by gut hormones (including secretin, cholecystokinin (CCK) and gastrin) and the autonomic nervous system.

The Endocrine Pancreas

The major role of the endocrine pancreas is the tight regulation of blood glucose levels and related energy sources, such as glycogen stores in liver and muscle. Representing only 1–1.5 g (<2 %) of the pancreas mass and scattered throughout the pancreas, with a higher concentration in the tail, are approximately a million clusters of endocrine, called the islets of Langerhans (after the German anatomical pathologist, Paul Langerhans, who discovered them in 1869) (Sakula 1988). Each islet is around 0.2 mm or 200 μm in diameter and encapsulated in a thin connective tissue capsule and is surrounded and infiltrated by a capillary network. This vascular network is particularly dense and contains highly fenestrated capillaries, such that islets have tenfold the blood flow than the exocrine pancreas and are able to respond rapidly to circulating nutrient levels, and similarly for the secreted hormones, importantly including insulin, to diffuse into the circulation (Guyton and Hall 1996; Greenstein and Wood 2011). It is these islets, isolated from human cadaver donors, that now form the basis of islet cell transplantation for a very small subset of adults with T1D, usually those with life-threatening recurrent severe hypoglycemia.

There are five known endocrine cell types in human islets of Langerhans:

(i) *β-cells*, which produce insulin and amylin and account for ~60 % of islet cells;
(ii) *α-cells*, which produce glucagon and account for ~30 % of islet cells;
(iii) *δ-cells*, about 5 % of islet cells, which produce somatostatin;
(iv) *γ-cells (also known as PP cells)*, about 5 % of islet cells, which produce pancreatic polypeptide; and
(v) *ε-cells*, <1 % of islet cells, which produce ghrelin (Eissa and Ghia 2015; Granata and Ghigo 2013; Horner and Lee 2015; Zigman et al. 2015).

The proportion as well as the architecture of islet cells is known to be different in mice (discussed by Manami Hara and colleagues in Chap. 2 of this volume). Mouse islets contain ~80 % β-cells that form a central core of the islet and are surrounded by non-β-islet cells, including ~10 % α-cells (Cabrera et al. 2006; Kilimnik et al. 2012; Kim et al. 2009). The function of each of these hormones is briefly described later.

Normal Glucose Homeostasis and the Importance of Hormonal Regulation

In non-diabetic subjects, blood glucose levels are tightly maintained within narrow limits, usually between 3.5 and 5.5 mmol/L in the fasted state and 5.0–7.5 mmol/L after meals. The body utilizes 180–260 g of glucose/day, of which about 50 % is taken up by the brain. Some tissues, such as the brain, red blood cells and the renal medulla, rely entirely on glucose for energy. Other tissues can additionally metabolize alternate substrates for their energy needs such as ketone bodies and fatty acids (Guyton and Hall 1996).

Glucose levels are controlled by the following hormones:

(i) insulin;
(ii) glucagon;
(iii) somatostatin (with i–iii arising from pancreatic islets);
(iv) incretin hormones [e.g., glucagon-like peptide-1 (GLP-1)] from the gut;
(v) catecholamines [e.g., adrenaline (epinephrine)] from the adrenal medulla;
(vi) cortisol from the adrenal cortex; and
(vii) growth hormone (GH) from the pituitary gland.

Fasting blood glucose is determined by the balance between the rate of endogenous glucose production (EGP), mainly from hepatic glycogenolysis and gluconeogenesis, and its utilization by tissues, in particular essential use by the brain, which is insulin independent. EGP prevents hypoglycemia and is supported by a low plasma insulin/glucagon ratio. Some insulin is required to maintain normoglycemia during fasting, as EGP, by default, is high in its absence. Glucose is spared for brain use during fasting through the provision of non-glucose nutrients (e.g., free fatty acids from adipose tissue lipolysis) to other tissues such as heart and skeletal muscle (Fig. 1.2) (Nolan et al. 2011; Guyton and Hall 1996).

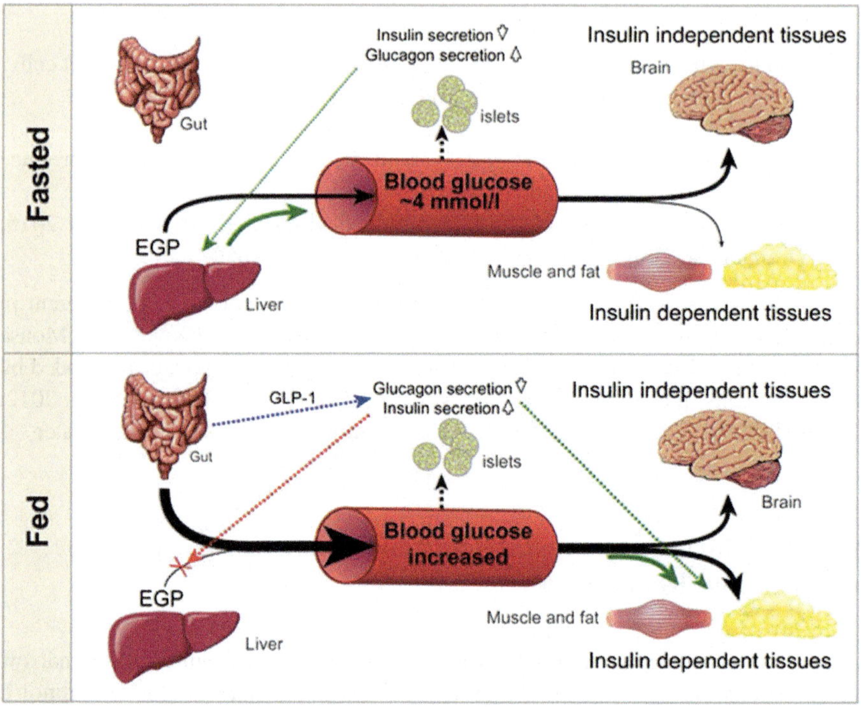

Fig. 1.2 Overview of normal glucose homoeostasis. In the fasting state, blood glucose concentration is determined by the balance between EGP production, mainly through hepatic glycogenolysis and gluconeogenesis, and use by insulin-independent tissues, such as the brain. EGP prevents hypoglycemia and is supported by a low insulin-to-glucagon ratio in plasma. In the fed state (meal with carbohydrate), glucose concentrations in the blood rise because of absorption in the gut, which stimulates insulin secretion by islet β-cells and suppresses glucagon secretion from α-cells. EGP is suppressed (which helps to curtail total glucose input into blood), and uptake into insulin-sensitive peripheral tissues, such as the heart, skeletal muscle, and adipose tissue, is activated (which increases the rate of glucose disposal). Neurohormonal processes include the release of the incretin hormones, such as GLP-1, which increases glucose-stimulated insulin secretion and glucose suppression of glucagon secretion. *GLP-1* glucagon-like peptide-1, *EGP* endogenous glucose production (Nolan et al. 2011)

In the fed state (e.g., meal with carbohydrate), the rate of glucose appearance into the blood increases due to gut glucose absorption. An elevation in blood glucose stimulates insulin secretion and suppresses glucagon secretion from, respectively, pancreatic islet β- and α-cells. This results in the suppression of EGP (helps to curtail the total rate of glucose appearance) and the activation of glucose uptake into insulin-sensitive peripheral tissues such as heart, skeletal muscle and adipose tissue (increases the rate of glucose disposal) (Fig. 1.2) (Nolan et al. 2011; Guyton and Hall 1996).

Other complex neurohormonal processes are also involved, including the release of the incretin hormones glucagon-like peptide-1 (GLP-1) and glucose-dependent insulinotropic peptide (GIP). Incretin hormones enhance both glucose-stimulated insulin secretion (GSIS) and glucose suppression of glucagon secretion (discussed in more detail below). These normal physiological responses prevent excessive rises in plasma glucose after meals and, in most circumstances, return plasma glucose levels close to the fasting level within 2 h. At the same time, the fed-state neurohormonal and metabolite mix suppresses adipose tissue lipolysis and promotes a general state of anabolic metabolism (Fig. 1.2) (Nolan et al. 2011; Guyton and Hall 1996; Steele et al. 2004; Vega-Monroy and Fernandez-Mejia 2011).

We now provide an overview of each of the major glucose-influencing hormones, with an emphasis on those produced by the pancreas.

Insulin

Insulin Structure and Synthesis

The structure of insulin is shown in Fig. 1.3. Mature, active insulin is a 5.8-kDa, 51-amino acid polypeptide with two (A and B) chains of 21 and 30 amino acid residues, respectively, joined by disulfide bridges and another disulfide bridge within the A chain, all between cysteine residues. There are two inactive precursors, pre-proinsulin (of 110 amino acids) and proinsulin (of 86 amino acids) (Fu et al. 2013).

Insulin synthesis is regulated at both transcriptional and translational levels. In the β-cell nucleus insulin coding (*pro*-ins) gene is transcribed to mRNA, which is then transferred into the cytoplasm, where translation occurs on ribosomes. Pre-proinsulin contains a hydrophobic N-terminal signal peptide, which interacts with cytosolic ribonucleoprotein signal recognition particles that facilitate translocation of the signal sequence, mRNA and ribosomes to the rough endoplasmic reticulum (RER) in the cytoplasm, which is responsible for protein assembly (Fu et al. 2013). The N-terminal signal sequence translocated to the RER membrane is elongated, to form pre-proinsulin. With removal of the N-terminal signal sequence, proinsulin is formed, folded and stabilized by the formation of the three disulfide bonds within the RER lumen and is then transported to the cytosolic Golgi complex. In the Golgi apparatus, the proinsulin enters immature secretory granules inside of which C-peptide is cleaved from the proinsulin (Fu et al. 2013). The insulin and C-peptide are then stored within secretory granules, along with amylin and other products including zinc, until their release by fusion with the plasma membrane and exocytosis in response to a range of stimuli (discussed below), with the most potent stimulus being rising glucose levels. Glucose metabolism activates insulin gene transcription and mRNA translation, having a particularly important role in stabilizing pre-proinsulin mRNA (Fu et al. 2013).

Insulin has a half-life in the circulation of about 6 min, while that of C-peptide is approximately 30 min. Circulating insulin is predominantly removed by the liver, and C-peptide is excreted by the kidneys (Gale 2015).

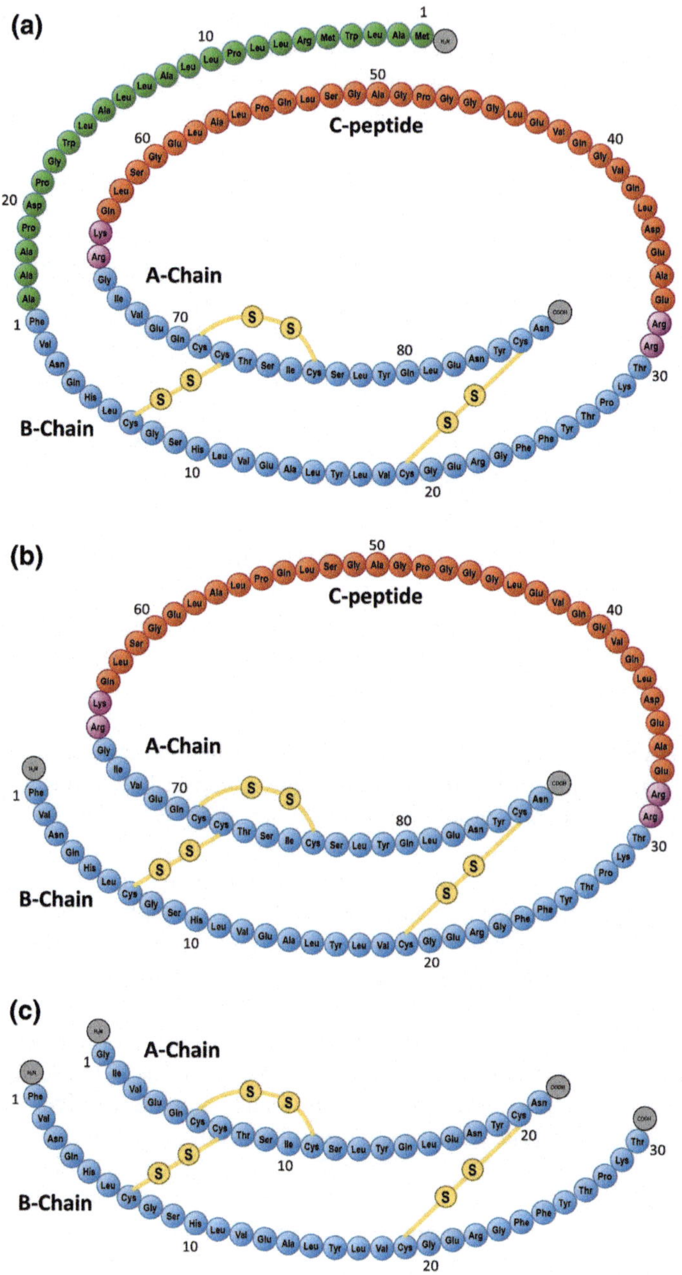

Fig. 1.3 Amino acid structures of pre-proinsulin (**a**), proinsulin (**b**) and insulin (**c**)

Signaling of Insulin Secretion

The islet β-cell senses multiple nutrient and neurohormonal inputs and accordingly secretes an appropriate amount of insulin for the requirements of the body at that time. Essential for insulin secretion is β-cell metabolic activation in response to an increased mixed nutrient supply, of which glucose is the most important (Fig. 1.4) (Nolan and Prentki 2008; Prentki et al. 2013).

In order to couple glucose sensing to insulin release, islet β-cell glucose metabolism is essential, and this is via three pathways with the production of metabolic coupling factors:

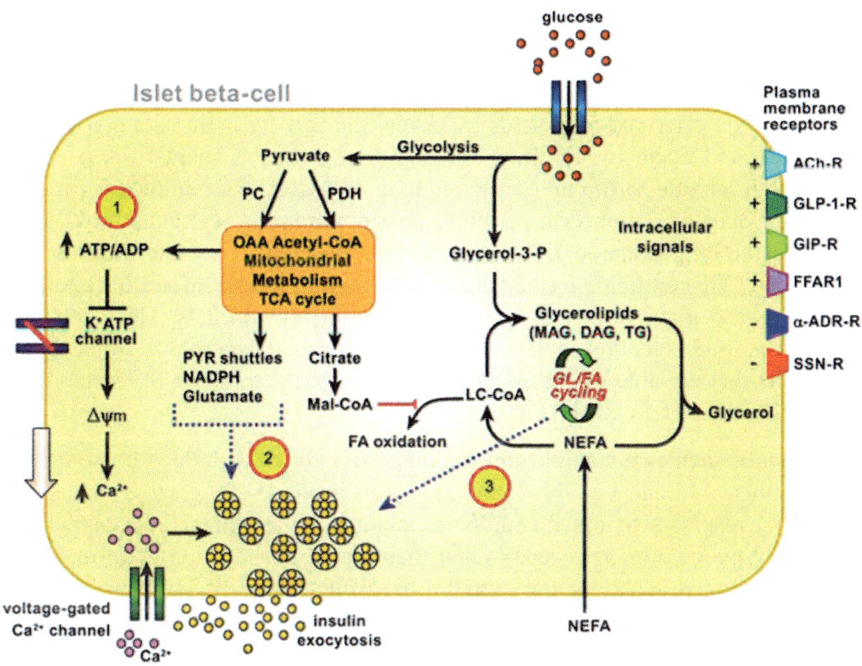

Fig. 1.4 Role of islet β-cell metabolic activation by fuels and neurohormonal agonists in insulin secretion. Islet β-cell glucose metabolism is essential for glucose to stimulate insulin secretion via three key pathways that produce metabolic coupling factors: *1* the $K_{ATP}{}^{+}$ channel-dependent pathway of GSIS. *2* The anaplerosis amplification pathway of GSIS. *3* The glycerolipid/fatty acid cycling amplification pathway (refer to text for details). Amino acids, such as glutamine and leucine, also interact with the glucose metabolism pathways to increase the coupling signals produced by glucose alone. The β-cell also responds to other neurohormonal and metabolic extracellular signals via various plasma membrane receptors. *PC* pyruvate carboxylase, *PDH* pyruvate dehydrogenase. *Ach-R* acetylcholine receptor, *FA* fatty acid, *GIP-R* gastric inhibitory polypeptide receptor, *GLP-1-R* glucagon-like peptide-1 receptor, *FFAR1* free fatty acid receptor 1, *α ADR-R* α2-adrenergic receptor, *SSN-R* somatostatin receptor. OAA, oxaloacetate, *CoA* coenzyme A, *MAG* monoacylglycerides, *DAG* diacylglycerides, *TG* triacylglycerides, *Δψm* change in plasma membrane potential, *Mal-CoA* malonyl-CoA, *LC-CoA* long-chain acyl-CoA, *GL* glycerolipid, *FA* fatty acid, *NEFA* non-esterified fatty acids (Nolan et al. 2011)

(i) *The K_{ATP}^{+} channel-dependent pathway of GSIS.* Glucose is metabolized via glycolysis to pyruvate and then acetyl-CoA via pyruvate dehydrogenase with subsequent oxidation in the tricarboxylic acid cycle. This gives rise to an increased cytosolic ATP/ADP ratio, which closes ATP-sensitive K^{+} channels, depolarizes the plasma membrane potential, opens voltage-gated Ca^{2+} channels, and Ca^{2+} influx activates insulin granule exocytosis (Fig. 1.4) (Nolan and Prentki 2008; Prentki et al. 2013).

(ii) *The anaplerosis amplification pathway of GSIS.* Pyruvate from glucose can also be metabolized via pyruvate carboxylase into the anaplerosis/cataplerosis pathway, which can impact on insulin secretion by increasing levels of cataplerosis-derived metabolic coupling molecules such as NADPH from the malate-, citrate- and isocitrate/α-ketoglutarate-pyruvate shuttles, as well as malonyl-CoA and glutamate (Fig. 1.4) (Nolan and Prentki 2008; Prentki et al. 2013).

(iii) *The glycerolipid/fatty acid cycling amplification pathway.* Glucose interacts with non-esterified fatty acids by promoting activity in an islet glycerolipid fatty acid cycle by elevating malonyl-CoA, via the anaplerosis pathway, which inhibits partitioning of long-chain acyl-CoA to the mitochondrion for fatty acid oxidation (via carnitine palmitoyltransferase-1 inhibition), such that the long-chain acyl-CoA are then more available for esterification processes. Glycerolipids formed are rapidly hydrolyzed by lipases back to fatty acids and glycerol creating a cycle of newly formed lipids. This cycle produces lipid signaling molecules such as monoacylglycerols and diacylglycerols that are able to enhance GSIS (Nolan and Prentki 2008; Prentki et al. 2013; Zhao et al. 2014) (Fig. 1.4).

Amino acids, such as glutamine and leucine, also interact with the glucose metabolism pathways to enhance the coupling signals produced by glucose alone. The β-cell also responds to other neurohormonal/metabolic extracellular signals by various plasma membrane receptors and their signal transduction pathways (e.g., G-protein-coupled signaling for example by cyclic AMP). Relevant to nutrient-induced insulin secretion, islet β-cells have cell surface receptors for fatty acids (e.g., FFAR1) that can modulate GSIS. Effector metabolic coupling factors interact with the insulin granule exocytosis machinery to cause insulin secretion (Nolan and Prentki 2008; Prentki et al. 2013) (Fig. 1.4).

Phases of Insulin Secretion

In healthy people, about half of the total daily insulin secretion is in the basal state, with pulses about every 10–15 min, and the remainder is postprandial (Lang et al. 1979). Postprandial Insulin secretion in healthy humans follows a biphasic response, shown in Fig. 1.5 (Curry et al. 1968; Henquin 2009; Seino et al. 2011).

First-phase insulin release starts in humans as plasma glucose levels rise above fasting levels about 2 min after food ingestion and lasts for about 10 min. This secretion is due to release of about 1 % of all insulin granules that are 'readily releasable,' called the readily releasable pool (RRP). Once the RRP is depleted, it is replaced by insulin from the 'reserve pool' (RP). While it was generally believed that the RRP consisted predominantly of docked granules, this has been questioned with the use of new methods of studying the exocytosis, which suggest that insulin granules released from the RRP in first-phase insulin release are mostly recruited to the plasma membrane (i.e., restless newcomers) (Seino et al. 2011). Recent studies have identified novel regulatory factors for first-phase insulin secretion and glucose homeostasis, including hypoxia inducible factor (HIF) 1α, von Hippel-Lindau, factor inhibiting HIF, nicotinamide phosphoribosyltransferase, and sirtuins (Cheng et al. 2013).

This first phase is followed by a longer second phase, utilizing the RP of insulin granules, that lasts for one to 2 h until the blood glucose level is returned to baseline (Fig. 1.5). While first-phase insulin secretion was believed to be a consequence of triggering of secretion by the K^{+}_{ATP} channel-dependent pathway of GSIS and the second phase a consequence of metabolic and neurohormonal amplification, it is now generally accepted that both first and second phases of insulin secretion are regulated by triggering (mostly via increasing intracellular Ca^{++}) and amplification processes (Prentki et al. 2013; Henquin 2009). This is why glucose control treatments that amplify secretory responses to glucose, such as the incretin-related agents, can partly restore first-phase insulin release; hence, this potentially reversible feature of insulin secretion is a worthwhile therapeutic target.

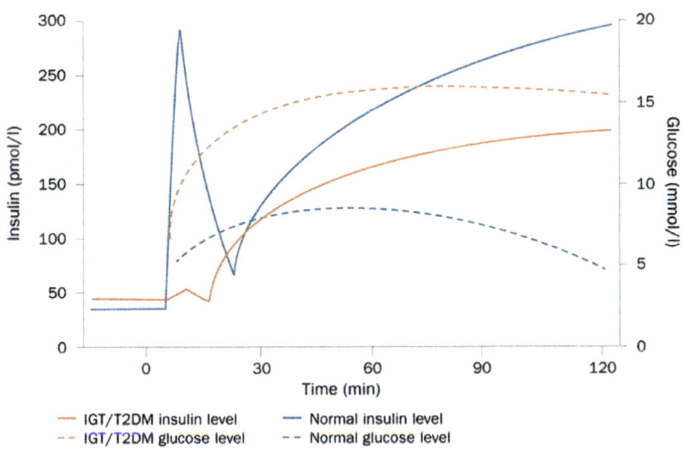

Fig. 1.5 First- and second-phase insulin release and corresponding glycemia after an intravenous glucose challenge in normal and impaired glucose tolerance/type 2 diabetic subjects (Jenssen and Hartmann 2015)

Non-nutrient Modulators of Insulin Secretion

Islet β-cells sense multiple neurohormonal stimuli that can modulate nutrient-stimulated secretion. These can be via endocrine and neural inputs from distant sites or can occur locally by paracrine and autocrine mechanisms, as well as at the cell-to-cell level through gap junctions and electrical activity. Of particular interest is the positive effect of the incretin hormones GLP-1 and GIP on insulin secretion. The autonomic nervous system can also modulate secretion, with the parasympathetic system being stimulatory and the sympathetic system being inhibitory (Wilcox 2005). Somatostatin and glucagon act in a paracrine fashion and are inhibitory (Fig. 1.4) (Wilcox 2005). Cortisol is also known to inhibit insulin secretion. In pregnancy, function and proliferation of islet β-cells can be modulated by gestational hormones such as prolactin, human placental lactogen and estrogen (Weinhaus et al. 2007; Tiano et al. 2011).

The Incretin Effect

The incretin effect refers to the phenomenon by which an oral glucose load induces a greater insulin response than the same amount of glucose delivered intravenously. The augmented insulin response with an oral glucose stimulus, shown in Fig. 1.6, is due to release of gut hormones (called incretins), including GLP-1 and GIP. Loss of the incretin effect occurs in T2D and is ameliorated by use of therapeutic drugs such as the dipeptidyl peptidase-4 (DPP-4) inhibitors and incretin mimetics (Holst et al. 2009) (Fig. 1.6).

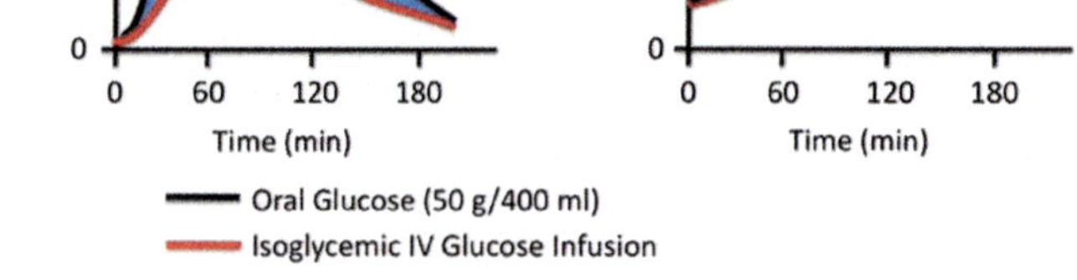

Fig. 1.6 Incretin effect on insulin release in normal and type 2 diabetic subjects (Nauck et al. 1986)

Altered Insulin Secretion in Obesity and in Diabetes

In *obesity*, a common risk factor for T2D, in the absence of abnormal glucose tolerance, β-cell mass and function is increased relative to lean non-diabetic subjects and insulin release (in both phases 1 and 2) is heightened (Polonsky 2000). In pre-diabetes and T2D, substantial β-cell mass has been lost, and both phases of insulin release are suppressed relative to lean (and obese) non-diabetic subjects (Fonseca 2009).

Type 2 diabetes: As shown in Fig. 1.5, in people with T2D, the first phase of insulin release is particularly low, with lower second-phase insulin release also. The first phase of insulin release is also blunted in people with pre-diabetes, in non-diabetic relatives of people with T2D, in gestational diabetes and in non-pregnant women with normal glucose tolerance but previous gestational diabetes. The incretin effect is also blunted in T2D (Seino et al. 2011; Holst et al. 2009).

Type 1 diabetes: By the time when T1D is clinically diagnosed, there has been substantial (but usually not total) β-cell loss. Abnormalities of insulin secretion occur prior to and at T1D diagnosis. This has been discerned by evaluation of subjects with autoantibodies, which place them at high risk of developing T1D, and of newly diagnosed T1D patients. Steele et al. (Steele et al. 2004) evaluated 42 newly diagnosed subjects. Basal insulin secretion was approximately half that of non-diabetic subjects and, in response to a mixed meal test, the first-phase insulin peak was markedly reduced and delayed. The second-phase insulin secretion was prolonged and failed to return blood glucose to baseline in many (Steele et al. 2004). On 1-year follow-up of the T1D subjects, C-peptide levels fell below the detectable in 47 % of the subjects. Over the 2-year follow-up, the insulin response to the mixed meal test predicted residual insulin secretion.

Insulin Actions

Insulin is an anabolic hormone with widespread actions encompassing carbohydrate, lipid and protein metabolism, but also affects on other cellular functions, such as cell proliferation and differentiation. Insulin signals via the insulin receptor, insulin receptor substrates and then the phosphoinositide 3-kinase-AKT/protein kinase B (PKB) pathway (responsible for most of the metabolic effects) or the mitogen-activated protein kinase (MAPK) pathway (responsible for the cell proliferation effects) (Taniguchi et al. 2006). The major sites of insulin action on glucose homeostasis are liver, muscle (skeletal and cardiac) and adipose tissue (the 'insulin-responsive' tissues). Tissues unresponsive to insulin for glucose uptake include renal tissue, the cornea, lens, red and white blood cells, the choroid plexus and gut epithelium. Insulin is essential for the transport of glucose into the insulin-responsive tissues. The various tissues of the body have different glucose transporter proteins (the GLUT proteins) that have different chromosome locations of their genes and varying affinity for glucose (Wilcox 2005). The insulin-responsive tissues express GLUT4 which is translocated from intracellular vesicles to the plasma membrane in response to insulin stimulation to promote glucose uptake (Wilcox 2005; Bryant et al. 2002). Insulin-stimulated glucose uptake is accompanied by tissue uptake of ions such as potassium and phosphate. Insulin promotes glycogen synthesis in the liver and in muscle, as well as protein and lipid synthesis, and inhibits protein breakdown, lipolysis and ketogenesis. Insulin suppresses gluconeogenesis and glycogenolysis and, therefore EGP, via direct and less well-understood indirect mechanisms in the periphery (e.g., reducing supply of gluconeogenic substrate by inhibiting lipolysis and protein catabolism) (Wilcox 2005; Gaw et al. 1999) and via the brain (Rojas and Schwartz 2014).

History of Exogenous Insulin Therapy

In 1921 and 1922, an orthopedic surgeon Frederick Banting, a medical student Charles Best, a professor of physiology John Macleod, and biochemist James Collip isolated insulin from dogs and tested it in pancreatectomized dogs. They then (in January 1922) tested it in severely ill people with T1D, a previously universally fatal condition, and showed it to be effective. Banting and Macleod were awarded the 1923 Nobel Prize in Physiology or Medicine 'for the discovery of insulin.' An excellent account of the history of the discovery of insulin and this award is that by medical historian Bliss (2007). Their work complemented and built upon that of other researchers, including Nicolae Paulescu, a Romanian professor who isolated insulin (which he called pancreatine) about the same time as the Canadian group (Bliss 2007). The pharmaceutical company, Lilly, which is still a major producer of insulin today, rapidly partnered with the University of Toronto to improve and scale up the production and bring to market bovine, followed by porcine insulin. Lilly and other companies now use recombinant DNA technology to produce human insulin to save, ease and prolong the lives of people around the world with T1D. Bovine and porcine insulin differ from human insulin by three and one amino acid, respectively. Since the 1980s, an increasing range of rapid-, short-, intermediate- and long-acting genetically engineered human analogue insulins have become available for the treatment of T1D and insulin-requiring T2D patients. Tragically, even today, over 90 years after Banting and Best's discovery and the development of insulin for human use, many people in less affluent countries, particularly those with T1D, develop early severe complications and die prematurely due to lack of access to affordable insulin. On the cover of *The Lancet* in 2006, UK Diabetologist Edwin Gale quoted, 'What is the commonest cause of death in a child with diabetes? The answer—from a global perspective—is lack of access to insulin' (Gale 2006).

Amylin

Amylin, also known as islet amyloid polypeptide (IAPP) was only discovered as a hormone in 1987. It is a 37-amino acid peptide that is co-secreted with insulin (as a 67-amino acid prohormone) from the β-cells of the islets of Langerhans at a ratio of approximately 1–25 (Westermark et al. 1987a, b, 2011). Amylin is broken down by peptidases in the kidney, but is not detected in urine. Amylin deposits are commonly found in the islets of people with T2D or insulinoma pancreatic tumors, where it is thought to modulate β-cell apoptosis; however, it is as yet unclear if this is causative or an epiphenomenon (Tomita 2011). Amylin is thought to modulate glycemia via slowing gastric emptying, increasing production of gastric acid, bile from the liver, and pancreatic enzymes and increasing satiety. Amylin release is controlled by the same factors as control insulin production and secretion and, in addition, is also activated by TNFα and by fatty acids (Westwell-Roper et al. 2014; Miegueu et al. 2013). A synthetic modified form of amylin (pramlintide) is in clinical use for pre-meal injection in people with T1D or T2D to reduce postprandial hyperglycemia (Riddle et al. 2015; Tran et al. 2015).

Glucagon

Glucagon, a 29-amino acid peptide hormone produced in and secreted by the α-cells of the islets of Langerhans is initially produced as proglucagon, which is then cleaved by pro-hormone convertase-2 in the α-cells. Glucagon has a counterregulatory effect on blood glucose compared to insulin, important in the prevention of hypoglycemia. Glucagon has a circulating half-life of 3–6 min (Greenstein and Wood 2011; Lebovitz and American Diabetes Association 2009).

Glucagon's main site of action is the liver, where it acts against insulin to:

- ↑ Glycogenolysis (the breakdown of glycogen, which is predominantly stored in liver and skeletal muscle).
- ↑ Gluconeogenesis (the synthesis of glucose from non-carbohydrate sources such as lactate, glycerol and amino acids). (Gluconeogenesis can also occur to a lesser extent in the kidney.)
- ↑ Ketogenesis. Ketone bodies can be used by some tissues as an alternate energy source (Valente-Silva et al. 2015).

Secretion of glucagon is *increased* by low glucose levels (in a non-diabetic person, and this release is impaired in people with diabetes), increased levels of amino acids (in particular alanine and arginine), increased gastrointestinal hormones (e.g., CCK), increased 'stress' hormones (catecholamines and glucocorticoids) and sympathetic and parasympathetic nervous stimulation. Glucagon secretion is *inhibited by* high levels of glucose, insulin, free fatty acids and the incretin hormone GLP-1 (Sharma et al. 2015).

Glucagon is used clinically as a deep subcutaneous or intramuscular injection to rapidly increase blood glucose levels in people with insulin-treated diabetes who are severely hypoglycemic and unable to take oral glucose-containing food or drink (Rowe et al. 2015), or in lower doses as part of their sick-day management plan (Chung and Haymond 2015), or in the acute management of people who have taken an overdose of insulin or insulin secretagogue oral hypoglycemic agents (White et al. 2014). Doctors, nurses, paramedics and lay community members with suitable training can administer glucagon, which is a lyophilized powder and requires suspension in liquid pre-injection. Research using glucagon in a 'bihormonal' insulin pump is underway (Bakhtiani et al. 2015; Shah et al. 2014), but hampered by the relatively poor stability of glucagon solutions. Intranasal glucagon is also likely to be available for clinical use soon (Pontiroli 2015).

Somatostatin

Somatostatin is a cyclic polypeptide secreted by the δ cells of the islets, and also by the hypothalamus, stomach and intestines. It has a circulating half-life of several minutes (Rai et al. 2015).

Somatostatin acts:
within the pancreas and the gut to

- ↓ Insulin secretion by the β-cells of the islets of Langerhans.
- ↓ Glucagon secretion by the α-cells of islets of Langerhans.
- ↓ Pancreatic exocrine secretion.
- ↓ Gastric emptying time and acid production.
- ↓ Splanchnic blood flow.
- ↓ Gastrin, CCK and other gut hormone levels; and

in the brain to

- ↓ GH release and
- ↓ Decrease release of other pituitary hormones, including thyroid-stimulating hormone (TSH) and prolactin.

Stimuli for the *release of somatostatin* tend to be the same as those for insulin and include glucose, arginine and gastrointestinal hormones (Arimura and Fishback 1981; Liu et al. 2010; Gahete et al. 2010).

Pancreatic Polypeptide

Pancreatic polypeptide (PP), a 36-amino acid peptide produced by both pancreatic gamma islet cells (also known as PP cells) and acinar cells, inhibits pancreatic secretion after a meal. Release of PP by a meal, primarily protein, occurs in a biphasic manner, with an initial rapid release in response to vagal stimulation and a more prolonged rise in response to hormonal stimulation, predominantly CCK (Lonovics et al. 1981; Batterham et al. 2003).

Ghrelin

While mainly produced by cells in the stomach, and also by adipose tissue, this 28-amino acid peptide is also produced by the few ε cells in the islets of Langerhans. Ghrelin-producing cells are more abundant in the developing fetal pancreas than in the adult pancreas and likely play a role in pancreas development and β-cell apoptosis and survival. Ghrelin stimulates food intake (hence, it is sometimes referred to as 'the hunger hormone'), GH release, adipogenesis and glucose uptake, and inhibits lipolysis (Granata and Ghigo 2013). Ongoing research is testing ghrelin-derived fragments on human β-cell and pancreas islet survival and on insulin sensitivity and glucose control (Favaro et al. 2012; Poykko et al. 2003).

Incretins

Incretin hormones are a family of hormones, including GLP-1 and GIP, that are released from gut endocrine cells, predominantly in response to food and/or gut distension, which markedly influence islet insulin and glucagon secretion. As discussed above, incretins potentiate glucose-induced insulin secretion in a glucose-responsive manner, i.e., the greater the glucose load, the greater the incretin and hence insulin release. Incretins are thought to be responsible for 50–70 % of postprandial insulin secretion (Holst et al. 2009). While in the presence of elevated blood glucose, GLP-1 suppresses glucagon secretion, GIP most likely has a glucagonotrophic effect, thus differentiating itself in its actions from GLP-1 (Lund et al. 2014). Incretins also have other beneficial effects such as delaying gastric emptying, promoting satiety and weight loss and reduced risk of hypoglycemia (Holst et al. 2009).

Two types of incretin-based therapies are now used for glucose control in people with T2D and sometimes in T1D. There are synthetic GLP-1 agonists and inhibitors of the peptidase enzyme DPP-4, which rapidly break down GLP-1 in the circulation. GLP-1 has a short half-life of only 1–2 min in blood. Modified, more resistant to degradation, forms of GLP-1 agonists can be given by injection. The first such GLP-1 agonist (exenatide) is injected subcutaneously once or twice a day pre-meals, but longer-acting (weekly and monthly) forms are becoming available (Zhang et al. 2015). Major advantages in diabetes care are a low risk of hypoglycemia and weight loss, but nausea and vomiting are not infrequent side effects of this drug therapy. More well-tolerated orally active once daily DPP-4 inhibitors ('the gliptins,' e.g., sitagliptin) prolong the half-life of endogenous GLP-1 and are already often used in combination with other oral hypoglycemic agents (Sujishi et al. 2015; Furuhashi et al. 2015).

Hormones produced by the adrenal gland and by the pituitary gland have effects that counteract insulin's effects.

Catecholamines

Two catecholamines—adrenaline (also called epinephrine) and noradrenaline (or norepinephrine)—are secreted by the adrenal medulla. Noradrenaline is also produced in the central and peripheral nervous system. The catecholamines have a circulating half-life of several minutes and are water soluble, hence are excreted in urine (Guyton and Hall 1996; Greenstein and Wood 2011). Catecholamines are commonly called 'stress hormones,' as stressful physical or emotional situations stimulate their release, as does low blood glucose levels, particularly in healthy people.

Catecholamines cause peripheral vasoconstriction and increased heart rate, blood pressure and cardiac output, which form part of the body's 'fight-or-flight response' to acute stress. Catecholamines also have several actions which promote a rise in blood glucose levels, including:

- ↓ Insulin secretion by the β-cells of the islets of Langerhans.
- ↑ Glycogenolysis in liver and muscle.
- ↑ Lipolysis in adipose tissue (releasing fatty acids and glycerol into the circulation) (Tran et al. 1981).

Cortisol

Cortisol is another 'stress hormone' that counteracts insulin's blood glucose-lowering effects. This steroid hormone is synthesized and secreted by the adrenal cortex under the control of hormones from the brain's hypothalamus (corticotropin-releasing hormone, CRF) which then stimulates release of adreno-corticotrophic hormone (ACTH) from the anterior pituitary gland. Cortisol is secreted in a continuous manner with a diurnal pattern, peaking in the morning and being lowest between midnight and 4 a.m. Cortisol levels rise rapidly, within minutes of stress or hypoglycemia (Guyton and Hall 1996).

With regard to glucose control, cortisol causes:

(i) ↑ Protein breakdown and ↓ protein synthesis.
(ii) ↑ Gluconeogenesis in liver by:
   (a) Increasing relevant enzyme activities.
   (b) Increasing the hepatic response to glucagon and catecholamines.
   (c) Mobilizing substrate from muscle.

(iii) ↑ Glycogenolysis in liver.
(iv) ↑ Lipolysis in liver and adipose tissue.
(v) ↓ Insulin secretion.
(vi) ↑ Insulin resistance, decreasing glucose uptake from blood by muscle and adipose tissue (Dimitriadis et al. 1997).

Cortisol also has widespread other effects on bone, muscle, skin and mood, not discussed herein (Morelius et al. 2005).

Growth Hormone

Growth hormone (GH), a 191-amino acid polypeptide hormone, is synthesized and secreted by cells in the anterior pituitary gland under the positive and negative control of growth hormone-releasing hormone (GHRH) and growth hormone inhibitory hormone (GHIH) or somatostatin.

GH release is stimulated by:

- Fasting,
- Hypoglycemia,
- Sleep (the GH rise during sleep is thought to contribute to the period of relative insulin resistance occurring at about 5 or 6 a.m. (the so-called dawn phenomenon), and
- Exercise.

GH release is suppressed by:

- Hyperglycemia and
- Glucocorticoids.

GH causes:

- ↑ Insulin resistance and ↑ glucose levels via ↑ gluconeogenesis and ↓ muscle glucose uptake.
- ↑ Lipolysis.
- ↑ Protein synthesis (Manson et al. 1988).

When present in excessive amounts, such as in acromegaly (due to a GH-producing pituitary tumor) or due to self-administration, as sometimes is used by bodybuilders or athletes, these actions can lead to glucose intolerance and diabetes mellitus. GH injections, including recombinant human GH in recent decades, are available for therapeutic use for some children with growth retardation, short stature, or for GH replacement therapy after pituitary tumor or for GH deficiency (Guyton and Hall 1996; Greenstein and Wood 2011).

Changes in these hormones in diabetes, in particular of those produced by the pancreatic islets, are discussed later in this chapter.

Types of Diabetes Mellitus

Tables 1.1 and 1.2 summarize and compare the major and some minor types of diabetes. The common types of diabetes are T1D, T2D and GDM. Diabetes, in particular T2D and GDM, can be asymptomatic. The classic symptoms are thirst, frequency of urination, including overnight (nocturia and sometimes bed-wetting) and, particularly in T1D, weight loss and hunger (in spite of increased food intake (Lebovitz and American Diabetes Association 2009). Due to low sensitivity and accuracy, fingerprick blood glucose meter readings should not be used to diagnose or exclude diabetes, but these devices are valuable aids for the home management of those diagnosed with diabetes, particularly if treated by exogenous insulin. In some countries, HbA1c levels, which reflect mean glucose levels over the preceding 2–3 months, are now approved for diagnostic purposes (Use of Glycated

Table 1.1 Types of diabetes mellitus

Type 2 diabetes
Type 1 diabetes
Latent autoimmune diabetes of adults (LADA)
Gestational diabetes mellitus (GDM)
Monogenic forms of diabetes (e.g., MODY)
Related to genetic syndromes (e.g., DIDMOAD)
Secondary to (acute or chronic) pancreatitis
Related to other endocrine disorders (e.g., acromegaly, Cushing's disease)
Drug-induced diabetes (e.g., corticosteroids, some antipsychotic drugs, HMG-CoA reductase inhibitors, some anti-HIV drugs)

Table 1.2 General characteristics of type 1 diabetes, type 2 diabetes and LADA

	Type 1 diabetes	LADA	Type 2 diabetes
Percentage of diabetes	5–15	1–2	85–90
Common age of onset	Youth, usually before 30 years	After 25–30 years	Usually middle age or older
Rapid clinical onset	+++	+/−	−
Family history of diabetes	Only in 10 %	+	+++
Personal or family history of autoimmune disease	++	++	−
Weight/BMI	Low, normal or increased	Normal or increased	Usually increased
C-peptide/insulin levels	Very low	Low	Normal or high
Autoantibodies	+++	++	−
Immediate need for exogenous insulin	+++	−	rarely
Acute complication risk	+++	+++	+++
Chronic complication risk	+++	+++	+++

Haemoglobin (HbA1c) in the Diagnosis of Diabetes Mellitus 2011), as well as for monitoring glucose control of those diagnosed with diabetes. However, some people may have diabetes based on fasting and postprandial blood glucose levels, or even on the 'gold' standard oral glucose tolerance test (oGTT), yet have a normal HbA1c level. Similarly, HbA1c levels may not accurately reflect blood glucose levels, such as due to renal failure, anemia or in hemoglobinopathies (National Glycohemoglobin Standardization Program 2013).

An excellent (online and free) resource related to the incidence and prevalence of diabetes globally is the International Diabetes Federation (IDF) atlas (International Diabetes Federation 2013).

Type 1 diabetes T1D usually accounts for 5–15 % of people with known diabetes and can occur at any age, but most commonly has its onset in childhood, presenting with days to several weeks of marked thirst, frequent urination and weight loss, and frequently with diabetic ketoacidosis (DKA) (a life-threatening state of marked hyperglycemia, ketosis, acidosis and dehydration) (Lebovitz and American Diabetes Association 2009). The incidence of T1D varies greatly between countries and ethnicities. The highest rates are in the Scandinavian countries, at approximately 35 new cases per 100,000 people per year; the lowest rates are in Asia. For example, in Japan and China the incidence is approximately 1 person per 100,000 per year. Australia, North America and Northern Europe incidence rates range between 8 and 17 new cases per 100,000 per year (International Diabetes Federation 2013).

T1D is usually due to the autoimmune destruction of the insulin-producing β-cells in the islets of Langerhans due to invasion of cytotoxic immune cells. Antibodies to islet proteins are usually present for years prior to clinical presentation. By the time a person presents clinically with T1D, it is estimated that they have lost about 50–60 % of their β-cell mass (Battaglia and Atkinson 2015). Sometimes soon after T1D diagnosis and after initiation of life-saving exogenous insulin injections (or delivery by an insulin pump), patients can experience what is commonly called a 'honeymoon' period (Lebovitz and American Diabetes Association 2009). The honeymoon period is a time in which people with recently diagnosed T1D can maintain normal or near-normal (non-diabetic) range blood glucose levels on low, and sometimes even no exogenous insulin therapy, for weeks to months, and sometimes more than a year. This relates to residual insulin production by surviving β-cells, which may improve in the presence of less hyperglycemia and elevated free fatty acid exposure ('glycotoxicity' and 'lipotoxicity') due to the exogenous insulin. In addition, the obvious symptoms of diabetes, leading to its diagnosis, may be precipitated by an intercurrent illness, which increases the body's stress hormones and insulin resistance. Once this temporary illness resolves, insulin requirements may be less. Sometimes people will attribute their T1D diagnosis to the related illness or injury, but it is almost certain that, had the temporary illness not occurred, they would still have been diagnosed with diabetes not long after. It is recommended that some exogenous insulin be continued during the honeymoon period as it helps preserve residual β-cell mass and also can lessen patient distress when insulin injections inevitably need to be restarted (if exogenous insulin was ceased during the honeymoon phase).

The autoimmune destruction of the insulin-producing islet cells begins years before and continues years after T1D diagnosis, particularly in the first two to three years after clinical presentation, often resulting in a need for a higher dose of injected insulin (Battaglia and Atkinson 2015). It is now recognized that some residual insulin production (reflected by circulating C-peptide levels) can remain, even in adults with over 50 years of T1D (Keenan et al. 2010). Residual endogenous insulin production has been associated with better clinical outcomes such as better glycemic control, including a lower risk of severe hypoglycemia and of vascular complications (The Diabetes Control and Complications Trial Research Group 1998). Postmortem examinations of the pancreata of people with T1D have also demonstrated the presence of some insulin within islets (Keenan et al. 2010). These findings support that the body can protect insulin-producing β-cells from destruction, such as by autoimmunity, glucotoxicity and lipotoxicity, and that β-cells may be able to regenerate. This naturally raises hope and potential for prevention of T1D. There are many national and international efforts, such as those led by Trialnet (https://www.diabetestrialnet.org/), directed at the early identification of people at high risk of T1D, usually based on family history, genetic and antibody profiles, and trials testing diets, supplements, vaccines and immune modulating therapies to retard or prevent T1D. As yet, none have proven clinically effective, but some major trials are still in progress, and some studies show promising changes in surrogate endpoints, such as antibody positivity. Ongoing research is merited, including the further development of the 'artificial pancreas,' an insulin pump linked with a glucose sensor, a controller and algorithm that is able to control a person's blood glucose usually better than by multiple daily insulin injection therapy (Battelino et al. 2015; Schmidt et al. 2015; Russell 2015).

While T1D is thought by most as being solely a disorder of the endocrine pancreas, there is also evidence of some subclinical exocrine gland dysfunction (Battaglia and Atkinson 2015; Atkinson 2005; Sun et al. 2015).

Latent Autoimmune Diabetes of Adulthood (LADA) A slower form of antibody-positive T1D is latent autoimmune diabetes of adulthood (LADA). LADA is usually diagnosed after the age of 25–40 years, but does not require exogenous insulin therapy for at least 6 months after diagnosis (Fourlanos et al. 2005). People with LADA usually require exogenous insulin within 5–12 years of diagnosis. It has been estimated that up to 10 % of people diagnosed with T2D have LADA.

Type 2 Diabetes The majority (85–90 %) of people with diabetes have T2D, which was previously regarded to be a condition of middle or older age onset, but now can occur earlier in life as obesity rates increase. Major risk factors for T2D include family history and ethnicity (e.g., Indigenous Australians, Indians, Inuits, Hispanics, Native American Indians, Maoris, Asians), as well as the modifiable risk factors of adiposity and physical inactivity (Lebovitz and American Diabetes Association 2009). With previously shorter life expectancies and less widespread screening for diabetes, lack of sharing of medical histories, previously higher 'cut-points' for diabetes diagnosis, and the possibility that T2D can be asymptomatic early in its natural history, many people are not be aware of their inherited risk for

T2D. Modulating factors such as age, weight, diet, smoking and physical activity will determine whether genetic risk becomes manifest as T2D, or its precursor of pre-diabetes (in which blood glucose levels are intermediate between normal and the elevated levels diagnostic of diabetes). Unfortunately, the modern lifestyle has increased obesity rates in youth, with 25 % of Australian children in 2011–2012 (Australian Health Survey: updated results 2011) and 4.9–8.5 % of children globally in 2010 (de Onis et al. 2010) being overweight or obese, and this has led to increased rates of pre-diabetes and T2D in youth, particularly those in high-risk ethnic (often insulin-resistant) groups. In countries with high rates of ethnic diversity, such as the USA, T2D now accounts for 18 % of diabetes in youth (Dabelea et al. 2014). As glucose tolerance declines with age, the proportion of people with T2D increases with age, with 14.8 and 14.2 % of adults over 65 and 75 years of age, respectively, having T2D, compared with 4.2 % of adults aged 45–54 years (Australian Health Survey: updated results 2011).

Gestational Diabetes The third most common form of diabetes is GDM which currently affects 5–8 % of pregnant women (Valente-Silva et al. 2015), including 4.6 % of pregnant women in Australia (Templeton MP-C 2008). The rates of gestational diabetes have increased several fold over recent decades related to increasing rates of obesity in women of child-bearing age, increased case ascertainment due to uptake of routine screening, and lower diagnostic cut-points (Nankervis et al. 2014). GDM commonly commences in the second or third trimester of pregnancy and usually resolves post-pregnancy, but it does signal that the mother is at increased risk of T2D (Di Cianni et al. 2003). Women with prior GDM have a 40 % risk of developing T2D at some stage later in life (Conrad Stöppler 2014). Such women should pay particular attention to maintaining a healthy diet, weight and exercise program and should be regularly screened for T2D. While not associated with the chronic vascular complications of diabetes, GDM, if untreated, is associated with increased risks to both mother and fetus/offspring, including increased rates of macrosomia (large babies) which increases rates of birth injury and need for caesarian section, pregnancy-induced (maternal) hypertension and pre-eclampsia, and in the child, increased risk of obesity, the metabolic syndrome, T2D and cardiovascular disease later in life.

The major features of these common types of diabetes are summarized and compared in Table 1.2. About 1–2 % of people with diabetes have a different form of diabetes than those described above, which are sometimes referred to as secondary forms of diabetes.

Secondary forms of diabetes have a different pathogenesis than that of T1D, LADA, T2D and GDM. Sometimes these secondary forms of diabetes can be reversed, such as if they are due to an underlying acquired illness or drug, and that illness resolves or the drug is removed. If not, the effects, such as those related to hyperglycemia, the long-term complications and the treatment and monitoring needed are similar to that of the more common T1D and T2D.

Secondary forms of diabetes include monogenic forms of diabetes, which may affect insulin secretion or insulin resistance, those associated with genetic syndromes, acquired forms due to pancreatic damage, other endocrine diseases and drug-induced diabetes (Lebovitz and American Diabetes Association 2009) (Table 1.1).

Monogenic Diabetes Some forms of diabetes are monogenic in origin, unlike T2D, which is thought to be of polygenic origin. These forms are due to single-gene defects that are autosomal dominantly inherited. An excellent review article is that by Tallapragada et al. (Tallapragada et al. 2015). The first and more common types of monogenic diabetes (of the current eight known) to be identified as the causes of 'maturity-onset diabetes of the young' (MODY) are due to mutations in the glucokinase gene (MODY2) and in the hepatocyte nuclear factor 1α gene (MODY3). Another group of monogenic diabetes relate to the genetic defects in the potassium-ATP channel on the β-cell membrane. Single-gene defects can also induce diabetes by increasing insulin resistance. Examples include type A insulin resistance syndrome and lipoatrophic diabetes.

Diabetes may also be part of *other genetic syndromes*. Examples include 'DIDMOAD' or Wolfram syndrome, which includes diabetes insipidus, diabetes mellitus, optic atrophy and deafness. People with chromosomal disorders, such as Down's syndrome (Trisomy 21) and Turner's syndrome, are also at increased risk of diabetes (Lebovitz and American Diabetes Association 2009).

Acute and chronic forms of damage to the pancreas can lead to diabetes. Acute pancreatitis, such as due to excess alcohol, gall-stones, or severe hypertriglyceridemia, can induce sufficient β-cell loss so as to necessitate exogenous insulin therapy, which is often also associated with pancreatic exocrine deficiency requiring oral replacement of digestive enzymes with food in addition to micronutrient supplements (Fieker et al. 2011). Pancreatic cancer, trauma and the surgical removal of a major part of the pancreas can each lead to diabetes. Hemochromatosis can also damage the pancreas (and other organs such as the liver) due to iron overload. Fibrocalculous pancreatopathy, such as related to malnutrition and infection, is not a uncommon cause of diabetes in impoverished regions.

Other endocrine disorders can induce diabetes, usually related to an overproduction of hormones that act against insulin. Examples include Cushing's syndrome (corticosteroid excess), glucagonoma (glucagon excess), phaeochromocytoma (catecholamine excess), thyrotoxicosis (thyroid hormone overproduction) and very rarely somatostatinoma (somatostatin-producing tumor) (Lebovitz and American Diabetes Association 2009).

Diabetes may also be *drug induced*, which often resolves with drug cessation or the use of non-diabetogenic alternates if available. Commonly needed drugs that can induce (pre-diabetes and type 2) diabetes are the immunosuppressive, anti-inflammatory corticosteroids, such as used in organ transplant, connective tissue diseases (e.g., rheumatoid arthritis and temporal arteritis) and severe asthma exacerbations. Some antidepressant and antipsychotic drugs, as well as protease inhibitors for the treatment of HIV, have significant diabetogenic effects, while other very commonly used drugs such as thiazide diuretics and HMG-CoA reductase inhibitors have weak, but substantiated effects, on potentiating pre-diabetes or T2D (Lebovitz and American Diabetes Association 2009).

Diabetes Complications

All forms of diabetes, with the exception of GDM, increase the susceptibility of those affected, to the acute complications of diabetes and, if the diabetes duration is long enough, also to the chronic microvascular and macrovascular complications, discussed below and summarized in Tables 1.3 and 1.4.

Acute Glycemia-Related Complications of Diabetes

In people without diabetes, glucose levels are normally tightly regulated within a narrow range. People with diabetes, however, particularly those requiring exogenous insulin, are at increased risk of severe hypoglycemic and hyperglycemic crises, both of which can be life-threatening.

Table 1.3 Complications of diabetes

Acute complications
Dehydration and electrolyte imbalance
Hyperglycemia including diabetic ketoacidosis and hyperosmolar non-ketotic coma
Increased risk of sepsis
Poor wound healing
Mental health issues, e.g., anxiety, depression, diabetes distress
Pregnancy related
Increased risk of reduced fertility
Increased risk of miscarriage
Increased risk of pre-eclampsia
Increased risk of growth retardation or macrosomia
Increased risk of congenital malformations in offspring
Increased risk of diabetes and of cardiovascular disease in offspring
Chronic complications
Microvascular complications
Diabetic retinopathy
Diabetic nephropathy
Diabetic neuropathy—peripheral neuropathy
Macrovascular complications
Coronary artery disease (CAD)
Cerebrovascular disease
Peripheral vascular disease including foot ulcers and amputations
Cardiomyopathy (independent of hypertension and/or coronary artery disease)
Diabetes dementia

Table 1.4 Other health problems more common in diabetes

Other health problems more common in diabetes
Cardiovascular disease
Hypertension
Congestive cardiac failure
Sudden death, most likely cardiac arrhythmia or CAD related
Transient ischemic attacks and cerebrovascular event
Peripheral vascular disease
Cardiomyopathy (due to hypertension, CAD and diabetes per se)
Eye
Glaucoma
Cataracts
Gastrointestinal tract
Nonalcoholic fatty liver disease (NAFLD)
Peptic ulcer disease
Cancer (except for decreased incidence of prostate cancer)
Increased risk infections including TB and fungal infections
Periodontal disease
Hearing loss (mild)—some debate

Hypoglycemia

Hypoglycemia is usually defined as a low blood glucose (<4.0 mmol/L). It usually causes symptoms and signs related to catecholamine release (e.g., anxiety, trembling, sweating and palpitations) and/or neuroglycopenia, i.e., related to lack of glucose to the brain (e.g., poor concentration, in-coordination, slurred speech, emotional outbursts, coma and seizures). The symptoms of sympathetic system activation usually occur, but not always, prior to life-threatening neuroglycopenic consequences of hypoglycemia, thus providing the person with diabetes early awareness a 'hypoglycemia warning system' (American Diabetes Association 2012, 2015; Cryer and American Diabetes Association 2013; Fox et al. 2009; Peters et al. 2013).

Mild to moderate hypoglycemia is defined as that which can be still self-managed by appropriate food or drink ingestion. Severe hypoglycemia is defined as that which causes marked impairment in consciousness, and even seizures, such that assistance by another person is essential for recovery, usually with administration of a glucagon injection or intravenous glucose.

Most people with T1D experience a mild hypoglycemic event on average once a week and a severe hypoglycemic event once per annum (American Diabetes Association 2012; Cryer and American Diabetes Association 2013).

Common precipitants of hypoglycemia are inappropriate (for subsequent carbohydrate intake and physical activity) insulin dosage, insulin administration issues, excess alcohol intake, recreational drug use, undiagnosed or uncontrolled medical conditions such as food malabsorption (e.g., related to celiac disease) or adrenal insufficiency, which are more common in people with T1D (Lebovitz and American Diabetes Association 2009). Recurrent severe hypoglycemia can have adverse psychosocial and socioeconomic impacts (including accidents-, family-, schooling- or employment-related issues and loss of driving rights), as well as cause seizures, permanent brain damage/cognitive impairment and death. Also frequent or severe hypoglycemia can impair warning responses to hypoglycemia, called 'hypoglycemia unawareness' which increases the risk of recurrent severe hypoglycemia (Lebovitz and American Diabetes Association 2009; Hendrieckx et al. 2014). People who have experienced severe hypoglycemia often opt for less acceptable (high) blood glucose control to prevent recurrence, even though they know this will increase their own risk of long-term vascular and neurological complications.

About one in three or four people with T1D will experience reduced hypoglycemia awareness at some stage of their life. In many, but not all cases, hypoglycemia awareness can be improved by several weeks to months of avoidance of hypoglycemia. In many cases, recurrent severe hypoglycemia can be lessened by identification and treatment of the precipitating factors, frequent home blood glucose monitoring, adjustment of insulin type (to modern insulin analogues), change in insulin delivery dose or mode (e.g., insulin pump use, particularly linked with continuous glucose monitors which can suspend insulin delivery) and additional education regarding insulin dosage adjustment, nutrition and physical activity. For adults with T1D with particularly debilitating recurrent severe hypoglycemia, islet cell transplantation (discussed elsewhere in this book) is a treatment option that can be successful in eliminating hypoglycemia risk (O'Connell et al. 2013).

Hyperglycemia

Hyperglycemia in diabetes is usually due to a relative lack of insulin (and/or excess glucagon). In T1D, hyperglycemia can be associated with ketosis and a metabolic acidosis (DKA), which is a medical emergency, usually requiring IV insulin, fluids and electrolytes and treatment of the precipitating cause which can be an intercurrent illness. In people with T2D, there is usually enough insulin to prevent severe ketosis, but the resultant hyperglycemia due to a relative lack of insulin and/or hyperglucagonemia can still reach extreme levels. At its worst, there can be marked hyperglycemia (even up to 100 mmol/L, severe dehydration and metabolic disarray (hyperosmolar non-ketotic coma or hyperglycemic hyperosmolar syndrome) that is at least as serious and life-threatening as DKA (Lebovitz and American Diabetes Association 2009). Even with milder hyperglycemia, there is impaired immune function, resistance to sepsis and tissue healing and, if chronic, increased risk of vascular complications.

The Vascular Complications of Diabetes

Microvascular Complications of Diabetes

The classically described microvascular complications include diabetic retinopathy, nephropathy and neuropathy. As a consequence of retinopathy, diabetes is a most common cause of vision loss in working-age adults (Facts About Diabetic Eye Disease 2015). Over 40 % of renal dialysis or transplant patients have renal failure due to diabetes (Collins et al. 2012). The various types of neuropathy cause various morbidities, but the most common would be peripheral neuropathy which contributes to diabetes being the most common cause for non-traumatic lower limb amputation. These major microvascular complications usually take at least 5 years to become manifest clinically, though subclinical damage can occur before this. Some people with T2D have diabetic microvascular complications at diagnosis, thought to be related to years of undiagnosed hyperglycemia. Not all people with diabetes will develop vascular complications, and in those (with T1D) who do not, their longevity is not substantially reduced (Mäkinen et al. 2008) relative to that of non-diabetic subjects. Most people with over 20 years of diabetes will have some evidence of diabetes complications, but in many it may not be at a degree to cause disability. For example, many will have 'background' diabetic retinopathy (King 2002) which does not threaten vision, but nonetheless needs regular review, including attention to risk factors (discussed below), so as to reduce the risk of progression to vision-threatening 'proliferative' disease. About a quarter of people with T1D who develop microalbuminuria will have spontaneous regression to normal levels of albumin excretion without loss of renal function. Good vascular risk factor control is thought to be of assistance in regression. It would seem prudent to assume so at this stage. As yet, it is unclear if these people who develop early microvascular damage and regress are at an increased risk of recurrence and progression of microvascular complications later on. If detected early, tight glucose, blood pressure, lipid and weight control, non-smoking, and the use of angiotensin-converting enzyme (ACE) inhibitor drugs (even if 'normotensive'), microvascular complications can be reversed or their progression slowed (Fowler and Vasudevan 2010). The onset of T2D in youth has recently been shown to be associated with high rates of early and severe vascular complications. Compared to people with T1D of similar diabetes duration, early onset T2D also has higher age-adjusted mortality rates (Craig et al. 2009; Wong et al. 2008a, b).

In general, if a person with diabetes develops one microvascular complication, they are at a higher risk of developing the other microvascular complications, as well as accelerated atherosclerosis and the related macrovascular complications (Rhee and Kim 2015; Bowling et al. 2015). We shall now briefly describe the major microvascular complications.

Diabetic Retinopathy

Diabetic retinopathy is typically divided into background diabetic retinopathy (which can be further subdivided into mild, moderate and severe background retinopathy) and proliferative retinopathy. There is no doubt that a preclinical stage, prior to the appearance of the first microaneurysms or hemorrhages, also exists. Figure 1.7 shows an image of a normal retina and of the various clinical stages of diabetic retinopathy. Nonproliferative retinopathy includes microaneurysms, soft exudates (retinal infarcts) and hard exudates (lipid deposits). The advanced-stage, proliferative retinopathy involves new fragile blood formation. These vessels are prone to leaking, which can cause sudden severe vision loss. Retinal hemorrhages can also induce fibrous tissue formation which can contract and cause retinal detachment.

Diabetic retinopathy is also associated with dysfunction and damage of the neural retina. The relative time course of the retinal neural and vascular damage is debated.

Diabetic Nephropathy

There are two aspects of renal function that can become abnormal in diabetes, and either one or both aspects can be abnormal. One aspect is the leakage of albumin (and other proteins) into the urine, and the other is failure of the filtration process. In periods of poor glucose control, particularly at T1D diagnosis, the glomerular filtration rate (GFR) is above normal (termed hyperfiltration). With progressive renal impairment, the GFR returns to within 'the normal range' and then declines toward, and may reach the level of needing renal replacement therapy of dialysis (often initially by peritoneal dialysis, then by hemodialysis) or by kidney transplant. Chronic kidney disease (CKD) is usually divided into five stages shown in Table 1.5. Normal renal function can be described as stage 0.

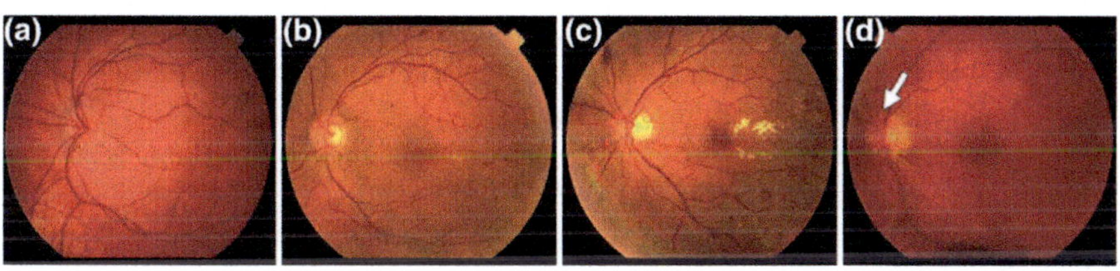

Fig. 1.7 Normal retina and stages of diabetic retinopathy. Fundus photographs showing the clinical stages of diabetic retinopathy: **a** a normal retina; **b** mild nonproliferative diabetic retinopathy, with hemorrhages, microaneurysms and hard exudates; **c** nonproliferative retinopathy; **d** proliferative diabetic retinopathy, with the optic disk (*white arrow*) and pre-retinal hemorrhage in the inferior retina. Taken from our publication—Farr et al. (2015). Licensed under CC BY 4.0 via http://www.ncbi.nlm.nih.gov/pmc/articles/PMC4649912/figure/f3/

Table 1.5 Stages of chronic renal disease

Stage	eGFR (mL/min/1.73 m^2)	Description (renal structure and function)	Predominant AER status
0	≥90	No damage/normal kidney function	
1	≥90	Kidney damage with normal/high GFR	Micro
2	60–89	Kidney damage with mild reduction in GFR	Micro
3	30–59	Kidney damage with moderate reduction in GFR	Macro
4	15–29	Kidney damage with severe reduction in GFR	Macro
5	<15	Kidney failure	

AER albumin excretion rate, *eGFR* estimated glomerular filtration rate

In persons with diabetes, *Stage 1* CKD, which is asymptomatic, is characterized by renal hypertrophy, hyperfiltration, normal serum creatinine levels, and often by increased urinary albumin loss, using in the microalbuminuric range. This albuminuria is usually worsened during and shortly after physical exercise or by poor glycemic control. Blood pressure is usually normal. This stage may regress spontaneously or by risk factor reduction and ACE inhibitor treatment.

Stage 2 CKD, also called incipient diabetic nephropathy, is also silent and characterized by normal GFR and abnormal renal morphology (though biopsies are not often performed unless there are atypical features, such as hematuria). Albuminuria is likely to be more permanent in stage 2 CKD, but will still be in the microalbuminuric range.

Stage 3 CKD is the initial stage of overt diabetic nephropathy. Its main manifestations are albuminuria (15–300 μg/min) and an early increase in serum creatinine (and fall in GFR). A slow, gradual increase over years in the amount of albuminuria is usual, and blood pressure usually also rises.

Stage 4 CKD, established overt diabetic nephropathy, is defined as persistent proteinuria (>0.5 g/24 h) and a fall in the eGFR to <60 mL/min/1.73 m^2, which is usually associated with hypertension. In the absence of antihypertensive agents (usually starting with an ACE inhibitor), GFR declines at a mean rate of about 1 mL/min/month. Long-term antihypertensive treatment reduces the rate of fall by about 60 %, substantially delaying renal failure.

Stage 5 CKD is end-stage renal failure, characterized by rising serum creatinine levels and an eGFR falling below 15 mL/min/1.73 m^2. Renal replacement therapy (dialysis or transplantation) is usually required (Lebovitz and American Diabetes Association 2009).

For people with T1D and end-stage renal failure, combined kidney–pancreas transplantation is an option when cadaver donors are available, resulting in improved survival (Kaku et al. 2015; Lindahl et al. 2014a, b; Light and Tucker 2013). The pancreas transplant is usually, but not always, performed at the same time as the renal transplant.

Diabetic Neuropathy

Neuropathy in diabetes is thought to have both a vascular and metabolic etiology. Several types of neuropathy may occur: peripheral neuropathy, autonomic neuropathy, 'plexopathy' or mononeuritis multiplex. Diabetes is also thought to be associated with a 'diabetes dementia' (Hatanaka et al. 2015; Shiue 2015).

T1D and T2D are common causes of *peripheral neuropathy*, which usually leads to sensory loss affecting the feet and hands in a 'glove and stocking' distribution, which may progress proximally. The major loss is usually sensory, but motor loss and muscle wasting can also occur.

Diabetes may also affect the *autonomic nervous system*, which can cause postural hypotension, abnormal cardiac reflexes, no pain (silent) or atypical pain during myocardial ischemia, erectile dysfunction, delayed gastric emptying, alternating diarrhea (particularly nocturnal) and constipation, incomplete bladder emptying and abnormal sweating (such as in response to eating) (Vinik et al. 2003).

An uncommon and painful cause of (often sudden onset unilateral) quadriceps, hip and buttock muscle wasting and weakness is a neuropathy affecting the lumbosacral plexus. This condition is also known as diabetic amyotrophy, proximal diabetic neuropathy, diabetic lumbosacral plexopathy or diabetic polyradiculopathy (Lebovitz and American Diabetes Association 2009).

Macrovascular Disease in Diabetes

In general, relative to their non-diabetic peers, people with diabetes are at a two- to four-fold increased risk of coronary artery disease, cerebrovascular disease and peripheral vascular disease. With regard to lower limb amputations, diabetes is associated with a 15-fold greater risk than non-diabetic subjects (Markakis et al. 2016; Wu et al. 2007; Young et al. 2003).

Accelerated Atherosclerosis

In diabetes, atherosclerosis is accelerated relative to that of similar aged non-diabetic subjects. While there are many similarities in the pathology of atheroma in diabetic and non-diabetic subjects, diabetes atheroma usually occurs prematurely with a greater plaque burden extending more distally. The plaques tend to be more lipid rich with a greater degree of inflammation and calcification and are therefore inherently more unstable in people with diabetes. Arterial collateral formation and wound healing are also more often impaired in diabetes subjects, and due to the more severe and distal disease, revascularization interventions are less often feasible for advanced vascular disease (Markakis et al. 2016). For these same reasons, clinical outcomes of cardiovascular disease events (e.g., myocardial infarction) are often worse with greater mortality in people with diabetes (Lima et al. 2013).

Peripheral Vascular Disease

Lower limb amputation in a person with diabetes is most often due to a combination of macrovascular, microvascular damage, neuropathy, tissue infection and impaired wound healing (Markakis et al. 2016). Macrovascular disease usually leads to a major above or below knee amputation, while microvascular disease usually leads to a non-healing foot ulcer and/or amputation of a toe or forefoot.

Cardiomyopathy

The incidence of congestive cardiac failure is also increased several fold in people with compared to without diabetes and has a poorer prognosis. While coronary artery disease, small arterial vessel disease (sometimes called syndrome little x), hypertension and renal failure, which are common in diabetes, can cause a cardiomyopathy, it is now accepted that diabetes per se can cause a metabolic cardiomyopathy (Felicio et al. 2015; Liu et al. 2014; Letonja and Petrovic 2014). This problem is thought to be more common in women with diabetes and to involve a hormonal basis.

The Pathology of Diabetes Vascular Damage

Endothelial dysfunction in both large and small blood vessels is a feature of diabetic vascular damage (Jenkins et al. 2004a). The vascular endothelium is more than an inert lining of blood vessels. As summarized in Table 1.6, the endothelium has a barrier function, modulates vascular tone and blood pressure, plays roles in inflammation, thrombosis and fibrinolysis, lipoprotein metabolism and angiogenesis.

Table 1.6 Endothelial dysfunction

Endothelial function	Feature in diabetes
Structural	Thick basement membranes
Barrier	↑ Permeability
Cell growth/angiogenesis (VEGF)	Cell proliferation/death/angiogenesis
Modulate thrombosis/fibrinolysis, platelets	↑ Thrombosis, ↓ tpa, ↑ PAI-1, platelet activation
Influences inflammation	↑ CAMs/monocyte adhesion
Modulate vascular tone (ET-1, NO, ACE)	Altered blood flow/↑ BP and capillary pressure
Lipid metabolism (LPL)	Dyslipidemia

ACE angiotensin-converting enzyme, *BP* blood pressure, *CAMs* cell adhesion molecules, *ET-1* endothelin 1, *LPL* lipoprotein lipase, *NO* nitric oxide, *PAI-1* platelet activator inhibitor 1, *tpa* tissue plasminogen activator, *VEGF* vascular endothelial growth factor

Common Mediators

Hyperglycemia together with disordered lipid and protein metabolism contributes to atherogenesis. Hyperglycemia results in a cascade of both intracellular and extracellular perturbations, including quantitative and qualitative changes in lipoproteins, increased inflammation and oxidative stress, including the formation of a family of compounds called advanced glycation end-products (AGEs) (Fig. 1.8) (Jenkins et al. 2004a).

Michael Brownlee has suggested a common intracellular pathway within endothelial cells that links hyperglycemia and vascular complications via induction of increased mitochondrial oxidative stress which then activates the polyol, hexosamine, protein kinase C and methylglyoxal/AGEs pathways. His excellent article and related American Diabetes Association Banting Best lecture are well worth reading/viewing (Atkinson 2005). This 'unifying hypothesis' explains why inhibition of single pathways, such as the PKC pathway (Deissler and Lang 2016; Tuttle et al. 2015) or AGE formation (Brownlee 2001, 2005) only partially reduces vascular complications, as glucose still activates the other pathways. Inhibition of common proximal modulators (such as improving glucose levels) and/or of common distal mediators, such as increased mitochondrial oxidative stress, as proposed and tested by Brownlee, is required. The Diabetes Control and Complications Trial (DCCT) (an intervention study) and its longitudinal observational follow-up study

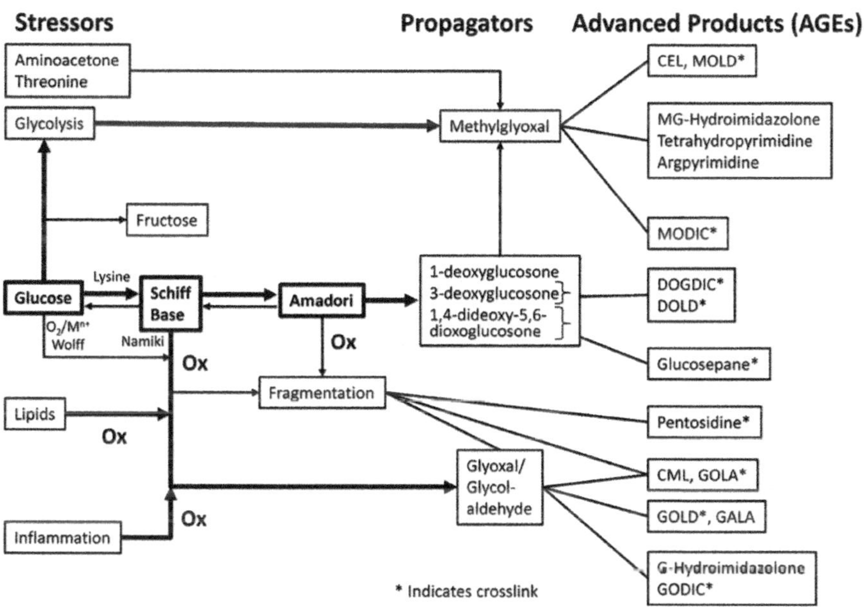

Fig. 1.8 Pathways of AGEs formation. Reproduced with permission, from Monnier and Wu (2003)

Epidemiology of Diabetes Interventions and Complications (EDIC) demonstrate efficacy of improved glucose control in reducing vascular complications in people with T1D (The Diabetes Control and Complications Trial Research Group 1998; White et al. 2008). The United Kingdom Prospective Diabetes Study (UKPDS) demonstrated similar benefit in people with T2D (Holman et al. 2008a, b).

Risk Factors for the Vascular Complications of Diabetes

Beta-Cell dysfunction and the resultant hyperglycemia and glycemic variability bring about and/or are compounded by multiple risk factors that can promote vascular damage. The risk factors, which are often inter-related, are usually divided into traditional and novel risk factors, as summarized in Table 1.7 (Jenkins et al. 2015). Risk factors may be further subdivided into those that are unmodifiable and modifiable.

Table 1.7 Risk factors for the chronic complications of diabetes

Traditional risk factors
Age (including age of onset)
Diabetes duration
Glycemic control
–hyperglycemia
–glycemic variability
–frequent hypoglycemia
Obesity
Increased waist circumference
Smoking
Hypertension
Dyslipidemia
High triglycerides
High LDL cholesterol
Low HDL cholesterol
Family history
Novel risk factors
Qualitative and quantitative changes in lipoproteins
E.g., lipoprotein glycation and/or oxidation
Inflammation
Oxidative distress
Advanced glycation end-product (AGE) formation
Insulin resistance
Angiogenesis-related growth factor disturbances
Genetics
Epigenetics
Changes in telomeres/telomerase-modifying enzymes
Histone modification
Unfavorable microRNA profiles

Traditional risk factors which are unmodifiable include age of diagnosis, which can be delayed for T2D, but not as yet for T1D, longer diabetes duration, gender, and an adverse family history. With regard to gender, both men and women are at high risk of macrovascular disease. Men tend to be at higher risk of microvascular complications, though the subgroup of females with prepubertal onset of T1D tend to have a higher risk than their male counterparts. The reasons for this are not fully elucidated. With regard to family history, the risk of diabetes complications is higher in subjects with other family members with diabetes complications, or in T1D patients with hypertension or T2D in their non-diabetic family members.

Modifiable traditional risk factors, which are also influenced by genetic and environmental factors, include hypertension, dyslipidemia and obesity. Obesity, in particular central obesity or an increased waist circumference, may compound hypertension, dyslipidemia and insulin resistance, but is also thought to act independently (The Diabetes Control and Complications Trial Research Group 1998; White et al. 2008). This may be via increased levels of inflammation and also by increased insulin resistance.

Most studies relate to factors associated with diabetes complications, but importantly there are several studies evaluating people with T1D who do not develop clinically significant complications.

Factors associated with a very favorable outcome of T1D, as reflected by living 50 or more years with it, but with no or only early vascular damage, include a normal weight, non-smoking status, a favorable lipid profile characterized by a high HDL cholesterol level, and a family history of parental longevity. There is marked concordance in the characteristics of these 50 or more year survivors with T1D who comprise the British 'Golden Years' Cohort and the Joslin Medalist's (Bain et al. 2003). Studies of the genetics and epigenetics of such patients are underway. Such knowledge may provide insight into treatments that can reduce the devastating complications of diabetes, while the 'cure' for diabetes is awaited.

Novel Risk Factors. Hyperglycemia in addition to increasing inflammation, oxidative stress and AGE formation can cause non-enzymatic glycation of many short-lived proteins (such as lipoproteins) and long-lived proteins (such as basement membranes and skin collagen). We have previously reviewed the adverse effects of hyperglycemia-induced modifications of lipoproteins (Jenkins et al. 2004b), which include adverse effects on arterial, retinal and renal cell survival and function.

Other novel risk factors include genetic factors, which require very large studies to explore, and more recently, epigenetic factors. As yet, no genetic or epigenetic markers are used in clinical practice. There are many excellent review articles.

Insulin resistance, while commonly associated with T2D, is also a feature of non-diabetic subjects (Festa et al. 2004) and people with T1D (Greenbaum 2002). Insulin resistance has been associated with increased risk of cardiovascular disease in non-diabetic subjects and of vascular complications in people with T1D (Bjornstad et al. 2015). As discussed later in this chapter, insulin-sensitizing drugs are often used in pre-diabetes (type 2), T2D and GDM, and sometimes in T1D.

Metabolic Memory for Glucose Control

The impacts of hyperglycemia and hypoglycemia on higher cerebral function and mood are evident immediately, but the effects on the vasculature are not evident for years. Importantly, the level of glycemic control in one decade can influence vascular complications a decade or even more later, irrespective of the level of control in the second decade. This phenomenon, called 'metabolic memory' or the 'legacy effect' (which are comparable terms), was initially coined in relationship to T1D studies in the DCCT/EDIC cohort (White et al. 2008) and in the 10-year follow-up of the UKPDS, respectively (Holman et al. 2008a).

Susceptible tissues for metabolic memory include retina, kidneys, nerves and arteries. A non-glucose-related example of a legacy effect is the persistence of a smoking-related increase in heart disease and cancer risk that persists for years after smoking cessation. A legacy effect for lipid and blood pressure control for the vasculature has also been demonstrated, but predominantly in non-diabetic subjects (O'Neal and Jenkins 2009). As yet, there are no specific lipid-related studies in diabetes for which the legacy effect has been evaluated, and given the major benefit of lipid-lowering drugs in diabetes, it will now be difficult to conduct such a study prospectively.

The timeframe of metabolic memory in T1D is best evidenced by the DCCT/EDIC study, in which $\approx$5.9 years of intensive versus conventional diabetes management (HbA1c 9 vs. 7 %) lowered vascular complication rates for at least 8–12 years (Aiello et al. 2015). The T2D UKPDS data demonstrate a similar legacy effect of glycemia that lasts for 10 years after having a HbA1c $\approx$ 7.0 % for 10 years (Manley 2003). Due to the low number of long-term glycemic control intervention studies with adequate follow-up, it is not yet clear if there is a threshold level for metabolic memory and how long the effect is maintained for a given time at each HbA1c level across the full HbA1c spectrum.

Potential mediators of glucose metabolic memory are epigenetic changes, which are acquired changes in DNA function without changes in DNA sequence (Okabe et al. 2012; Thomas 2014; Keating and El-Osta 2015; Mathiyalagan et al. 2015). Epigenetic changes include DNA and histone methylation, histone acetylation and telomere shortening. Glucose- and AGE-induced modifications of long-lived tissues such as vascular basement membranes may also mediate metabolic memory and also be a therapeutic target (Jenkins et al. 2004a). Some currently used drugs, such as ACE inhibitors, 'statins' and metformin also have anti-AGE and DNA-protective effects, as well as their primary actions related to blood pressure, lipid and glucose lowering, respectively. Novel drugs such as histone deacetylase inhibitors, which reduce epigenetic damage, are currently in human cancer clinical trials and protect against diabetic nephropathy in an animal model (West and Johnstone 2014).

Diabetes Treatments

Current research strategies for both T1D, T2D and GDM also include approaches related to diabetes prediction, diabetes prevention, diabetes control (using a range of drugs and insulin delivery devices), and the ultimate, but currently elusive, cure. Other chapters in this book, by experts in their area, include potential preventative strategies for islet damage that leads to T1D, to pancreas or islet cell transplantation (which are clinically available in some affluent countries for a very small proportion of people with T1D), to islet β-cell replacement with stem cell therapy.

Other drugs related to blood pressure, lipid control and novel modalities to retard or treat the vascular complications of diabetes, such as anti-VEGF agents for diabetic retinopathy and dialysis for renal failure, are also relevant to the care of many people with diabetes, but of relevance to the book topic, in this section we focus on glucose control agents.

Most existent treatments for glucose control in diabetes have mechanisms of action that are linked to pancreatic islet hormones, as described earlier in this chapter. Drug classes for glucose control are summarized in Table 1.8 and reviewed in more detail in other excellent books such as by the American Diabetes Association (Lebovitz and American Diabetes Association 2009).

Broadly, the drugs for glycemic control are mostly glucose lowering for daily use, but drugs to elevate glucose are used on an occasional basis for the treatment of hypoglycemia. There is an increasing number of insulin formulations available for T1D or for insulin-requiring GDM or T2D. On average, within 10 years of T2D diagnosis, many people will need exogenous insulin due to β-cell failure. Insulin delivery devices, also an active area of research, include syringes, disposable and reusable syringes, as well as external and implantable insulin pumps, including the 'artificial pancreas,' also known as a closed loop insulin pump. Ongoing research aims to improve exogenous insulin therapy, with approaches including faster onset of action, longer duration and 'smart insulins' that have responsiveness to glucose levels (so as to reduce the risk of hypoglycemia). Insulin secretagogues are usually used mainly in T2D, with the exception of non-sulfonylurea drugs, which are also approved for T1D use (always in conjunction) with insulin.

Insulin sensitizers, incretin-based therapies, glycosuria-inducing drugs and drugs used to delay dietary carbohydrate absorption are mainly used for T2D, but some have shown benefit in people with T1D, as an adjunct to insulin therapy. This is an active area of clinical research. Combination therapies of oral agents or of injectable and oral agents, always including insulin, are commonly needed for individuals with T1D. Many diabetes associations recommend treatment algorithms for T2D (Gunton et al. 2014), when lifestyle measures alone are inadequate to achieve optimal glucose control.

Glucagon, which can only be delivered by injection, is usually used to treat severe hypoglycemia by trained carers or healthcare staff to raise blood glucose levels. Insulin pump-related research is exploring glucagon administration, but

Table 1.8 Glucose control drugs used in diabetes

Drug class	Type 1 diabetes	Type 2 diabetes	Gestational diabetes
Glucose lowering			
Insulin[a]			
Rapid-acting	X	X	X
Short-acting	X	X	X
Intermediate-acting	X	X	X
Long-acting	X	X	X
Pre-mixed (short and intermediate)	X	X	X
Insulin secretagogues			
Sulfonylureas		X	
Non-sulfonylurea related			
Meglitinides (e.g., repaglinide)		X	
Insulin sensitizers			
Metformin	–[b]	X	X
Thiazolidinediones	–[b]	X	
Incretin-based therapies			
GLP-1 agonists[a]	–[b]	X	
DPP-IV inhibitors	–[b]	X	
Amylin related[a]	X	X	
Glycosuria-inducing drugs			
SGLT2 inhibitors	–[b]	X	
Agents to delay complex carbohydrate absorption			
α-Glucosidase inhibitors	–[b]	X	X
Hypoglycemia treatment			
Glucagon injection[a]	X	X	X
Glucose tablets or gels	X	X	X
IV glucose	X	X	X

[a]Injectable drug
[b]Emerging or off-label use

is hampered by poor glucagon stability, currently necessitating a fresh glucagon solution daily. Other treatments for hypoglycemia, apart from food, include oral glucose tablets or gels or intravenous glucose fluids. Much important drug-related research continues alongside endeavors to protect, regenerate and replace the hormones produced by the pancreatic islets.

Glucose Monitoring

To guide the type and amount of glycemic control therapies used by the person with diabetes, doctors commonly measure the HbA1c level in blood. The HbA1c is a measure of the amount of non-enzymatic glycation of hemoglobin, and this reflects mean blood glucose control over the previous 2–3 months. People with insulin-treated diabetes usually test their own blood glucose levels regularly (often four times a day or more if they have T1D). Special portable blood glucose testing meters and strips have been available for this over the past four decades. The glucose meters are becomingly increasingly smaller, more accurate and with added features such as memory, glucose profile recognition and the capacity (with healthcare professional setup) to calculate optimal insulin doses for the individual person according to their glucose level at that time and what they are about to eat (i.e., 'smart meters').

Home urine glucose monitoring, used clinically since the early 1900s, is now infrequently used. Some patients with diabetes, usually with T1D, have access to continuous glucose monitoring (CGM); however, the cost is prohibitive to most patients for regular use. In some countries, the government or health insurers will cover CGM. CGM requires a disposable glucose sensor (each lasts 5–7 days) to be placed into the subcutaneous fat for quantification of interstitial fluid glucose levels every few minutes using enzymatic glucose oxidase chemistry, as for blood glucose testing (Davis et al. 2015; DeSalvo and Buckingham 2013). If linked with a compatible insulin pump, a CGM can trigger the pump to suspend insulin delivery in the setting of a low or predicted low glucose level and to restart insulin delivery when the glucose level has risen to a safe level or the user restarts insulin delivery. Earlier versions, still available, can suspend insulin delivery at a low glucose level for up to 2 h (Choudhary et al. 2015; Prazny 2015; Tauschmann and Hovorka 2014; Agrawal et al. 2011; Pickup 2011). The use of CGM communicating with pumps and using the 'low glucose suspend' option can significantly improve glucose control, quality of life and reduce risk of severe hypoglycemia (Buckingham et al. 2015; Thabit et al. 2015). Excellent progress is being made with successful clinical trials in the home use of the 'closed loop' system, particularly for nocturnal use. This system includes CGM and an insulin pump with inbuilt control algorithm software for insulin adjustment according to interstitial fluid glucose levels (Tauschmann et al. 2015). All-day closed loop systems with insulin or with insulin and glucagon (van Bon et al. 2014) are also progressing well in clinical trials, so it is expected that fully closed loop insulin pumps (i.e., 'artificial' or 'bionic' pancreases) will be available for clinical use in our lifetime (Battelino et al. 2015; Russell 2015; Malchesky 2015).

Conclusion

Diabetes was known to physicians from about 2500 BC, and the term 'diabetes' was first used by the Greeks about 250 BC, yet it is only since 1922 when Banting, Best, McLeod, Collip and their colleagues first injected insulin into humans that long-term survival for people with T1D has greatly improved, and the lives of those with insulin-requiring T2D or with GDM has improved. In the last few decades, we are still discovering hormones produced by the pancreatic islets. Pleasingly, we also have learnt how to mass-produce and safely use drugs that replace or replicate the actions of the pancreatic islets, improve peripheral insulin sensitivity or modulate gut or renal tract glucose handling to improve glucose control. We have also learnt lifestyle and medical and surgical means to reduce the risk of the many potential acute and chronic complications of diabetes. Whole pancreas or pancreatic islet transplantation is also clinically available to some people with T1D, and this is also an exciting area of ongoing research. While much has been learnt about pancreatic islets and the effects and interactions of the body's various glucose control hormones and mechanisms, much remains to be learnt and translated into clinical practice for even better outcomes for those with or at risk of diabetes.

Acknowledgments Authors are grateful for funding support from the University of Sydney, Sydney Medical School Foundation, and to Associate Professor Anand Hardikar and Dr. Mugdha Joglekar for helpful discussions.

References

Agrawal P et al (2011) Usage and effectiveness of the low glucose suspend feature of the Medtronic Paradigm Veo insulin pump. J Diabetes Sci Technol 5(5):1137–1141

Aiello LP et al (2015) Intensive diabetes therapy and ocular surgery in type 1 diabetes. N Engl J Med 372(18):1722–1733

American Diabetes Association (2012) Diabetes and hypoglycemia: topical and important articles from the American Diabetes Association scholarly journals. American Diabetes Association, Alexandria, 266 p

American Diabetes Association (2015) Practical insulin: a handbook for prescribing providers, 4th edn. American Diabetes Association, Alexandria

Arimura A, Fishback JB (1981) Somatostatin: regulation of secretion. Neuroendocrinology 33(4):246–256

Atkinson MA (2005) ADA Outstanding Scientific Achievement Lecture 2004. Thirty years of investigating the autoimmune basis for type 1 diabetes: why can't we prevent or reverse this disease? Diabetes 54(5):1253–1263

Australian Health Survey: updated results, 2011–2012 (2013) Canberra

Bain SC et al (2003) Characteristics of type 1 diabetes of over 50 years duration (the golden years cohort). Diabet Med 20(10):808–811

Bakhtiani PA et al (2015) A novel, stable, aqueous glucagon formulation using ferulic acid as an excipient. J Diabetes Sci Technol 9(1):17–23

Battaglia M, Atkinson MA (2015) The streetlight effect in type 1 diabetes. Diabetes 64(4):1081–1090

Batterham RL et al (2003) Pancreatic polypeptide reduces appetite and food intake in humans. J Clin Endocrinol Metab 88(8):3989–3992

Battelino T, Omladic JS, Phillip M (2015) Closed loop insulin delivery in diabetes. Best Pract Res Clin Endocrinol Metab 29(3):315–325

Bjornstad P et al (2015) Insulin sensitivity and complications in type 1 diabetes: new insights. World J Diabetes 6(1):8–16

Bliss M (2007) The discovery of insulin, 25th edn. The University of Chicago Press, Chicago

Bowling FL, Rashid ST, Boulton AJ (2015) Preventing and treating foot complications associated with diabetes mellitus. Nat Rev Endocrinol 11(10):606–616

Brownlee M (2001) Biochemistry and molecular cell biology of diabetic complications. Nature 414(6865):813–820

Brownlee M (2005) The pathobiology of diabetic complications: a unifying mechanism. Diabetes 54(6):1615–1625

Bryant NJ, Govers R, James DE (2002) Regulated transport of the glucose transporter GLUT4. Nat Rev Mol Cell Biol 3(4):267–277

Buckingham BA et al (2015) Predictive low-glucose insulin suspension reduces duration of nocturnal hypoglycemia in children without increasing ketosis. Diabetes Care 38(7):1197–1204

Cabrera O et al (2006) The unique cytoarchitecture of human pancreatic islets has implications for islet cell function. Proc Natl Acad Sci USA 103(7):2334–2339

Cheng K, Andrikopoulos S, Gunton JE (2013) First phase insulin secretion and type 2 diabetes. Curr Mol Med 13(1):126–139

Choudhary P et al (2015) Evidence-informed clinical practice recommendations for treatment of type 1 diabetes complicated by problematic hypoglycemia. Diabetes Care 38(6):1016–1029

Chung ST, Haymond MW (2015) Minimizing morbidity of hypoglycemia in diabetes: a review of mini-dose glucagon. J Diabetes Sci Technol 9(1):44–51

Collins AJ et al (2012) United States Renal Data System 2011 Annual Data Report: Atlas of chronic kidney disease and end-stage renal disease in the United States. Am J Kidney Dis, 2012. 59(1 Suppl 1): A7, e1–420

Conrad Stöppler M (2014) Diabetes Prevention 2014. Available from http://www.medicinenet.com/diabetes_prevention/page4.htm

Craig ME et al (2009) Delayed referral of new-onset type 1 diabetes increases the risk of diabetic ketoacidosis. Med J Aust 190(4):219

Cryer PE, American Diabetes Association (2013) Hypoglycemia in diabetes: pathophysiology, prevalence, and prevention, 2nd ed. American Diabetes Association, Alexandria, xi, 236 p

Curry DL, Bennett LL, Grodsky GM (1968) Dynamics of insulin secretion by the perfused rat pancreas. Endocrinology 83(3):572–584

Dabelea D et al (2014) Prevalence of type 1 and type 2 diabetes among children and adolescents from 2001 to 2009. JAMA 311(17):1778–1786

Davis T et al (2015) Automated insulin pump suspension for hypoglycaemia mitigation: development, implementation and implications. Diabetes Obes Metab 17(12):1126–1132

de Onis M, Blossner M, Borghi E (2010) Global prevalence and trends of overweight and obesity among preschool children. Am J Clin Nutr 92(5):1257–1264

Deissler HL, Lang GE (2016) The protein kinase C inhibitor: ruboxistaurin. Dev Ophthalmol 55:295–301

DeSalvo D, Buckingham B (2013) Continuous glucose monitoring: current use and future directions. Curr Diabetes Rep 13(5):657–662

Di Cianni G et al (2003) Intermediate metabolism in normal pregnancy and in gestational diabetes. Diabetes Metab Res Rev 19(4):259–270

Dimitriadis G et al (1997) Effects of glucocorticoid excess on the sensitivity of glucose transport and metabolism to insulin in rat skeletal muscle. Biochem J 321(Pt 3):707–712

Eckel RH et al (2011) Obesity and type 2 diabetes: what can be unified and what needs to be individualized? Diabetes Care 34(6):1424–1430

Eissa N, Ghia JE (2015) Immunomodulatory effect of ghrelin in the intestinal mucosa. Neurogastroenterol Motil 27(11):1519–1527

Facts About Diabetic Eye Disease. 2015 [cited 2015]. Available from https://nei.nih.gov/health/diabetic/retinopathy

Faerch K et al (2014) Cardiovascular risk stratification and management in pre-diabetes. Curr Diab Rep 14(6):493

Farr RJ et al (2015) A comparative analysis of high-throughput platforms for validation of a circulating microRNA signature in diabetic retinopathy. Sci Rep 5:10375

Favaro E et al (2012) The ghrelin gene products and exendin-4 promote survival of human pancreatic islet endothelial cells in hyperglycaemic conditions, through phosphoinositide 3-kinase/Akt, extracellular signal-related kinase (ERK)1/2 and cAMP/protein kinase A (PKA) signalling pathways. Diabetologia 55(4):1058–1070

Felicio JS et al (2015) Present insights on cardiomyopathy in diabetic patients. Curr Diabetes Rev Sep 14. Epub ahead of print

Festa A et al (2004) Differences in insulin resistance in nondiabetic subjects with isolated impaired glucose tolerance or isolated impaired fasting glucose. Diabetes 53(6):1549–1555

Fieker A, Philpott J, Armand M (2011) Enzyme replacement therapy for pancreatic insufficiency: present and future. Clin Exp Gastroenterol 4:55–73

Fonseca VA (2009) Defining and characterizing the progression of type 2 diabetes. Diabetes Care 32(Suppl 2):S151–S156

Fourlanos S et al (2005) Latent autoimmune diabetes in adults (LADA) should be less latent. Diabetologia 48(11):2206–2212

Fourlanos S, Harrison LC, Colman PG (2008) The accelerator hypothesis and increasing incidence of type 1 diabetes. Curr Opin Endocrinol Diabetes Obes 15(4):321–325

Fowler GC, Vasudevan DA (2010) Type 2 diabetes mellitus: managing hemoglobin A(1c) and beyond. South Med J 103(9):911–916

Fox LA, Weber SL, American Diabetes Association (2009) Diabetes 911: how to handle everyday emergencies. American Diabetes Association, Alexandria, viii, 117 p

Fu Z, Gilbert ER, Liu D (2013) Regulation of insulin synthesis and secretion and pancreatic β-cell dysfunction in diabetes. Curr Diabetes Rev 9(1):25–53

Furuhashi M et al (2015) Reduction of serum concentration of FABP4 by sitagliptin, a dipeptidyl peptidase-4 inhibitor, in patients with type 2 diabetes mellitus. J Lipid Res 56(12):2372–2380

Gahete MD et al (2010) Somatostatin and its receptors from fish to mammals. Ann N Y Acad Sci 1200:43–52

Gale EA (2002) The rise of childhood type 1 diabetes in the 20th century. Diabetes 51(12):3353–3361

Gale EA (2006) Dying of diabetes. Lancet 368(9548):1626–1628

Gale EA (2015) Diapedia, the living textbook of diabetes. 2015 [cited 11 November 2015]. Available from www.diapedia.org

Gaw A, Cowan RA, O'Reilly DSJ, Stewart MJ, Shepherd J (1999) Clinical biochemistry an illustrated colour textbook 2nd edition pp 165, Harcourt Brace, Edinburgh

Granata R, Ghigo E (2013) Products of the ghrelin gene, the pancreatic β-cell and the adipocyte. Endocr Dev 25:144–156

Greenbaum CJ (2002) Insulin resistance in type 1 diabetes. Diabetes Metab Res Rev 18(3):192–200

Greenstein B, Wood DF (2011) The endocrine system at a glance, 3rd edn. At a glance series. Wiley-Blackwell, Chichester, 143 p

Gregg EW et al (2014) Changes in diabetes-related complications in the United States, 1990–2010. N Engl J Med 370(16):1514–1523

Gunton JE et al (2014) A new blood glucose management algorithm for type 2 diabetes: a position statement of the Australian Diabetes Society. Med J Aust 201(11):650–653

Guyton AC, Hall JE (1996) Textbook of medical physiology, 9th edn. W.B. Saunders, Philadelphia xliii, 1148 p

Hatanaka H et al (2015) Peripheral oxidative stress markers in diabetes-related dementia. Geriatr Gerontol Int. doi: 10.1111/ggi.12645. Epub ahead of print

Hendrieckx C et al (2014) Severe hypoglycaemia and its association with psychological well-being in Australian adults with type 1 diabetes attending specialist tertiary clinics. Diabetes Res Clin Pract 103(3):430–436

Henquin JC (2009) Regulation of insulin secretion: a matter of phase control and amplitude modulation. Diabetologia 52(5):739–751

Holman RR et al (2008a) 10-Year follow-up of intensive glucose control in type 2 diabetes. N Engl J Med 359(15):1577–1589

Holman RR et al (2008b) Long-term follow-up after tight control of blood pressure in type 2 diabetes. N Engl J Med 359(15):1565–1576

Holst JJ, Vilsboll T, Deacon CF (2009) The incretin system and its role in type 2 diabetes mellitus. Mol Cell Endocrinol 297(1–2):127–136

Horner K, Lee S (2015) Appetite-related peptides in childhood and adolescence: role of ghrelin, PYY, and GLP-1. Appl Physiol Nutr Metab 40(11):1089–1099

Insulin-treated diabetes in Australia 2000–2007, in Diabetes series. 2009. Australian Institute of Health and Welfare, Canberra

International Diabetes Federation (2013) IDF Diabetes Atlas, 6th edn. International Diabetes Federation, Brussels

Jenkins AJ (2015) The world I want—a world with less diabetes. Med J Aust 202(2):3

Jenkins AJ et al (2004a) Lipoproteins, glycoxidation and diabetic angiopathy. Diabetes Metab Res Rev 20(5):349–368

Jenkins AJ et al (2004b) Lipoproteins and diabetic microvascular complications. Curr Pharm Des 10(27):3395–3418

Jenkins AJ, Januszewski AS, O'Neal DN (2015) Addressing vascular risk factors in diabetes. Endocrinol Today 4(4):4

Jenssen T, Hartmann A (2015) Emerging treatments for post-transplantation diabetes mellitus. Nat Rev Nephrol 11(8):465–477

Kaku K et al (2015) Living donor kidney transplantation preceding pancreas transplantation reduces mortality in type 1 diabetics with end-stage renal disease. Transplant Proc 47(3):733–737

Keating ST, El-Osta A (2015) Epigenetics and metabolism. Circ Res 116(4):715–736

Keenan HA et al (2010) Residual insulin production and pancreatic β-cell turnover after 50 years of diabetes: Joslin Medalist Study. Diabetes 59(11):2846–2853

Kilimnik G et al (2012) Quantification of islet size and architecture. Islets 4(2):167–172

Kim A et al (2009) Islet architecture: a comparative study. Islets 1(2):129–136

King GL (2002) Diabetic retinopathy. Diabetes Care 25(Suppl 1):4

Letonja M, Petrovic D (2014) Is diabetic cardiomyopathy a specific entity? World J Cardiol 6(1):8–13

Light J, Tucker M (2013) Simultaneous pancreas kidney transplants in diabetic patients with end-stage renal disease: the 20-yr experience. Clin Transplant 27(3):E256–E263

Lang DA et al (1979) Cyclic oscillations of basal plasma glucose and insulin concentrations in human beings. N Engl J Med 301(19):1023–1027

Lebovitz HE, American Diabetes Association (2009) Therapy for diabetes mellitus and related disorders, 5th edn. American Diabetes Association, Alexandria, vii, 665 p

Lima EG et al (2013) Impact of diabetes on 10-year outcomes of patients with multivessel coronary artery disease in the Medicine, Angioplasty, or Surgery Study II (MASS II) trial. Am Heart J 166(2):250–257

Lindahl JP et al (2014a) In patients with type 1 diabetes simultaneous pancreas and kidney transplantation preserves long-term kidney graft ultrastructure and function better than transplantation of kidney alone. Diabetologia 57(11):2357–2365

Lindahl JP, Jenssen T, Hartmann A (2014b) Long-term outcomes after organ transplantation in diabetic end-stage renal disease. Diabetes Res Clin Pract 105(1):14–21

Liu Y et al (2010) The evolution of somatostatin in vertebrates. Gene 463(1–2):21–28

Liu Q, Wang S, Cai L (2014) Diabetic cardiomyopathy and its mechanisms: role of oxidative stress and damage. J Diabetes Investig 5(6):623–634

Lonovics J et al (1981) Pancreatic polypeptide. A review. Arch Surg 116(10):1256–1264

Lund A et al (2014) Glucagon and type 2 diabetes: the return of the α cell. Curr Diab Rep 14(12):555

Malchesky PS (2015) Artificial organs 2014: a year in review. Artif Organs 39(3):260–287

Manley S (2003) Haemoglobin A1c—a marker for complications of type 2 diabetes: the experience from the UK Prospective Diabetes Study (UKPDS). Clin Chem Lab Med 41(9):1182–1190

Mäkinen V-P et al (2008) Metabolic phenotypes, vascular complications, and premature deaths in a population of 4,197 patients with type 1 diabetes. Diabetes 57(9):2480–2487

Manson JM, Smith RJ, Wilmore DW (1988) Growth hormone stimulates protein synthesis during hypocaloric parenteral nutrition. Role of hormonal-substrate environment. Ann Surg 208(2):136–142

Markakis K, Bowling FL, Boulton AJ (2016) The diabetic foot in 2015: an overview. Diabetes Metab Res Rev 32(Suppl 1):169–78

Mathiyalagan P et al (2015) Epigenetic-mediated reprogramming of pancreatic endocrine cells. Antioxid Redox Signal 22(16):1483–1495

Miegueu P et al (2013) Amylin stimulates fatty acid esterification in 3T3-L1 adipocytes. Mol Cell Endocrinol 366(1):99–107

Monnier VM, Wu X (2003) Enzymatic deglycation with amadoriase enzymes from Aspergillus sp. as a potential strategy against the complications of diabetes and aging. Biochem Soc Trans 31(Pt 6):1349–1353

Morelius E, Theodorsson E, Nelson N (2005) Salivary cortisol and mood and pain profiles during skin-to-skin care for an unselected group of mothers and infants in neonatal intensive care. Pediatrics 116(5):1105–1113

Nankervis AM, HD Moses R, Ross GP, Callaway L, Porter C, Jeffries W, Boorman C, De Vries B, McElduff A (2014) ADIPS Consensus Guidelines for the Testing and Diagnosis of Hyperglycaemia in Pregnancy in Australia and New Zealand Australian Diabetes in Pregnancy Society: Sydney

National Glycohemoglobin Standardization Program 2013 01 November 14]. Available from http://www.ngsp.org/interf.asp and http://www.ngsp.org/factors.asp

Nauck M, Stöckmann F, Ebert R, Creutzfeldt W (1986) Reduced incretin effect in type 2 (non-insulin-dependent) diabetes. Diabetologia 9(1):46–52

Nolan CJ, Prentki M (2008) The islet β-cell: fuel responsive and vulnerable. Trends Endocrinol Metab 19(8):285–291. doi:10.1016/S0140-6736(11)60614-4

Nolan CJ, Damm P, Prentki M (2011) Type 2 diabetes across generations: from pathophysiology to prevention and management. Lancet 378(9786):169–181

O'Connell PJ et al (2013) Multicenter Australian trial of islet transplantation: improving accessibility and outcomes. Am J Transplant 13(7):1850–1858

Okabe J et al (2012) Distinguishing hyperglycemic changes by Set7 in vascular endothelial cells. Circ Res 110(8):1067–1076

O'Neal DN, Jenkins AJ (2009) Metabolic memory in type 1 diabetes. Diabetes Manag 27:18

Polonsky KS (2000) Dynamics of insulin secretion in obesity and diabetes. Int J Obes Relat Metab Disord 24(Suppl 2):S29–S31

Peters AL et al (2013) American Diabetes Association/JDRF Type 1 Diabetes sourcebook. JDRF, American Diabetes Association, Alexandria, xx, 624 p

Pontiroli AE (2015) Intranasal glucagon: a promising approach for treatment of severe hypoglycemia. J Diabetes Sci Technol 9(1):38–43

Poykko SM et al (2003) Low plasma ghrelin is associated with insulin resistance, hypertension, and the prevalence of type 2 diabetes. Diabetes 52(10):2546–2553

Pickup JC (2011) Semi-closed-loop insulin delivery systems: early experience with low-glucose insulin suspend pumps. Diabetes Technol Ther 13(7):695–698

Prazny M (2015) Progress in the development of insulin pumps and their advanced automatic functions. Vnitr Lek 61(4):355, 357, 359–60

Prentki M, Matschinsky FM, Madiraju SR (2013) Metabolic signaling in fuel-induced insulin secretion. Cell Metab 18(2):162–185

Rai U et al (2015) Therapeutic uses of somatostatin and its analogues: current view and potential applications. Pharmacol Ther 152:98–110

Riddle MC et al (2015) Fixed ratio dosing of pramlintide with regular insulin before a standard meal in patients with type 1 diabetes. Diabetes Obes Metab 17(9):904–907

Rhee SY, Kim YS (2015) Peripheral arterial disease in patients with type 2 diabetes mellitus. Diabetes Metab J 39(4):283–290

Rojas JM, Schwartz MW (2014) Control of hepatic glucose metabolism by islet and brain. Diabetes Obes Metab 16(Suppl 1):33–40

Rowe BH et al (2015) Acute management and outcomes of patients with diabetes mellitus presenting to Canadian emergency departments with hypoglycemia. Can J Diabetes 39(Suppl 4):9–18

Russell SJ (2015) Progress of artificial pancreas devices towards clinical use: the first outpatient studies. Curr Opin Endocrinol Diabetes Obes 22(2):106–111

Sakula A (1988) Paul Langerhans (1847–1888): a centenary tribute. J R Soc Med 81(7):414–415

Schmidt S et al (2015) An artificial pancreas for automated blood glucose control in patients with type 1 diabetes. Ther Deliv 6(5):609–619

Seino S, Shibasaki T, Minami K (2011) Dynamics of insulin secretion and the clinical implications for obesity and diabetes. J Clin Invest 121(6):2118–2125

Shiue I (2015) Impaired day-to-day activities worsened but diabetes control improved self-rated health: The UK diabetes survey, 2006. Diabetes Metab 41(6):516–519

Shah VN et al (2014) Closed-loop system in the management of diabetes: past, present, and future. Diabetes Technol Ther 16(8):477–490

Sharma A et al (2015) Therapeutic stimulation of GLP-1 and GIP protein with DPP-4 inhibitors for type-2 diabetes treatment. J Diabetes Metab Disord 14:15

Shaw J, Tanamas S. (2012) Diabetes: the silent pandemic and its impact on Australia. Baker IDI Heart and Diabetes Institute, Melbourne, p 27 https://static.diabetesaustralia.com.au/s/fileassets/diabetes-australia/e7282521-472b-4313-b18e-be84c3d5d907.pdf accessed May 2016

Steele C et al (2004) Insulin secretion in type 1 diabetes. Diabetes 53(2):426–433

Sujishi T et al (2015) Sitagliptin can inhibit the development of hepatic steatosis in high-fructose diet-fed ob/ob mice. J Clin Biochem Nutr 57(3):244–253

Sun J et al (2015) Pancreatic β-cells limit autoimmune diabetes via an immunoregulatory antimicrobial peptide expressed under the influence of the gut microbiota. Immunity 43(2):304–317

Tauschmann M, Hovorka R (2014) Insulin pump therapy in youth with type 1 diabetes: toward closed-loop systems. Expert Opin Drug Deliv 11(6):943–955

Tallapragada DSP, Bhaskar S, Chandak GR (2015) New insights from monogenic diabetes for “common” type 2 diabetes. Front Genet 6:251

Taniguchi CM, Emanuelli B, Kahn CR (2006) Critical nodes in signalling pathways: insights into insulin action. Nat Rev Mol Cell Biol 7(2):85–96

Tauschmann M et al (2015) Factors associated with glycemic control during free-living overnight closed-loop insulin delivery in children and adults with type 1 diabetes. J Diabetes Sci Technol 9(6):1346–1347

Thabit H et al (2015) Unsupervised overnight closed loop insulin delivery during free living: analysis of randomised cross-over home studies in adults and adolescents with type 1 diabetes. Lancet 385(Suppl 1):S96

Thomas MC (2014) Glycemic exposure, glycemic control, and metabolic karma in diabetic complications. Adv Chronic Kidney Dis 21(3):311–317

Templeton MP-C I (2008) Gestational diabetes mellitus in Australia, 2005–06. In: Diabetes series. Australian Institute of Health and Welfare, Canberra

The Diabetes Control and Complications Trial Research Group (1998) Effect of intensive therapy on residual β-cell function in patients with type 1 diabetes in the diabetes control and complications trial. A randomized, controlled trial. Ann Intern Med 128(7):517–523

Tiano JP et al (2011) Estrogen receptor activation reduces lipid synthesis in pancreatic islets and prevents β cell failure in rodent models of type 2 diabetes. J Clin Invest 121(8):3331–3342

Tomita T (2011) Immunocytochemical staining for islet amyloid polypeptide in pancreatic endocrine tumors. Islets 3(6):344–351

Tran MA, Dang TL, Berlan M (1981) Effects of catecholamines on free fatty acid release from bone marrow adipose tissue. J Lipid Res 22(8):1271–1276

Tran L et al (2015) Pharmacologic treatment of type 2 diabetes: injectable medications. Ann Pharmacother 49(6):700–714

Tuttle KR et al (2015) Effect of ruboxistaurin on albuminuria and estimated GFR in people with diabetic peripheral neuropathy: results from a randomized trial. Am J Kidney Dis 65(4):634–636

Use of Glycated Haemoglobin (HbA1c) in the Diagnosis of Diabetes Mellitus (2011) Abbreviated report of a WHO consultation. WHO, Geneva

Valente-Silva P et al (2015) Ketone bodies effectively compete with glucose for neuronal acetyl-CoA generation in rat hippocampal slices. NMR Biomed 28(9):1111–1116

van Bon AC et al (2014) Feasibility of a portable bihormonal closed-loop system to control glucose excursions at home under free-living conditions for 48 hours. Diabetes Technol Ther 16(3):131–136

Vega-Monroy ML, Fernandez-Mejia C (2011) Beta-cell function and failure in type 1 diabetes, type 1 diabetes-pathogenesis, genetics and immunotherapy, Prof. David Wagner (Ed.), InTech, doi: 10.5772/22089. Available from: http://www.intechopen.com/books/type-1-diabetes-pathogenesis-genetics-andimmunotherapy/beta-cell-function-and-failure-in-type-1-diabetes accessed May 2016

Versini M et al (2015) Unraveling the hygiene hypothesis of helminthes and autoimmunity: origins, pathophysiology, and clinical applications. BMC Med 13:81

Vinik AI et al (2003) Diabetic autonomic neuropathy. Diabetes Care 26(5):1553–1579

Weinhaus AJ et al (2007) Regulation of glucokinase in pancreatic islets by prolactin: a mechanism for increasing glucose-stimulated insulin secretion during pregnancy. J Endocrinol 193(3):367–381

West AC, Johnstone RW (2014) New and emerging HDAC inhibitors for cancer treatment. J Clin Investig 124(1):30–39

Westermark P et al (1987a) Islet amyloid in type 2 human diabetes mellitus and adult diabetic cats contains a novel putative polypeptide hormone. Am J Pathol 127(3):414–417

Westermark P et al (1987b) Amyloid fibrils in human insulinoma and islets of Langerhans of the diabetic cat are derived from a neuropeptide-like protein also present in normal islet cells. Proc Natl Acad Sci USA 84(11):3881–3885

Westermark P, Andersson A, Westermark GT (2011) Islet amyloid polypeptide, islet amyloid, and diabetes mellitus. Physiol Rev 91(3):795–826

Westwell-Roper CY, Ehses JA, Verchere CB (2014) Resident macrophages mediate islet amyloid polypeptide-induced islet IL-1β production and β-cell dysfunction. Diabetes 63(5):1698–1711

White NH et al (2008) Prolonged effect of intensive therapy on the risk of retinopathy complications in patients with type 1 diabetes mellitus: 10 years after the Diabetes Control and Complications Trial. Arch Ophthalmol 126(12):1707–1715

White NH et al (2014) Intravenous glucagon in a deliberate insulin overdose in an adolescent with type 1 diabetes mellitus. Pediatr Diabetes

Wilcox G (2005) Insulin and insulin resistance. Clin Biochem Rev 26(2):19–39

Wong J et al (2008a) Timing is everything: age of onset influences long-term retinopathy risk in type 2 diabetes, independent of traditional risk factors. Diabetes Care 31(10):1985–1990

Wong J et al (2008b) Different accelerators to early-onset type 2 diabetes: a comparison of Anglo-Celtic and Chinese patients. J Diabetes Complicat 22(6):389–394

Wu SC et al (2007) Foot ulcers in the diabetic patient, prevention and treatment. Vasc Health Risk Manag 3(1):65–76

Young BA et al (2003) Effects of ethnicity and nephropathy on lower-extremity amputation risk among diabetic veterans. Diabetes Care 26(2):495–501

Zhang Y et al (2015) In vitro and in vivo sustained release of exenatide from vesicular phospholipid gels for type II diabetes. Drug Dev Ind Pharm, pp 1–8

Zhao S et al (2014) α/β-Hydrolase domain-6-accessible monoacylglycerol controls glucose-stimulated insulin secretion. Cell Metab 19(6):993–1007

Zigman JM, Bouret SG, Andrews ZB (2015) Obesity impairs the action of the neuroendocrine ghrelin system. Trends Endocrinol Metab 27(1):54–63

Chapter 2
Comparative Analysis of Islet Development

Ananta Poudel, Omid Savari, Zehra Tekin and Manami Hara

Introduction

Islet cells play a critical role in glucose homeostasis. The islet is a highly vascularized microorgan embedded in the exocrine pancreas, which mainly consists of endocrine hormone-secreting cells: beta cells (insulin), alpha cells (glucagon), delta cells (somatostatin), pancreatic polypeptide (PP) cells, and epsilon cells (ghrelin). Here, we review the evolutionary changes followed by a molecular hierarchy of genes involved in pancreas development and a model of islet formation that recent technological advances have made possible. Considering current concerns, species differences between humans and rodents are discussed in detail.

Fetal Endocrine Cell Development in Various Species

The development and organization of hormone-producing cells in the pancreatic islets of Langerhans is important and essential for normal islet function. Fetal pancreas and endocrine cell development is reviewed in this chapter, starting from vertebrates to large mammals.

A. Poudel · M. Hara (✉)
Department of Medicine, The University of Chicago,
5841 South Maryland Avenue, MC1027, Chicago, IL, USA
e-mail: mhara@uchicago.edu

O. Savari · Z. Tekin
Departments Surgery, The University of Chicago,
5841 South Maryland Avenue, MC1027, Chicago, IL, USA

A.A. Hardikar (ed.), *Pancreatic Islet Biology*, Stem Cell Biology
and Regenerative Medicine, DOI 10.1007/978-3-319-45307-1_2

Zebrafish

Zebrafish is a useful model for the study of vertebrate pancreas organogenesis, with advantages of external fertilization, large progeny sizes, rapid development, and optical clarity of its embryos (Driever et al. 1996; Ober et al. 2003; Ward et al. 2007; Tehrani and Lin 2015).

The zebrafish pancreas develops from the posterior foregut endoderm (Field et al. 2003). Origin of endocrine and exocrine structures is uniquely separated in zebrafish. While dorsal bud exclusively gives rise to endocrine pancreas, ventral bud forms mostly exocrine tissue besides very little endocrine cells. The posterior-dorsal bud evanginates by 24 h post-fertilization (hpf), and the antero-ventral bud is formed by 32–40 hpf. The two buds fuse by 52 hpf to form a single primary islet consisting of beta, alpha, delta and epsilon cells (Field et al. 2003; Argenton et al. 1999; Biemar et al. 2001). However, these early endocrine cells have limited proliferation and ultimately contribute very little to the adult endocrine cell mass (Hesselson et al. 2009). Ventral bud-derived endocrine precursors migrate in a dorsal and posterior direction to envelop the principal islet. After 120 hpf, ventral bud-derived beta cells expand forming secondary islets, including PP cells (Li et al. 2009; Wang et al. 2011).

Xenopus

In Xenopus, the dorsal rudiment evanginates at stage 35/36 (2 days, 2 h), followed by appearance of two small ventral rudiments at stage 37/38 (2 days, 5.5 h) (Nieuwkoop and Faber 1967; Maake et al. 1998; Kelly and Melton 2000). Dorsal and ventral anlagen fuse by stage 41 (3 days, 4 h). Clusters of beta cells and singly scattered alpha cells appear at stage 41, followed by delta cells at stage 43 and PP cells at stage 46 (Maake et al. 1998). In late tadpoles at stage 54 (26 days), segregation of beta cells and alpha cells is observed where beta cells form large groups of cells distributed throughout the pancreas, whereas small clusters of alpha cells reside in the periphery (Horb and Slack 2002). This arrangement is similarly observed in the frog pancreas. During metamorphic climax, the beta cell clusters increase in size by aggregation of pre-existing beta cells following exocrine remodeling (Pearl et al. 2009).

Guinea Pig

The gestation period for guinea pig is longer than other rodents (63 days). All four endocrine cells are observed at E25, where alpha cells are the most abundant, followed by PP and beta cells (Petersson 1966; Reddy et al. 1985, 1992). Endocrine cells are dispersed throughout primitive tubules where beta cell clusters are observed occasionally. Only a few delta cells are present. Between E35 and E40, endocrine cells are prominent in both the ductules and islet-like structures. As islets develop, beta cell mass increase and alpha cells segregate to the periphery.

Pig

The gestation period for pigs is ~114 days. Pancreas develops from ventral and dorsal buds. By embryonic day 19, all four endocrine cells are detected with no ducts or acini in the pancreatic primordial (Alumets et al. 1983; Zabel et al. 1995; Carlsson et al. 2010). Relative mass of beta cells to alpha cells increases gradually from days 25–34 (more alpha cells), ~52 (equal mass), to 60 (more beta cells). Delta cells first appear at day 31, but are rarely observed until day 89 (Carlsson et al. 2010). No islet formation, with a core of beta cells and other endocrine cells in the periphery, is observed before birth (Alumets et al. 1983).

Alligator

In *Alligator mississippiensis*, pancreas arises from dorsal and ventral pancreatic buds that develop at stages 8 and 13 (embryonic days 8–13), respectively (Jackintell and Lance 1994). Two anlagen fuse at 14th stage and become similar to adult pancreas at stage 17 (days 22–23). Number of cells in clusters increases during later stages, and pancreas become fully mature at stage 27 (days 60–63). In alligator and crocodile, alpha and beta cell development precedes by delta and PP cells, similarly in rat and mouse. Alpha cells appear as early as stage 10 (day 11.5), followed by beta cells at stage 11 (day 12.5), delta cells at stage 13 (day 15) and PP cells at stage 17 (day 22.5). Alpha cells are dominant during early stages, and then, uniquely delta cells become most abundant. This dominance of delta cells is also observed in American alligator and adult Nile crocodile and caiman (Rhoten 1987). The vertebrate endocrine pancreas shows an evolutionary trend that the dorsal pancreas contains larger islets as observed in some teleosts, crocodilians and birds (Epple and Brinn J 1987).

Cattle

In cattle, alpha cells are the first endocrine cells that appear at 26 d.p.c in dorsal pancreatic primordium and duodenum, followed by beta cells at 27 d.p.c and delta cells at 45 d.p.c (Carlsson et al. 2010). These endocrine cells are singly scattered until 50 d.p.c, and thereafter, beta and alpha cells start forming small clusters. At this stage, alpha cells are still dominant. Between 89 and 105 d.p.c, small groups of endocrine cells assemble into larger clusters. The number of beta cells increases and exceeds alpha cells after 110 d.p.c. Islet architecture resembles rodents with a beta cell core with alpha and delta cells in periphery.

Buffalo

The gestation period of buffalo is 308–312 days. Beta, alpha and delta cells are first observed as scattered small clusters in 2nd month, followed by PP cell one month later (Lucini et al. 1998). By 8th month, islets appear to be formed with a beta cell core with alpha and delta cells in periphery. Large clusters of islets observed in early fetuses decreases in number with the 1st year after birth and become absent in 4- to 5-year-old animals.

Genes Involved in Endocrine Cell Development

The transgenic mouse technology has advanced our understanding of endocrine cell development. These mouse studies have elucidated genes that have defects, which result in a wide range of malformation in the pancreas and islet structure.

Pancreatic duodenal homeobox 1 (Pdx1) is an essential transcription factor that determines pancreas formation. The early expression of Pdx1 occurs in pancreatic progenitor cells around E8.5 in mice (Guz et al. 1995). Pdx1 null mice completely lack pancreas and die few days after birth (Jonsson et al. 1994). Pancreatic agenesis in humans caused by a homozygous point deletion of the gene (termed insulin promoter factor (IPF)-1 in humans) has also been reported (Stoffers et al. 1997). PTF1-p48 (ptf1a) basic helix–loop–helix transcription factor (bHLH) is involved in exocrine cell differentiation (Krapp et al. 1998; Kawaguchi et al. 2002). PTF1-p48 knockout mice lacked exocrine pancreas and succumbed to neonatal death. Endocrine cells were localized in intra-mesenterial ducts at E12, which then migrated to the spleen around E16-E18. These cells were singly scattered in the spleen and did not form islets. Another bHLH transcription factor neurogenin (Ngn) 3 is involved in the formation of all endocrine cell types. During development, Ngn3-expressing cells are detected as early as E 9.5, and expression level peaks at E15.5 and declines thereafter (Gradwhol et al. 2000; Schwitzgebel et al. 2000). NeuroD is a bHLH transcription factor involved in islet formation and organization. NeuroD null mice died 3–5 days after birth and showed a marked reduction in beta cells. Islet morphogenesis was arrested in mid- to late embryonic period, and endocrine cells only formed small clusters (Naya et al. 1997; Miyata et al. 1999). NKx2.2 is a homeobox transcription factor, which is expressed in beta, alpha and PP cells but not in delta cells (Sussel et al. 1998). Nkx2.2-deficient mice developed severe hyperglycemia and died shortly after birth, mainly due to the absence of beta cells. Paired homeobox protein Pax4 is expressed in the early pancreas, but is later restricted to beta cells (Sosa-Pineda et al. 1997). Inactivation of Pax4 resulted in the absence of beta and delta cells and mice died within 3 days after birth. Arx-deficient mice developed hypoglycemia and died 2 days after birth with a loss of alpha cells concomitant with increased beta and delta cells (Collombat et al. 2003). The study further showed the opposite action of Arx and Pax4. Mice with compound deficiency of Arx and Nkx2.2 lacked alpha and beta cells with hyperplasia of epsilon cells (Kordowich et al. 2011). Mice lacking the glucagon receptor (GcgR) exhibited lower blood glucose levels, hyperglucagonemia and alpha and delta cell hyperplasia (Gelling et al. 2003; Vuguin et al. 2006). Glucagon-like peptide (GLP)-1 and GLP-1 amide content was increased in the pancreas as well as circulation (Gelling et al. 2003). GcgR null mice displayed impaired insulin secretion (Sørensen et al. 2006). However, blocking glucagon signaling led to improved insulin action as well as resistance to diet-induced obesity and streptozotocin-mediated beta cell loss and hyperglycemia (Sørensen et al. 2006; Conarello et al. 2007). Prohormone convertase 2 (PC2)-deficient mice similarly showed alpha and delta cell hyperplasia starting around 3 months of age (Furuta et al. 1997) and delayed beta cell differentiation (Vincent et al. 2003). These mice were further examined in the long term up to 18 months to see whether sustained glucagon deficiency would lead to islet tumorigenesis (Tehrani and Lin 2015). PC2-deficient mice displayed marked changes in islet morphology from alpha cell hypertrophy/hyperplasia to adenomas and carcinomas in 6–8 months (Jones et al. 2014). POU domain transcription factor Brn4 regulates the expression of proglucagon gene by its interaction with G1 promoter element (Hussain et al. 1997). However, Brn4 knockout mice showed completely normal appearances of all endocrine cells and transcription factors analyzed (Heller et al. 2004). Paired-box protein Pax6 is expressed in beta, alpha, delta and PP cells (St-Onge et al. 1997). Pax6 null mice lacked alpha cells. Double mutants lacking both Pax6 and Pax4 failed to develop any endocrine cells. In embryos homozygous for a mutant allele of the *Pax6* gene, *Small eye* (*Sey*^{Neu}), all four types of pancreatic endocrine cells were decreased, however, at the lesser degree compared to Pax6 null mice (Sander et al. 1997). Disruption of homeobox gene Nkx6.1 led to a loss of beta cell precursors, which profound defect was observed after the start of the secondary transition (Sander et al. 2000). Double-knockout mice of Nkx2.2 and Nkx6.1 had similar phenotype as Nkx2.2 mice, indicating that Nkx6.1 lies downstream of Nkx2.2 in the pathway of beta cell formation. Islet-1 (Isl1) is a LIM homeodomain transcription factor, which is expressed in mature pancreatic endocrine cells as well as in mesenchymal cells surrounding the dorsal bud. Isl1 null embryos exhibited no dorsal pancreatic mesenchyme with associated failure of exocrine cell differentiation in the dorsal but not in the ventral pancreas (Ahlgren et al. 1997). There was also a complete loss of differentiated islet cells, suggesting a dual role of Isl1 in the pancreas. MafA is a basic leucine zipper transcription factor that functions as a potent transactivator for the insulin gene (Matsuoka et al. 2003). MafA null mice showed glucose intolerance and developed diabetes with diminishing beta cell mass as they aged (Zhang et al. 2005).

Islet Formation

Pancreatic islets are formed in various sizes including single and small clusters of endocrine cells, which are unevenly distributed in the exocrine pancreas in mammals. Despite differences in the body/pancreas size and total beta cell mass, the islet size distribution falls into a similar range throughout various species, suggesting certain regulatory mechanisms of the islet size to maintain proper functional properties (Samols et al. 1986; Weir and Bonner-Weir 1990). Little is known about how pancreatic islets are formed during development in mammals. The widely accepted model of islet formation is by the local aggregation of endocrine cells that migrate from the ductal epithelium in the late embryonic stage (Pictet and Rutter 1972; Herrera et al. 1991; Bouwens and De 1996; Jensen 2004). This aggregation model is based on observations using pancreatic tissue sections from rats and mice. Two-dimensional analysis using thinly cut sections has certain limitations and could only capture part of larger structures (Hara et al. 2006), which potentially hampers deducing dynamic islet formation in the pancreas. Using transgenic mice with fluorescent-tagged beta cells combined with an automated computational analysis, we have proposed a new model for islet formation that occurs by a process of fission during neonatal development following contiguous endocrine cell proliferation (Miller et al. 2009). This model accounts for the morphological transformation from embryonic endocrine cord-like structures into distinct spherical islets of various sizes observed in the adult pancreas (Fig. 2.1). The entire distribution of beta cells in the intact pancreas at P7 shows ventral (v) and dorsal (d) pancreatic regions (Fig. 2.1A.a). Stretches of interconnected islets were identified, which are color-coded in *blue* together with spherical-shaped islets (green) and small clusters of beta cells (<10 cells, red; Fig. 2.1A.b). A closer view of the interconnected islets is shown in Fig. 2.1B. Such islet clustering is often found along large blood vessels (Fig. 2.1B.c). Immunohistochemical staining of insulin and glucagon depicts rows of alpha cells spanning each islet-like mass within a continuous elongated structure (Fig. 2.1C). Acinar cells surrounding these interconnected islet structures, which expand at a high rate during development, may also play a role in islet fission. A fission model of islet formation in the neonatal pancreas is detailed in Fig. 2.1D. Islet formation in the neonatal pancreas may occur by fission of elongated structures composed of beta cells and surrounding alpha cells, following the contiguous proliferation and branching of endocrine cells into cord-like structures in the fetal and newborn pancreas. The fission process appears to be random, which may explain the diversity of islet sizes seen in the adult pancreas. Islets including small clusters are coated with a layer of extracellular matrix (Fig. 2.1D.c) that stabilizes the structure. Beta cell mass expansion within an islet leads to an increase in islet volume and the formation of spherical-shaped islets with a reduced alpha cell ratio. This process of islet formation may also produce small isolated clusters of endocrine cells that persist throughout a lifetime.

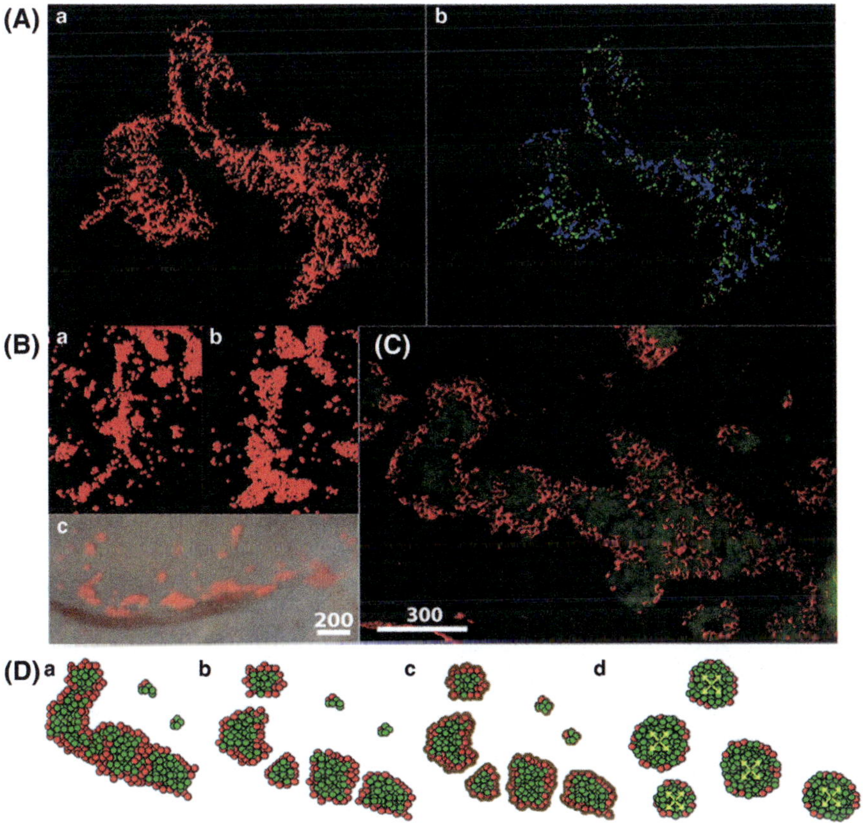

Fig. 2.1 Islet formation during neonatal development in mice. **A a** The entire distribution of beta cells in the intact pancreas at P7 (*d* dorsal, *v* ventral). **b** Three distinct populations of beta cell clusters are *color-coded. Blue* interconnected islets, *green spherical-shaped* islets and *red* small clusters of beta cells (<10 cells). **B** Interconnected islets. **a** P7; **b** P12; **c** a stretch of interconnected islets in the neonatal pancreas (P10). Merged image of fluorescent and bright field images of interconnected islets. Scale bar is 200 μm. **C** Immunohistochemical analysis of coalescent islets (P13). Insulin (*green*) and glucagon (*red*) staining is shown. *Scale bar* is 300 μm. **D** A fission model of islet formation in the neonatal pancreas. **a** Endocrine cells proliferate contiguously, forming branching cord-like structures in the fetal and newborn pancreas. **b** Islet formation in the neonatal pancreas may occur by fission of elongated structures composed of beta cells and surrounding alpha cells. The fission process appears to be random, producing islets of different size. **c** Each islet is subsequently coated with a layer of extracellular matrix that stabilizes the structure. This process of islet formation may also result in small isolated clusters of pancreatic endocrine cells that persist even in the adult pancreas. **d** Beta cell mass expansion within an islet leads to an alpha cell ratio of 5–10 %. Reproduced from (Miller et al. 2009)

Regional Differences in Islet Distribution and Composition in the Human Pancreas

The large size of the human pancreas challenges unbiased quantitative analyses that require a practical stereological approach. There is marked heterogeneity within an individual, where islet distribution/density is relatively low in the head and gradually increases through the body toward the tail region by >twofold (Wang et al. 2013). The head of human pancreas has unique characteristics in early development as well as the anatomical disposition that may lead to a preferential loss of beta cells in patients with type 2 diabetes and the development of pancreatic cancer (Savari et al. 2013). The head originates from the ventral pancreatic bud and exclusively contains the pancreatic polypeptide (PP) cell-rich area. We have shown that the PP cell-rich region is more narrowly segregated than previously reported (Adrian et al. 1982; Hazelwood 1993; Asakawa et al. 2003; Wierup et al. 2002; Andralojc et al. 2009) and largely restricted to the uncinate process (Wang et al. 2013). Regional distribution of the total endocrine cell area from a 50-year-old male is plotted from the head–body–tail region, which shows a narrowly restricted PP cell-rich area that does not cover the entire head region (Fig. 2.2A). In the inset, a deduced PP cell-rich region in the pancreas specimen (left) and in vivo (right) is illustrated. The PP cell-rich region contained a large number of irregularly shaped structures as well as islets with PP cells in the periphery (Fig. 2.2B.a), which are masked when only three major hormones are stained (adjacent section in Fig. 2.2B.b). Intra-specimen comparison of the PP cell-rich and PP cell-poor regions (Fig. 2.2C.a) revealed significantly reduced beta and alpha cell areas in the PP cell-rich region (Fig. 2.2C.b). The analysis on PP cell-rich and PP cell-poor areas in the same section (shown in Fig. 2.2C.c, d) highlights the differences in the distribution of islets/clusters and cellular composition.

Islet Plasticity in Mice and Humans

Human islets exhibit distinct islet architecture particularly in large islets that comprise of a relatively abundant fraction of alpha cells mixed with beta cells, whereas mouse islets show largely similar architecture of a beta cell core with alpha cells in the periphery. (Brissova and Fowler 2005; Cabrera et al. 2006; Kharouta et al. 2009). In human islets, islet architecture is size dependent. This size dependency of endocrine cellular composition in humans is quantitatively analyzed in comparison with mice (Fig. 2.3). Islet sizes are in a wide range from a single endocrine cell to a large islet consisting of several thousand cells. In mouse islets, the beta cell is the major component of islets throughout the size distribution (Fig. 2.3a, left). However, in human islets, while small islets show similar cellular composition with mice, in larger islets (>60 μm in diameter), the fraction of alpha and delta cells increases (Fig. 2.3a, right). These large islets typically exhibit relatively

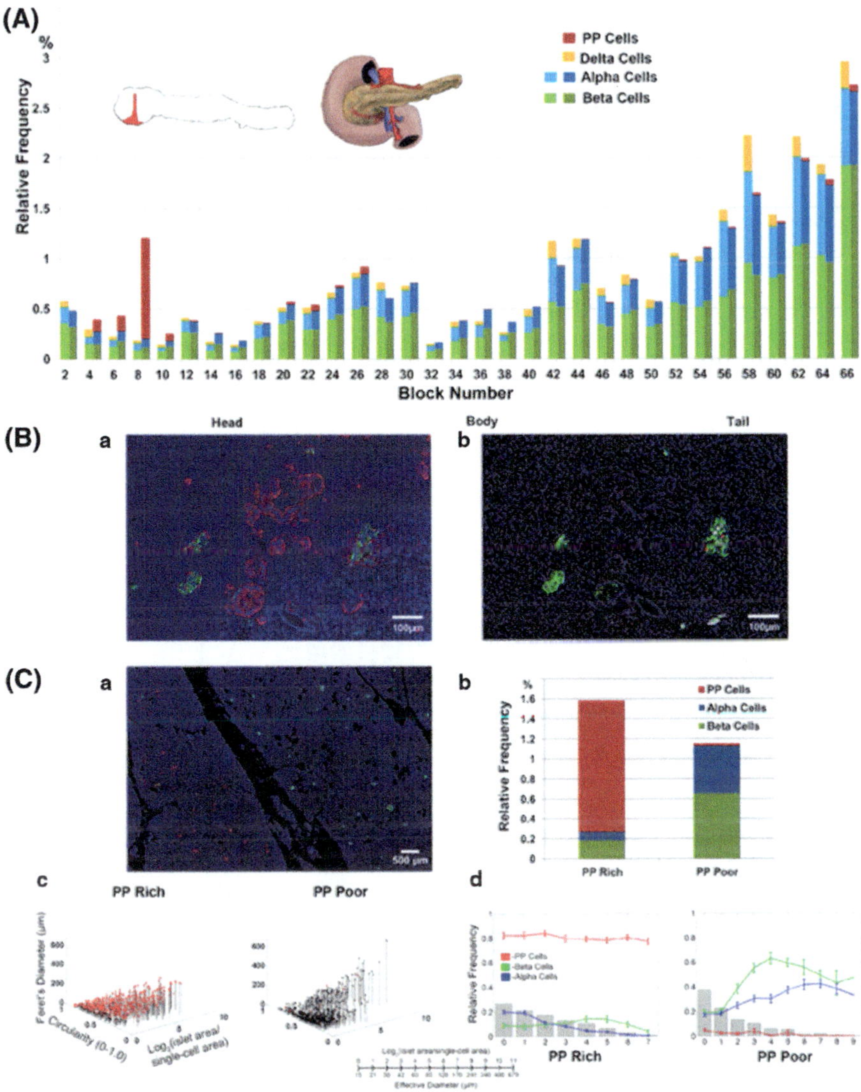

Fig. 2.2 Whole pancreas analysis of the PP cell distribution. **A** Regional distribution of PP, beta, alpha and delta cell mass. *Inset* (*left*) Restricted PP cell-rich area illustrated in *red*. (*Right*) deduced PP cell-rich region in vivo. **B a** Representative view of islets in the PP cell-rich area (PP in *red*, insulin in *green*, glucagon in white and nuclei in *blue*). **b** Adjacent section immunostained for insulin (*green*), glucagon (*red*), somatostatin (*white*) and nuclei (*blue*). **C** Intra-specimen comparison of the PP cell-rich and PP cell-poor regions. **a** A clear boundary between the PP cell-rich region (area in the *left*) and PP cell-poor region (*right*). **b** Total endocrine cell area in each region. **c** Three-dimensional plot of individual islet/cluster with PP cell containing clusters in *red*. PP cell-rich area (*left*) and PP cell-poor area (*right*). **d** Islet size distribution and cellular composition. PP cell-rich area (*left*) and PP cell-poor area (*right*). Reproduced from (Wang et al. 2013)

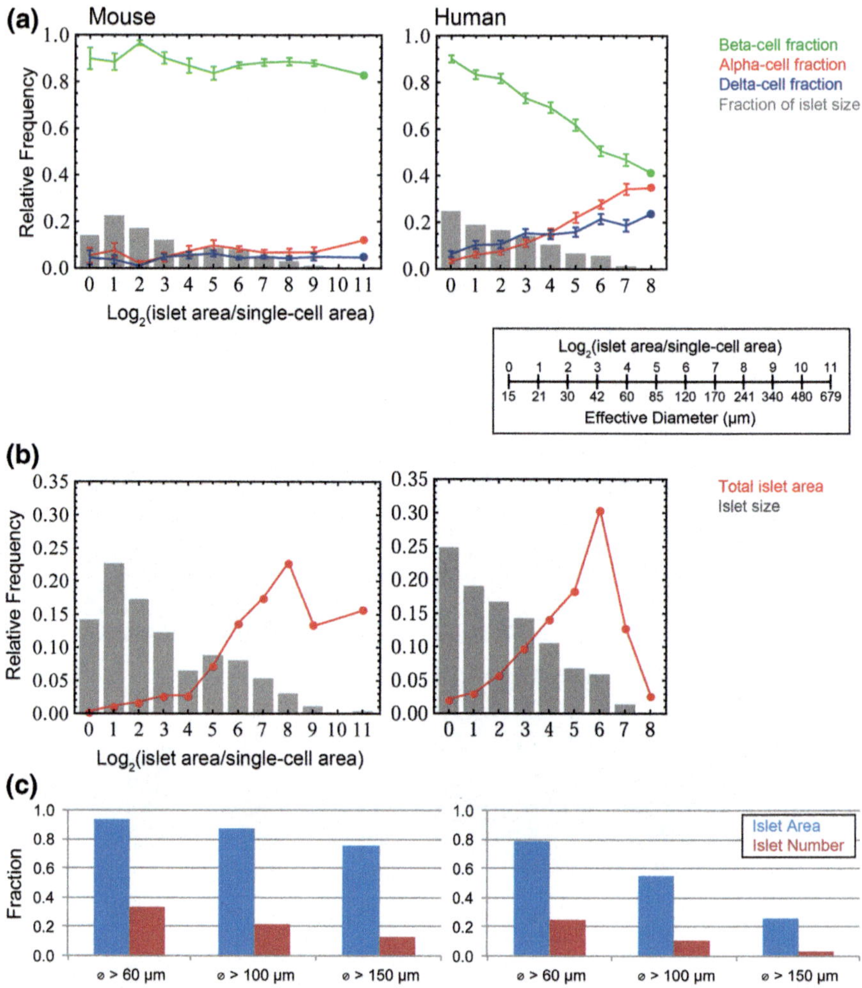

Fig. 2.3 Islet size-dependent changes in endocrine cellular composition in human islets. **a** *left*, mouse pancreas (a CD-1 female mouse at 3 month). Relative frequency of islet size (*gray bar*) and ratios of alpha (*red*), beta (*green*) and delta (*blue*) cells within islets are plotted against islet size; means ± SEM. Note that islet size is presented as a logarithmic scale considering the high number of small islets and the low number of large islets. In addition, islet area is divided by the single-cell area (178 μm^2; Hara et al. 2003) to make them as dimensionless values representing the number of cells in a given islet area. See the conversion between logarithmic islet area (logarithmic) and effective diameter (μm). *Right* human pancreas. **b** Fraction of islet size distribution (*gray bar*) and fraction of total islet area (*red line*). **c** The contribution of large islets to the total endocrine cell area is plotted with the cutoff points in islet size of >60, 100 and 150 μm in diameter (*left* mouse; *right* human corresponding to b. Reproduced from (Kilimnik et al. 2012)

intermingled architecture of beta and non-beta cells. (Kharouta et al. 2009; Kim et al. 2009) Note that such changes in large islets are not an intrinsic characteristic of human islets, but are also observed in mice under insulin resistance such

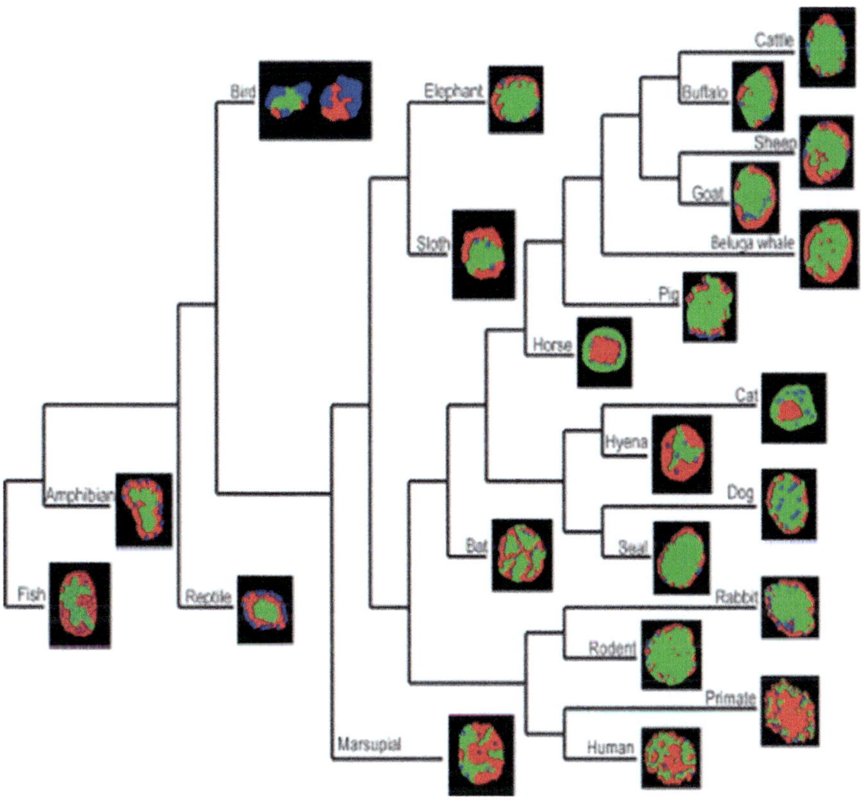

Fig. 2.4 Multi-species comparison of islet structure and composition. Islets from various species are organized into a phylogenetic tree. Representative islets are pseudo-colored models of actual islets based on immunohistochemical images composed of alpha cells (*red*), beta cells (*green*) and delta cells (*blue*). The following species are shown: mouse (*Mus musculus*); human (*Homo sapiens sapiens*); rhesus macaque (*Macaca mulatta*); cat (*Felis domesticus*); dog (*Canis lupus familiaris*); fur seal (*Arctocephalus pusillus*); Egyptian fruit bat (*Rousettus aegyptiacus*); horse (*Equus ferus caballus*); cattle (*Bos taurus*); domestic goat (*Capra aegagrus hircus*); domestic sheep (*Ovis aries*); domestic pig (*Sus domesticus*); European rabbit (*Oryctolagus cuniculus*); striped hyena (*Hyaena hyaena*); water buffalo (*Bubalus bubalis*); beluga whale (*Delphinapterus leucas*); African elephant (*Loxodonta africana*); three-toed sloth (*Bradypus variegatus*); common brushtail opossum (*Trichosurus vulpecula*); zebra finch (*Taeniopygia guttata*); European asp (*Vipera aspis*); northern leopard frog (*Rana pipiens*) and rainbow trout (*Salmo gairdneri*). Note that high-quality images were not available for the echidna and camel. No information on delta cells was available for the beluga whale. Reproduced from (Kim et al. 2009)

as pregnancy, obesity, diabetes and inflammation (Kim et al. 2009). Figure 2.3b shows the relative contribution of each bin of small to large islets to total endocrine cell area (left: mouse and right: human corresponding to Fig. 2.3a). The greater number of endocrine clusters and small islets (gray bars) does not markedly share the total area (plotted in a red line), but the fewer number of large islets mainly comprises the islet mass. The contribution of large islets to the total endocrine cell area is highlighted in Fig. 2.3c with the cutoff points in islet size of >60, 100 and 150 mm in diameter (left: mouse and right: human corresponding to Fig. 2.3b).

Islet Plasticity in Various Species

Lastly, there is great diversity of islet structure among various species as shown in Fig. 2.4 (Steiner et al. 2010). Most of species generally show rodent-like islet architecture with a core of beta cells surrounded by alpha and delta cells in the periphery. However, this arrangement is not "the default," since horses and cats conversely have islets with a core of alpha cells accompanied with a thick mantle of beta cells. In line with diverse islet plasticity, Nyman et al. have elegantly shown that blood flow is not coupled to the distinct islet cellular arrangement (Nyman et al. 2008).

Conclusion

Emerging studies have demonstrated the fundamental species differences between rodents and humans from morphology to molecular mechanisms of islet development. Challenges in studying the human pancreas are (1) inter-individual variability in endocrine cell mass; (2) intra-individual variability in endocrine cell mass (i.e., regional differences); and (3) intra-islet variability reflecting islet plasticity. It is critical to have a better understanding of human islet development, which has an important implication for the development of therapeutic surrogate beta cells for a cure for diabetes.

References

Adrian TE, Mitchenere P, Sagor G et al (1982) Effect of pancreatic polypeptide on gallbladder pressure and hepatic bile secretion. Am J Physiol 243:G204–G207

Ahlgren U, Pfaff SL, Jessell TM et al (1997) Independent requirement for ISL1 in formation of pancreatic mesenchyme and islet cells. Nature 385:257–260

Alumets J, Hakanson R, Sundler F (1983) Ontogeny of endocrine cells in porcine gut and pancreas, An immunocytochemical study. Gastroenterology 85:1359–1372

Andralojc KM, Mercalli A, Nowak KW et al (2009) Ghrelin-producing epsilon cells in the developing and adult human pancreas. Diabetologia 52:486–493

Argenton F, Zecchin E, Bortolussi M (1999) Early appearance of pancreatic hormone expressing cells in the zebrafish embryo. Mech Dev 87:217–221

Asakawa A, Inui A, Yuzuriha H et al (2003) Characterization of the effects of pancreatic polypeptide in the regulation of energy balance. Gastroenterology 124:1325–1336

Biemar F, Argenton F, Schmidtke R et al (2001) Pancreas development in zebrafish: early dispersed appearance of endocrine hormone expressing cells and their convergence to form the definitive islet. Dev Biol 230:189–203
Bouwens L, De BE (1996) Islet morphogenesis and stem cell markers in rat pancreas. J Histochem Cytochem 44:947–951
Brissova M, Fowler MJ (2005) Nicholson WE (2005) Assessment of human pancreatic islet architecture and composition by laser scanning confocal microscopy. J Histochem Cytochem 53:1087–1097
Cabrera O, Berman M, Kenyon NS (2006) The unique cytoarchitecture of human pancreatic islets has implications for islet cell function. PNAS 103:2334–2339
Carlsson GL, Heller RS, Serup P et al (2010) Immunohistochemistry of pancreatic development in cattle and pig. Anat Histol Embryol 39:107–119
Collombat P, Mansouri A, Sorensen JH et al (2003) Opposing actions of Arx and Pax4 in endocrine pancreas development. Genes Dev 17:2591–2603
Conarello SL, Jiang G, Mu J et al (2007) Glucagon receptor knockout mice are resistant to diet-induced obesity and streptozotocin-mediated beta cell loss and hyperglycaemia. Diabetologia 150:142–150
Driever W, Solnica-Krezel L, Schier AF et al (1996) A genetic screen for mutations affecting embryogenesis in zebrafish. Development 123:37–46
Epple A, Brinn JE (1987) The comperative physiology of the pancreatic islets. Springer, Berlin
Field HA, Dong PD, Beis D et al (2003) Formation of the digestive system in zebrafish. II. Pancreas morphogenesis. Dev Biol 261:197–208
Furuta M, Yano H, Zhou A et al (1997) Defective prohormone processing and altered pancreatic islet morphology in mice lacking active SPC2. Proc Natl Acad Sci USA 94:6646–6651
Gelling RW, Du XQ, Dichmann DS et al (2003) Lower blood glucose, hyperglucagonemia, and pancreatic cell hyperplasia in glucagon receptor knockout mice. Proc Natl Acad Sci USA 100:1438–1443
Gradwhol G, Dierich A, LeMeur M et al (2000) neurogenin3 is required for the development of the four endocrine cell lineages of the pancreas. Proc Natl Acad Sci USA 97:1607–1611
Guz Y, Montminy MR, Stein R et al (1995) Expression of murine STF-1, a putative insulin gene transcription factor, in β cells of pancreas, duodenal epithelium and pancreatic exocrine and endocrine progenitors during ontogeny. Development 121:149–161
Hara M, Wang X, Kawamura T et al (2003) Transgenic mice with green fluorescent protein-labeled pancreatic β-cells. Am J Physiol Endocrinol Metab 284:E177–E183
Hara M, Dizon RF, Glick BS et al (2006) Imaging pancreatic beta-cells in the intact pancreas. Am J Physiol Endocrinol Metab 290:E1041–E1047
Hazelwood RL (1993) The pancreatic polypeptide (PP-fold) family: gastrointestinal, vascular, and feeding behavioral implications. Proc Soc Exp Biol Med 202:44–63
Heller RS, Stoffers DA, Liu A et al (2004) The role of Brn4/Pou3f4 and Pax6 in forming the pancreatic glucagon cell identity. Dev Biol 268:123–134
Herrera PL, Huarte J, Sanvito F et al (1991) Embryogenesis of the murine endocrine pancreas; early expression of pancreatic polypeptide gene. Development 113:1257–1265
Hesselson D, Anderson RM, Beinat M (2009) Distinct populations of quiescent and proliferative pancreatic beta-cells identified by HOTcre mediated labeling. Proc Natl Acad Sci USA 106:14896–14901
Horb ME, Slack JM (2002) Expression of amylase and other pancreatic genes in Xenopus. Mech Dev 113:153–157
Hussain MA, Lee J, Miller CP et al (1997) POU domain transcription factor brain 4 confers pancreatic alpha-cell-specific expression of the proglucagon gene through interaction with a novel proximal promoter G1 element. Mol Cell Biol 17:7186–7194
Jackintell LA, Lance VA (1994) Ontogeny and regional distribution of hormone producing cells in embryonic pancreas of *Alligator Mississippienses*. Gen Comp Endocrinol 94:244–260
Jensen J (2004) Gene regulatory factors in pancreatic development. Dev Dyn 229:176–200

Jones HB, Reens J, Brocklehurst SR et al (2014) Islets of Langerhans from prohormone convertase-2 knockout mice show α-cell hyperplasia and tumorigenesis with elevated α-cell neogenesis. Int J Exp Pathol 95:29–48

Jonsson J, Carlsson L, Edlund T et al (1994) Insulin-promoterfactor 1 is required for pancreas development in mice. Nature 371:606–609

Kawaguchi Y, Cooper B, Gannon M et al (2002) The role of the transcriptional regulator Ptf1a in converting intestinal to pancreatic progenitors. Nat Genet 32:128–134

Kelly OG, Melton D (2000) Development of the pancreas in *Xenopus laevis*. Dev Dyn 218:615–627

Kharouta M, Miller K, Kim A et al (2009) No mantle formation in rodent islets—the prototype of islet revisited. Diabetes Res Clin Pract 85:252–257

Kilimnik G, Jo J, Periwal V et al (2012) Quantification of islet distribution and architecture. Islets 4:167–172

Kim A, Miller K, Jo J et al (2009) Islet architecture: a comparative study. Islets 1:129–136

Kordowich S, Collombat P, Mansouri A et al (2011) Arx and Nkx2.2 compound deficiency redirects pancreatic alpha- and beta-cell differentiation to a somatostatin/ghrelin co-expressing cell lineage. BMC Dev Biol 11:52–68

Krapp A, Knofler M, Ledermann B et al (1998) The bHLH protein PTF1-p48 is essential for the formation of the exocrine and the correct spatial organization of the endocrine pancreas. Genes Dev 12:3752–3763

Li Z, Wen C, Peng J et al (2009) Generation of living color transgenic zebrafish to trace somatostatinexpressing cells and endocrine pancreas organization. Differentiation 77:128–134

Lucini C, Castaldo L, Lai O et al (1998) Ontogeny, postnatal development and ageing of endocrine pancreas in *Bubalus bubalis*. J Anat 192:417–424

Maake C, Hanke W, Reinecke M (1998) An immunohistochemical and morphometric analysis of insulin, insulin-like growth factor I, glucagon, somatostatin, and PP in the development of the gastroentero-pancreatic system of *Xenopus laevis*. Gen Comp Endocrinol 110:182–195

Matsuoka T, Zhao L, Artner I et al (2003) Members of the large Maf transcription family regulate insulin gene transcription in islet cells. Mol Cell Biol 23:6049–6062

Miller K, Kim A, Kilimnik G et al (2009) Islet formation during neonatal development. PLoS ONE 4:e7739

Miyata T, Maeda T, Lee JE (1999) NeuroD is required for differentiation of the granule cells in the cerebellum and hippocampus. Genes Dev 13:1647–1652

Naya FJ, Huang HP, Qiu Y et al (1997) Diabetes, defective pancreatic morphogenesis and abnormal enteroendocrine differentiation in BETA2/NeuroD-deficient mice. Genes Dev 11:2323–2334

Nieuwkoop PD, Faber J (1967) Normal table of *Xenopus laevis* (Daudin). North-Holland Publishing Co, Amsterdam

Nyman LR, Wells KS, Head WS et al (2008) Real-time, multidimensional in vivo imaging used to investigate blood flow in mouse pancreatic islets. J Clin Investig 118:3790–3797

Ober EA, Field HA, Stainier DY (2003) From endoderm formation to liver and pancreas development in zebrafish. Mech Dev 120:5–18

Pearl EJ, Bilogan CK, Mukhi S et al (2009) Xenopus pancreas development. Dev Dyn 238:1271–1286

Petersson B (1966) The two types of alpha cells during the development of the guinea pig pancreas. 2. Zellforsch. 75:371–380

Pictet R, Rutter WJ (1972) Development of the embryonic endocrine pancreas. In: Geiger SR (ed) Handbook of physiology section 7: endocrinology. American Physiological Society, Washington, DC, pp 25–66

Reddy SN, Bibby NJ, Elliott RB (1985) Cellular distribution of insulin, glucagon, pancreatic polypeptide hormone and somatostatin in the fetal and adult pancreas of the guinea pig: A comparative immunohistochemical study. Eur Cell Biol 38:301–305

Reddy S, Bibby NJ, Elliott RB (1992) An immunocytochemical study of endocrine cell development in the early fetal guinea pig pancreas. Gen Comp Endocrinol 86:275–283

Rhoten WB (1987) Quantitative immunocytohemical analysis of the endocrine pancreas of the Nile Crocodile. Am J Anat 173:103–115

Samols E, Bonner-Weir S, Weir GC (1986) Intra-islet insulin-glucagon somatostatin relationships. Clin Endocrinol Metab 15:33–58

Sander M, Neubuser A, Kalamaras J et al (1997) Genetic analysis reveals that PAX6 is required for normal transcription of pancreatic hormone genes and islet development. Genes Dev 11:1662–1673

Sander M, Sussel L, Conners J et al (2000) Homeobox gene Nkx6.1 lies downstream of Nkx2.2 in the major pathway of β-cell formation in the pancreas. Development 127:5533–5540

Savari O, Zielinski MC, Wang X et al (2013) Distinct function of the head region of human pancreas in the pathogenesis of diabetes. Islets 5:226–228

Schwitzgebel VM, Scheel DW, Conners JR (2000) Expression of neurogenin3 reveals an islet cell precursor population in the pancreas. Development 127:3533–3542

Sørensen H, Winzell MS, Brand CL et al (2006) Glucagon receptor knockout mice display increased insulin sensitivity and impaired beta-cell function. Diabetes 55:3463–3469

Sosa-Pineda B, Chowdhury K, Torres M et al (1997) The *Pax4* gene is essential for differentiation of insulin-producing β cells in the mammalian pancreas. Nature 386:399–402

Steiner DJ, Kim A, Miller K et al (2010) Pancreatic islet plasticity—interspecies comparison of islet architecture and composition. ISLETS 2:135–145

Stoffers DA, Zinkin NT, Stanojevic V (1997) Pancreatic agenesis attributable to a single nucleotide deletion in the human IPF1 gene coding sequence. Nat Genet 15:106–110

St-Onge L, Sosa-Pineda B, Chowdhory K et al (1997) Pax6 is required for differentiation of glucagon-producing α-cells in mouse pancreas. Nature 387:406–409

Sussel L, Kalamaras J, Hartigan-O'Connor DJ et al (1998) Mice lacking the homeodomain transcription factor Nkx2.2 have diabetes due to arrested differentiation of pancreatic beta cells. Development 125:2213–2221

Tehrani Z, Lin S (2015) Endocrine pancreas development in zebrafish. Cell Cycle 10(20):3466–3472

Vincent M, Guz Y, Rozenberg M et al (2003) Abrogation of protein convertase 2 activity results in delayed islet cell differentiation and maturation, increased alpha-cell proliferation, and islet neogenesis. Endocrinology 144:4061–4069

Vuguin PM, Kedees MH, Cui L (2006) Ablation of the glucagon receptor gene increases fetal lethality and produces alterations in islet development and maturation. Endocrinology 147:3995–4006

Wang Y, Rovira M, Yusuff S et al (2011) Genetic inducible fate mapping in larval zebrafish reveals origins of adult insulin-producing β-cells. Development 138:609–617

Wang X, Zielinski MC, Misawa R (2013a) Quantitative analysis of the pancreatic polypeptide cell distribution in the human pancreas. PLoS One 8:e55501

Wang X, Misawa R, Zielinski MC (2013b) Regional differences in islet size distribution and architecture in human adult pancreas. PLoS One 8:e67454

Ward AB, Warga RM, Prince VE (2007) Origin of the zebrafish endocrine and exocrine pancreas. Dev Dyn 236:1558–1569

Weir GC, Bonner-Weir S (1990) Islets of Langerhans: the puzzle of intraislet interactions and their relevance to diabetes. J Clin Invest 85:983–987

Wierup N, Svensson H, Mulder H et al (2002) The ghrelin cell: a novel developmentally regulated islet cell in the human pancreas. Regul Pept 107:63–69

Zabel M, Surdyk-Zasada J, Lesisz I et al (1995) Immunocytochemical studies on endocrine cells of alimentary tract of the pig in the embryonic and fetal period of life. Folia Morphol (Warsz.) 54:69–80

Zhang C, Moriguchi T, Kajihara M et al (2005) MafA is a key regulator of glucose-stimulated insulin secretion. Mol Cell Biol 25:4969–4976

Chapter 3
Mathematical Modeling of Islet Generation, Degeneration and Regeneration

Junghyo Jo, Deborah A. Striegel, Manami Hara and Vipul Periwal

Introduction

Mathematical modeling has become an essential tool in biology as experimental observations become more quantitative (Mogilner et al. 2006; Tomlin and Axelrod 2007). It weaves together individual detailed observations, tests and weights different hypotheses, and makes non trivial predictions. Such predictions can then be tested through additional experiments and lead to strengthening or revision of our understanding. The islets of Langerhans in the pancreas are an attractive system to apply mathematical modeling. First, because they are physiologically important as the crucial micro-organs for controlling glucose homeostasis, its malfunction or destruction leads to diabetes, one of the most prevalent metabolic diseases (American Diabetes 2014). Second, research on diabetes has led to the accumulation of substantial information about islet structure and physiology. Finally, despite their complex function, islets have a rather simple structure composed of three major elements: alpha, beta, and delta cells. Together, these factors make the pancreatic islet an excellent system to tackle with mathematical modeling.

Humans have a few million islets scattered in the pancreas, while mice have a few thousands of islets. They have heterogeneous sizes from small clusters of a few cells to large ones of several thousand cells. The size range, however, is conserved between species including human, monkey, pig, and rabbit, regardless

J. Jo
Asia Pacific Center for Theoretical Physics, Pohang, Korea

D.A. Striegel · V. Periwal (✉)
Laboratory of Biological Modeling, NIDDK, NIH, Bethesda, MD, USA
e-mail: vipulp@mail.nih.gov

M. Hara
Department of Medicine, The University of Chicago, Chicago, IL, USA

A.A. Hardikar (ed.), *Pancreatic Islet Biology*, Stem Cell Biology and Regenerative Medicine, DOI 10.1007/978-3-319-45307-1_3

of their body size (Kim et al. 2009). The characteristic size distribution suggests that there may be an optimal size for islet function. The orchestrated functioning of alpha, beta, and delta cells is responsible for maintaining blood glucose levels within a narrow range. Under fasting conditions, alpha cells secrete glucagon to increase glucose levels, while under fed conditions, beta cells secrete insulin to decrease glucose levels and reduce lipolysis. Since the two reciprocal cells look sufficient to control glucose levels, the role of the third cell type, delta cells, is still mysterious. Furthermore, these endocrine cells do not work independently, but act in concert, stimulating or suppressing hormone secretions from their neighboring cells (Koh et al. 2012). Considering the interactions between alpha, beta, and delta cells, islet structures must have a functional significance. Indeed, islets have a special organization of the three cell types. In particular, rodent islets have beta cells in the core and non-beta cells on the periphery (Brissova et al. 2005; Cabrera et al. 2006). On the other hand, human islets lack a clear order in their spatial organization (Bosco et al. 2010; Brissova et al. 2005; Cabrera et al. 2006; Orci 1976). The development of these specialized islet structures and their species dependence has not been completely understood.

Diseases in biology can be represented as specific mathematical perturbations of a model of the biological system. The perturbed states, however, provide positive opportunities to deduce the structure–function relationship of the system. The degeneration of pancreatic islets results in diabetes leading to persistently elevated blood glucose levels. Type 1 diabetes (T1D) is an autoimmune disease where beta cells are specifically attacked by immune cells (Notkins and Lernmark 2001), while type 2 diabetes (T2D) is a metabolic disease where overstressed beta cells finally degenerate under insulin resistant conditions (Prentki and Nolan 2006). Once beta-cell number has been depleted by metabolic stress or aging, regeneration is very limited because beta cells rarely proliferate in adults (Teta et al. 2005). Therefore, diabetic states represent the specific destruction of beta cells from pancreatic endocrine cell mass. These perturbed states provide an opportunity to examine not only the functional role of beta cells, but also their functional relationship with alpha and delta cells in the islet. Unger has emphasized that diabetes originates from the imbalance between alpha and beta cells rather than the sole degeneration of beta cells (Unger and Orci 2010).

Research on islet biology and diabetes has led to the accumulation of substantial data. Mathematical modeling has already been applied to analyze some of the available quantitative information and to extract important information hidden in the raw data (Soria et al. 2000). As a good example, Finegood et al. analyzed postnatal beta-cell mass growth with a mathematical model and inferred the contribution of beta-cell death (Finegood et al. 1995). Later, they experimentally confirmed that beta-cell apoptosis indeed participates in the islet remodeling in the postnatal period (Scaglia et al. 1997). Recent advances in islet imaging have been providing more opportunities to examine the processes of islet generation (Jo et al. 2011, 2013; Miller et al. 2009), degeneration (Kilimnik et al. 2011), and regeneration (Lee et al. 2010). In this chapter, we review these studies and discuss some future perspectives for mathematical modeling of pancreatic islets.

Generation

In pancreatic development, the primitive endodermal epithelium differentiates into the pancreatic duct, exocrine, and endocrine cells (Slack 1995). Then, proximal endocrine cells aggregate to form cell clusters which later develop into the islets of Langerhans. For this micro-organ development, blood vessels play a pivotal role for providing not only oxygen and nutrients, but also inductive signals for endocrine cell differentiation (Lammert et al. 2001). On the other hand, endocrine cells produce vascular endothelial growth factor (VEGF) to attract endothelial cells of blood vessels. The organogenesis of pancreatic islets, thus, results from the mutual interaction between the endocrine cells and blood vessels. This process explains the colocalization of islets and blood vessels in the pancreas (Fig. 3.1). Hastings et al. have examined the spatial distribution of islets in the pancreas (Hastings et al. 1992; Schneider et al. 1996). They found that this distribution of pancreatic islets follows a fractal structure like a tree. Scale invariance is a central feature of fractals. This means that we can observe self-similar structures regardless of the scale at which we consider the given objects. Fractality is usually quantified by the fractal dimension, D. This is a useful parameter that tells us how many islets exist within a radius, r, from an arbitrary center. The number of islets is proportional to $N \propto r^D$. Thus, the log–log plot of N versus r becomes linear (Fig. 3.2). The fractal dimension has been measured as $D = 2.1$ based on the three-dimensional positions of entire islets in the mouse pancreas (Jo et al. 2013). A natural step in this line of inquiry is to ascertain the reason for such a spatial distribution. The answer may be found in the recent study by Magenheim et al. reporting that blood vessels reduce the branching and differentiation of epithelial cells (Magenheim et al. 2011). This

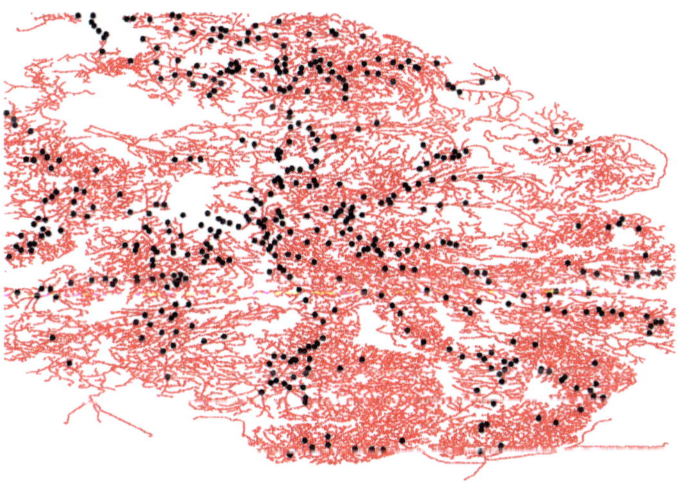

Fig. 3.1 Spatial distribution of blood vessels and islets in the pancreas. Captured were positions of blood vessels (*red dots*) and centroids of fluorescent islets (*black dots*) in the MIP-GFP mouse

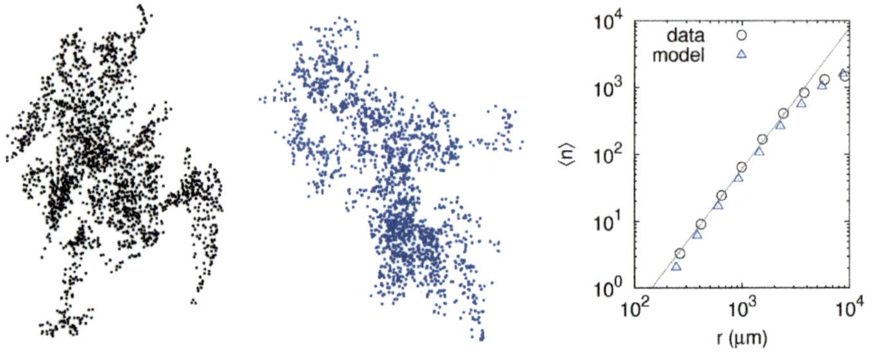

Fig. 3.2 Fractal spatial distribution of islets in the three-dimensional (3D) pancreas. Optical projection tomography determined 3D centroid positions (*black dots*) of individual islets in a splenic pancreas lobe of a C57BL/6 mouse at 8 weeks. A mathematical model of self-avoiding growth produced 3D positions (*blue dots*) of islets. The graph on the right plots the average number <*n*> of islets within a distance *r*. Both islet distributions show a power law with the slope, 2.1, corresponding to the fractal dimension of the spatial islet distribution. Note that the finite size of pancreata gives a cutoff distance above which power–law behavior disappears

implies that blood vessels can play the dual role for inducing and inhibiting endocrine cell differentiation. Therefore, the self-avoiding formation of islets on the branched epithelium can lead to their fractal distribution in the pancreas. Such a self-avoiding process has been simulated with mathematical modeling. The model was successful in reproducing the fractal distribution of islets in Fig. 3.2 (Jo et al. 2013).

In addition to prenatal islet formation, postnatal formation and growth of islets are important for islet maturation. More than 90 % of beta cells in adults are generated after birth, and postnatally formed small beta-cell aggregates contribute enormously to functional beta-cell mass in adults (Chintinne et al. 2010). Advanced imaging methods have provided detailed information about the postnatal islet formation and growth (Kilimnik et al. 2009; Kim et al. 2011; Miller et al. 2009). They measured the size of every islet at different postnatal periods. The changes in the islet size distribution (Fig. 3.3) contain dynamic information corresponding to islet growth. The dynamic information includes the islet neogenesis rate and the islet growth rate and its islet size dependence. These dynamic rates have been inferred from the changes in islet size distribution by using mathematical modeling (Jo et al. 2011, 2012). It has been demonstrated that new islets are not formed after 4 postnatal weeks, the weaning period, in mice. The islet growth rate diminishes and becomes negligible after weaning. In addition, the size dependence of this rate shows that small islets grow faster than large islets, implying cells in small islets proliferate more frequently. Furthermore, interconnected-islet-like structures are frequently observed during the postnatal period, but later disappear. This phenomenon can be explained by an islet fission process. When we included such a process in the mathematical model to explain the changes in

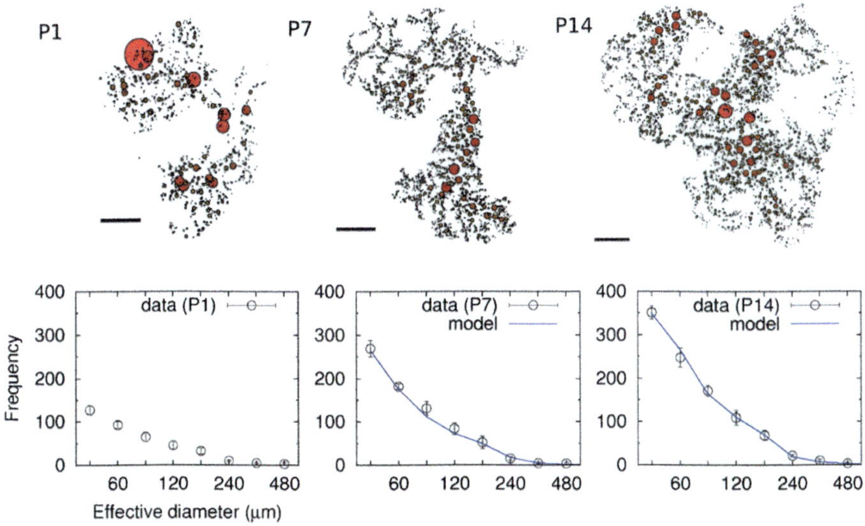

Fig. 3.3 Islet distributions in postnatal pancreata. Centroid positions and sizes of individual islets in the MIP-GFP mice were captured in postnatal day 1, 7, and 14. Islet size is represented by an effective diameter with which a *circle* gives the same area of an elongated islet. Note that the *circles* in the images do not represent real islet areas. *Scale bars* are 2 mm. Frequencies of islet sizes were counted (data, *black circles*) and predicted by a mathematical model (model, *blue lines*). *Error bars* represent standard deviations of data

islet size distributions, we could fit the data better. We also compared different models to distinguish whether islet fission occurs symmetrically or randomly, and confirmed that the random fission model is favored by the data. Large islets are susceptible to fission into smaller islets. The random fission model preference suggests that islet fission is a passive process of attaching and detaching through heterotypic contacts. Therefore, the mathematical modeling shows that islet formation and growth is a tightly regulated process involving preferential expansion of small islets and fission of large islets.

In fact, islets are not just simple aggregates of homogeneous cells. They have a particular organization of the three major cell types. Rodent islets have the mantle–core structure where beta cells are located in the islet core (Fig. 3.4a), while alpha and delta cells are mostly located on the islet periphery (Fig. 3.4b, c). However, human islets lack a clear order in the cellular organization. They have been described as random aggregates of cells (Brissova et al. 2005; Cabrera et al. 2006), assemblages (lobules, cloverleaf patterns, ribbon-like structures) of beta-cell-core subunits (Bonner-Weir and O'Brien 2008; Erlandsen et al. 1976; Grube et al. 1983; Orci 1976), or folded trilaminar plates (Bosco et al. 2010). Recently, advanced imaging methods enable the capture of the center positions of every cell in each islet (Hoang et al. 2014). These data were used quantitatively to judge whether human islets are random cell aggregates or not. First, we examined cell–cell contacts and counted contact number ratios of alpha–alpha ($P_{\alpha\alpha}$), beta–beta

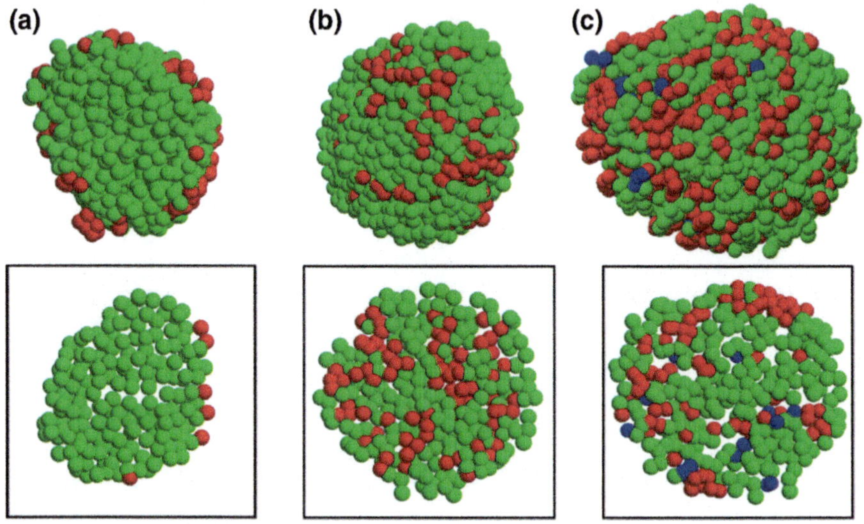

Fig. 3.4 Three-dimensional islet structures. Spatial distribution of alpha (*red*) and beta (*green*) cells in **a** mouse islet and **b** a human islet. In addition to alpha and beta cells, delta cells (*blue*) are also visualized in **c** a human islet. To show internal islet structures clearly, their corresponding two-dimensional sections are also shown in *boxes*

($P_{\beta\beta}$), and alpha–beta cells ($P_{\alpha\beta}$). If islets are random cell aggregates, the contact number should be coincident with the theoretical predictions, $P_{\alpha\alpha} \approx P_{\alpha}^2$, $P_{\beta\beta} \approx P_{\beta}^2$, and $P_{\alpha\beta} \approx 2\ P_{\alpha}P_{\beta}$ based on the fraction of alpha and beta cells, P_{α} and P_{β}. The mathematical analysis then clearly showed that human islets have more homotypic contacts ($P_{\alpha\alpha} > P_{\alpha}^2$ and $P_{\beta\beta} > P_{\beta}^2$) and fewer heterotypic contacts ($P_{\alpha\beta} < 2\ P_{\alpha}P_{\beta}$). Therefore, human islets are unlikely to be random cell aggregates. Interestingly, rodent islets give similar results for the preferential homotypic cell contacts, although they look different from human islets. This observation raises the following question. Can rodent and human islets have the same rule for their organization, but just different fractions of alpha and beta cells? Steinberg introduced the differential adhesion hypothesis that differences in stickiness between cell types result in special cellular organization (Steinberg 1963). We found that homotypic attractions are slightly, but significantly, stronger than heterotypic attractions in both mouse and human islets (Hoang et al. 2014). For the modeling, we assumed that islets have an equilibrium structure, rather than a non-equilibrium structure considering sequential development. Indeed the differential fractions of alpha and beta cells with a slight difference of their relative adhesiveness were sufficient to explain the structural differences between mouse and human islets (Fig. 3.5). Pseudo-islet formation supports our model. We propose that islets have a dynamic structure to flexibly alter their structures for various physiological demands.

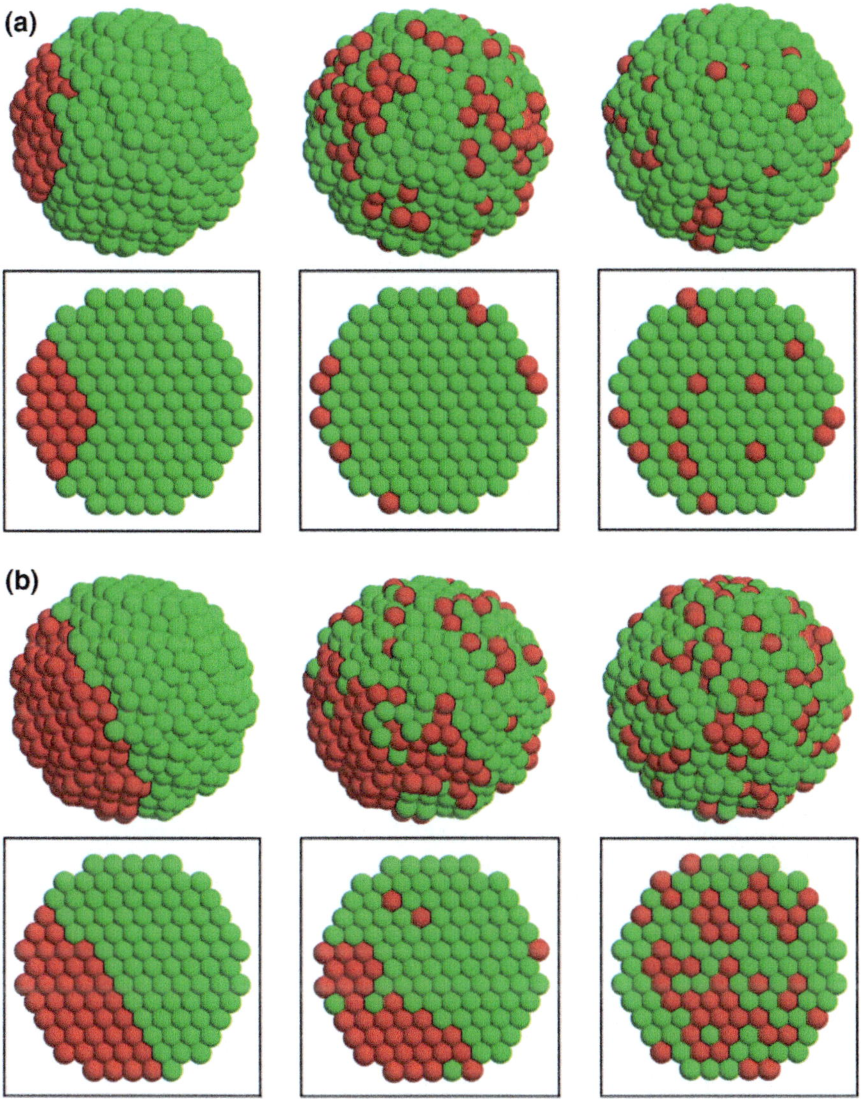

Fig. 3.5 Self-organization model for islet structures. Various organization patterns are generated by the differential adhesion models depending on cellular composition and relative adhesiveness between cell types in hexagonal close-packed lattices. Jaa = 1 is a reference adhesiveness. Jbb = 1 and Jab = 0.7, 0.93, and 0.98 (*left to right*) are used. Fractions of *green* cells are **a** 70 % and **b** 90 %. To show internal islet structures clearly, their corresponding two-dimensional sections are also shown in *boxes*. The mantle–core structure (*middle* in **a**) represents mouse islets, while the partial mixing structure (*right* in **b**) represents human islets

Degeneration

Pancreatic islets degenerate in diabetes. In particular, 70–100 % loss of beta-cell mass has been observed in the autoimmune disease, T1D, while 0–65 % loss of beta-cell mass has been observed in the slowly progressive metabolic disease, T2D (Matveyenko and Butler 2008). Persistent elevation of blood glucose levels and/or increased demand due to insulin resistance overload beta cells to secrete more insulin under diabetes. However, it is surprising that 50 % loss of beta-cell mass does not lead to impaired glucose tolerance, suggesting tight compensatory control of glucose homeostasis by the remaining beta cells (Butler et al. 2003). Thus, the adaptation of islets under diabetes may give new information about the compensation of islet function under perturbed conditions. One simple question is whether destruction and/or adaptation depend on islet size. Pancreatic islets have various sizes from small groups of a few cells to large ones of several thousand cells. On the one hand, extra-islet beta cells are scattered all over the exocrine tissue and they represent 15 % of all beta cells in human pancreas (Bouwens and Pipeleers 1998). On the other hand, more than 50 % of total beta-cell area in a pancreas comes from islets with diameters in the largest 2 % only (Jo et al. 2012). Therefore, it has been speculated that different sizes of islets may have different functions.

T1D is an autoimmune disease, characterized by loss of beta-cell mass, leading to lowered insulin secretion and hyperglycemia (Devendra et al. 2004). NOD mice provide an animal model of some aspects of T1D (Anderson and Bluestone 2005). Alanentalo et al. have measured islet size distributions during T1D progression in NOD mice at 6, 8, 12, and 16 weeks of age by optical projection tomography (Alanentalo et al. 2010). They report a preferential loss of smaller islets (<120 μm in diameter) and a possible regeneration process in the largest islets (>200 μm) to compensate for the ongoing beta-cell destruction. However, the process in humans may be more complex (Poudel et al. 2015).

Insulin resistance and partial beta-cell loss are hallmarks of type 2 diabetes. The additional metabolic demand for remaining beta cells can lead to regeneration and reorganization of islets in T2D. The recent development of a novel imaging method has permitted the scanning of each islet in an entire section of a human pancreas and the extraction of the location of every endocrine cell inside each islet (Kilimnik et al. 2011). The large-scale imaging analysis was performed with pancreata from brain-dead donors, including both T2D and non-diabetic subjects. This powerful method enables the automatical examination of several thousands of islets and quantification of changes in islet structures under T2D. A preferential loss of large islets has been observed. The death of islet cells may lead to the shrinkage of islets. The shrinkage of large islets may then contribute to an increase in the number of small islets. This hypothesis may explain the preferential loss of large islets even with an equal occurrence of cell death in every islet. Furthermore, in large islets, beta cells are preferentially lost among various endocrine cells. The preferential loss of beta cells results in the increase in alpha-cell fraction in

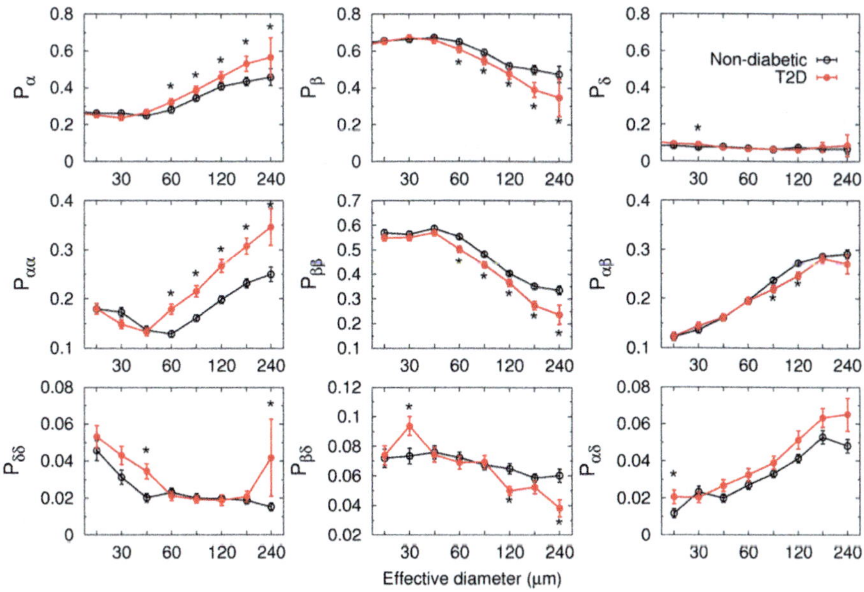

Fig. 3.6 Changes in islet structures under type 2 diabetes (T2D). Cellular compositions (alpha-, beta-, and delta-cell fractions depending on islet size in non-diabetic (8903 islets from $n = 14$ pancreata) and T2D (7929 islets from $n = 12$ pancreata) subjects. Student's t test compared the results between non-diabetic and T2D subjects at each size bin with $*p < 0.05$. Cell–Cell contact ratios are also plotted to represent the structural changes in islets under T2D

islets (Fig. 3.6). Note that delta-cell fraction is not altered under T2D. The fractional changes in islet cells are correlated with the cell–cell contact ratio in islets. The relative decrease in the beta-cell fraction leads to a decrease in the contacts of beta–beta cells, while the relative increase in alpha-cell fraction increases the contacts of alpha–alpha cells. However, the reciprocal change in alpha- and beta-cell fractions does not alter the contact ratio of alpha–beta cells. Here, it is an open question whether the structural change in islets under T2D is a result of passive degeneration or an active regeneration process.

Graph Theory of Islet Organization

As we have seen, the organization of islets shows more structure than just the islet size dependence of cell–cell contact frequencies for different endocrine cell types. The challenge is to first find a framework for the quantification of this organization. From a functional perspective, proximity to capillaries to monitor glucose levels and to release insulin, as well as contact with other beta cells, is crucial for beta cells. Graph theory is a mathematical framework that is well suited to quantifying such interactions between cells.

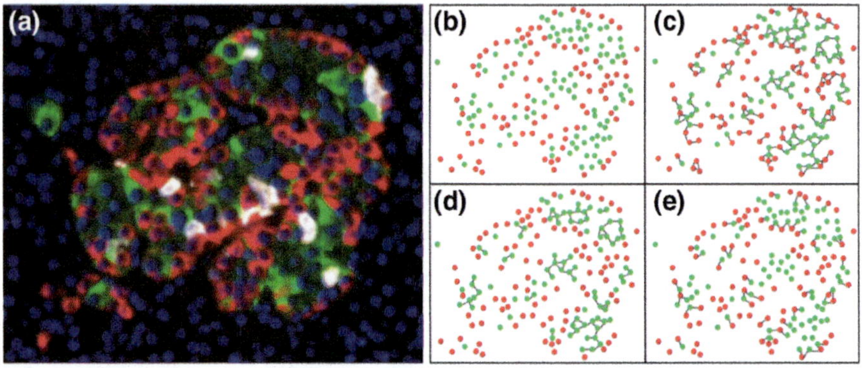

Fig. 3.7 Graph representations of islets. Graphs were created for each islet by representing the (x, y) coordinates of cells by vertices (**a**, **b**) and edges representing particular cell contacts. The alpha–beta–delta (**c**), beta–beta (**d**), and alpha–beta (**e**) contact graphs represent all cellular interactions, beta-cell clusters, and the alpha–beta interface, respectively

A graph, or network, consists of entities, called vertices, that display a certain connectivity represented by edges between vertices. In the context of islets, the vertices are the alpha, beta, and/or delta cells arranged by their (x, y) coordinates determined by the imaging method described above and the edges represent specific cell–cell contacts. Figure 3.7 shows the resulting alpha–beta, beta–beta, and alpha–beta–delta contact graphs for an islet. The accuracy of a graph representation of an islet depends on the determination of edges. Since the distance between contacting cells is not precisely known because an optical plane hits cells at a different height for each cell, the mathematical analysis described was performed for graphs with edges determined by cell–cell distances between 8 and 13 microns. This ensures that results obtained are robust to uncertainty in cell–cell distances.

Certain graph measures can be used to quantify the architectural structure: The *degree* of a vertex is the number of edges with that vertex as an endpoint. In the case of an islet, this describes the connectivity of a given cell. A *component* is a subset of vertices of a graph such that each vertex in the subset shares an edge with at least one other vertex in the subset. If a component of a graph has no edges, it is called a *singular component.* For islets, *non-singular components* represent beta-cell clusters. These graph measures and, in addition, the number of vertices per component were computed on beta–beta graphs for the human dataset described above. We found that T2D islets have a higher average degree, independent of islet size and cell–cell contact length (Fig. 3.8a–c). Computing the component measure showed that beta-cell clusters in large T2D islets have more cells per cluster (Fig. 3.8d–f); however, the number of clusters is greatly reduced (Fig. 3.8g–i). We believe the increase in average degree and number of cells per cluster is due to the initial increase in beta-cell mass observed with T2D (Rhodes 2005).

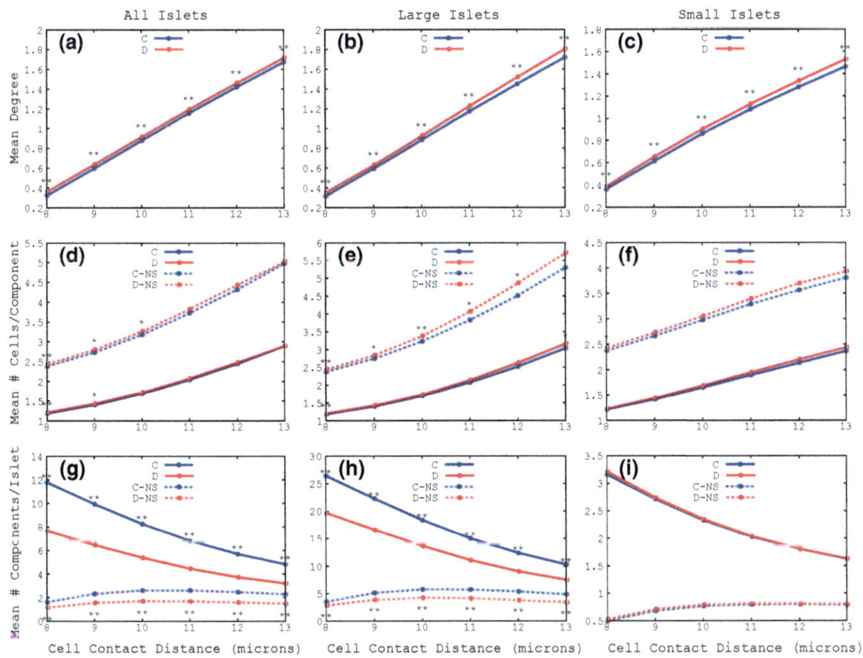

Fig. 3.8 Graph measures of control and T2D islets. Graphs were created and measures were calculated for each cell–cell distance between 8 and 13 microns where the cell–cell distance represents the maximal length of an edge between vertices. A statistically significant difference was found for the average degree between the control and T2D when all (**a**), large (**b**), and small (**c**) islets were analyzed. The number of cells per beta-cell cluster (**d–f**) showed a statistically significant difference for large non-singular islets only. Also, the number of components per islet (**g–i**) is reduced in singular and non-singular large T2D islets as compared to controls. *$p < 0.05$ between control and T2D groups; ** denotes a statistically significant difference after the Bonferroni correction (using $n = 64$)

While graph measures allow for the quantification of structural properties of a graph and give valuable insight into architectural changes that occur with T2D, these results are static and cannot capture the dynamic structural changes that occur throughout a lifespan of islet cell rearrangement, death and, to a lesser extent, replication processes. To address this, Monte Carlo simulations consisting of numerous iterations of vertex removals and additions were used to examine the dependence of the disappearance of a cell or appearance of a cell nearby on the degree, age, and component size of a cell. To illustrate how a simulation dependent on a given factor affects the graph of an islet, we applied the degree-dependent simulation procedure to an individual islet where the probability of the appearance or disappearance of a cell is highest for cells less than a given value (pp_a or pp_d, respectively). The graphs resulting after 2000 iterations for $pp_a = 0{,}3$, and 7 crossed with $pp_d = 0$, 3, and 7 are given in Fig. 3.9 and architecturally range from one giant cluster to many singular and non-singular clusters.

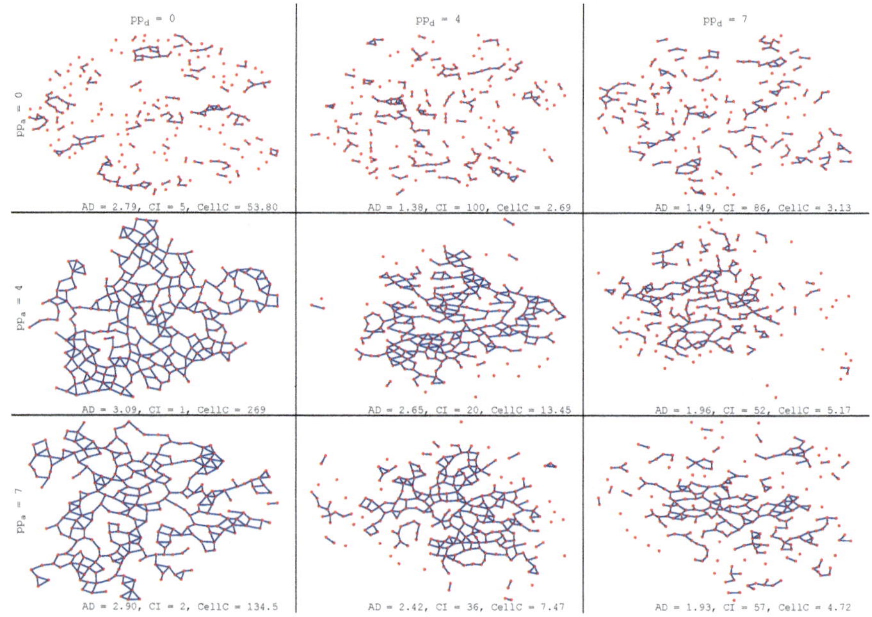

Fig. 3.9 Effects of degree-dependent appearance and disappearance on an islet. Simulations on an individual islet for $pp_a = 0,3$, and 7 crossed with $pp_d = 0$, 3, and 7 were run, and the resulting architectures and measures are shown. *AD* average degree, *CI* number of components in the islet, CellC = number of cells per component

Cells are known to undergo rearrangements throughout a lifespan. However, our calculations of the graph measures showed that control islets maintain a particular architecture that differs from T2D islets. To further examine which factors are responsible for maintaining the structural islet architecture observed in control and T2D islets, degree- and component-dependent rearrangement processes were applied to a subset of the control and diabetic large islets. The resulting graph measures were then compared to the original values of these measures to determine which factor(s), and associated parameter values, maintain the original values. The degree- and component-dependent simulation procedures maintained the original architecture only when cells in components larger than (or cells with degrees higher than) the probability parameter disappeared independent of the appearance parameter (results for the average degree are shown in Fig. 3.10a). For T2D islets, the architecture-preserving parameter value was shifted to a lower value when compared to control islets. This slight deviation results in the altered graph measures observed in T2D islets and further illustrates the sensitivity of the arrangement of beta cells in their clusters.

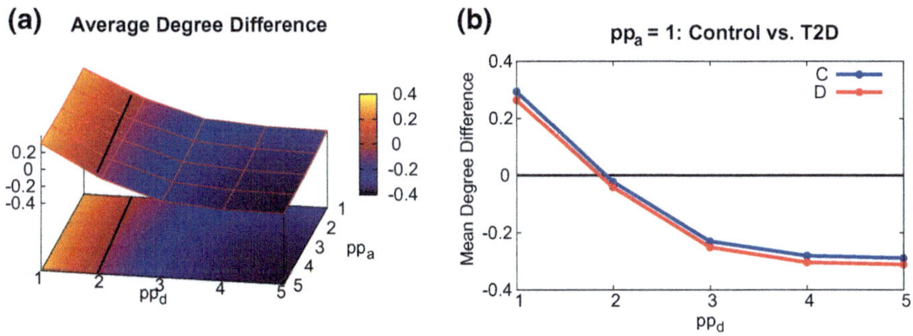

Fig. 3.10 Architecture-preserving parameter values for component-size-dependent simulations. Simulations of the appearance and disappearance of cells where the probability of such events is highest for cells located in clusters greater than a given value ($pp_a = 1,\ldots,5$ and $pp_d = 1,\ldots,5$, respectively) were run. The average degree differences between the resulting simulation and the original graph were calculated for each parameter combination for the control islets (**a**). The *black line* shows the parameter values where the simulated graphs have the same average degree as the original graph. The difference in average degree between the simulated and original control and simulated and original T2D architectures for $pp_a = 1$ is also shown (**b**)

Regeneration

It is known that beta-cell proliferation is very limited in adults. Nevertheless, weight gain and pregnant conditions may stimulate beta-cell proliferation. In addition, a non-physiological perturbation such as a pancreatectomy or ductal ligation could stimulate beta-cell regeneration. Although beta-cell proliferation is the main source for new beta cells (Dor et al. 2004), other sources such as ductal stem cells or transdifferentiation from alpha or acinar cells might contribute to the increase in beta-cell mass (Lysy et al. 2013). Thus, finding key factors for stimulating islet regeneration during the developmental period or stimulating conditions is an important issue for beta-cell regeneration.

One direct approach is to control cell cycles. Rane et al. have shown cyclin-dependent kinase 4 (Cdk4) exercises some control over beta-cell proliferation (Rane et al. 1999). The kinase can play a dual role for (1) recruiting quiescent cells (in 'dormancy') into the active cell cycle and (2) accelerating cell cycles. The distinct population of quiescent and active cells for proliferation were probed by using a DNA analog-based lineage-tracing method (Teta et al. 2007). A double-labeling experiment allowed the labeling of proliferating cells differently depending on time. Frequent proliferation of active cells in the mixture population of active and quiescent cells can generate the same number of proliferated cells with infrequent proliferation of purely active cells. However, the double-labeling experiment distinguished between these two cases (Fig. 3.11). The first case is more likely to have doubly labeled cells that proliferated during both labeling periods. A model was used to quantify the ratio of active and quiescent cells and proliferation

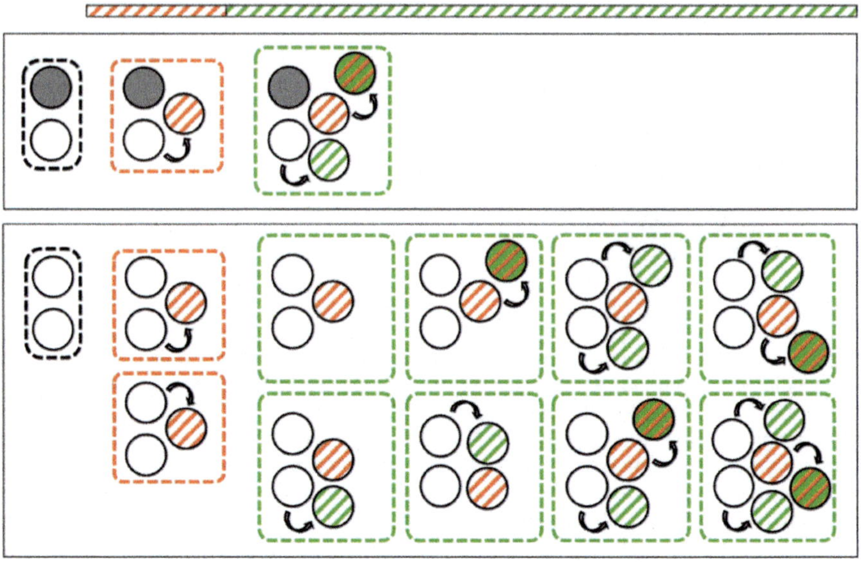

Fig. 3.11 Schematic description of cell proliferation under the double-labeling experiment. A DNA analog-based lineage-tracing technique using dual thymidine analogs allows to detect distinct rounds of cell division. Thus, the cells which proliferated during the first period with an orange analog exhibit an orange color, while the cells which proliferated during the second period with a *green* analog exhibit a *green color*. When a cell which proliferates during the first period, proliferates again in the second period, it expresses both *orange* and *green colors*. The first scenario (*top panel*) represents a heterogeneous population of quiescent (*black*) and active (*white*) cells, while the second scenario (*bottom panel*) represents a homogeneous population of active cells with half the proliferation rate (50 % per round) compared with the proliferation rate (100 % per round) of the active cells in the first scenario. Both scenarios produce the same result after the first round, but they diverge in the second round. *Each dashed square* represent one realized configuration among possible configurations

rate of active cells during regeneration (Lee et al. 2010). This mathematical analysis demonstrated that the recruitment of quiescent beta cells into the cell cycle and the acceleration of beta-cell proliferation contribute to the regeneration of beta cells under partial pancreatectomy.

Perspectives

As biological data become more and more quantitative, we have more chances to deduce design principles in biological systems. Here, mathematical modeling can be widely used to sharpen deductions from the analysis of quantitative data to integrate different types of data, to compare different hypotheses given incomplete information, and to provide testable hypotheses for the consequences of perturbations. This chapter is focused on explaining how mathematical modeling/analysis has contributed to our understanding of the generation, degeneration, and regeneration of pancreatic islets.

In addition to the structural aspect, however, mathematical modeling has also contributed to our understanding of the physiological aspects of islet function. As an excellent example, Sherman and his colleagues have modeled insulin secretion patterns of beta cells based on electrophysiological data and found that the electrical couplings between beta cells are essential to induce the specific firing pattern of insulin secretion (Sherman and Rinzel 1991). In the future, mathematical modeling may contribute to integrate the structure and the physiology of islets, and give a more comprehensive understanding of the control system for energy homeostasis. One gram of islets in the human pancreas (Meier et al. 2008) precisely controls our blood glucose levels with great precision. From an engineering perspective, these microscopic systems possess amazing performance, considering that the latest artificial pancreas weighs about 60 g just for the simple injection of exogenous insulin. We expect that mathematical modeling will be essential for the unraveling of the design principles and working algorithms of this natural system for glucose control.

References

Alanentalo T, Hornblad A, Mayans S, Karin Nilsson A, Sharpe J, Larefalk A, Ahlgren U, Holmberg D (2010) Quantification and three-dimensional imaging of the insulitis-induced destruction of beta-cells in murine type 1 diabetes. Diabetes 59:1756–1764

American Diabetes A (2014) Diagnosis and classification of diabetes mellitus. Diabetes Care 37(Suppl 1):S81–S90

Anderson MS, Bluestone JA (2005) The NOD mouse: a model of immune dysregulation. Annu Rev Immunol 23:447–485

Bonner-Weir S, O'Brien TD (2008) Islets in type 2 diabetes: in honor of Dr. Robert C. Turner. Diabetes 57:2899–2904

Bosco D, Armanet M, Morel P, Niclauss N, Sgroi A, Muller YD, Giovannoni L, Parnaud G, Berney T (2010) Unique arrangement of alpha- and beta-cells in human islets of Langerhans. Diabetes 59:1202–1210

Bouwens L, Pipeleers DG (1998) Extra-insular beta cells associated with ductules are frequent in adult human pancreas. Diabetologia 41:629–633

Brissova M, Fowler MJ, Nicholson WE, Chu A, Hirshberg B, Harlan DM, Powers AC (2005) Assessment of human pancreatic islet architecture and composition by laser scanning confocal microscopy. J Histochem Cytochem 53:1087–1097

Butler AE, Janson J, Bonner-Weir S, Ritzel R, Rizza RA, Butler PC (2003) Beta-cell deficit and increased beta-cell apoptosis in humans with type 2 diabetes. Diabetes 52:102–110

Cabrera O, Berman DM, Kenyon NS, Ricordi C, Berggren PO, Caicedo A (2006) The unique cytoarchitecture of human pancreatic islets has implications for islet cell function. Proc Natl Acad Sci USA 103:2334–2339

Chintinne M, Stange G, Denys B, Hellemans K, Pipeleers-Marichal M, Ling Z, Pipeleers D (2010) Contribution of postnatally formed small beta cell aggregates to functional beta cell mass in adult rat pancreas. Diabetologia 53:2380–2388

Devendra D, Liu E, Eisenbarth GS (2004) Type 1 diabetes: recent developments. BMJ 328:750–754

Dor Y, Brown J, Martinez OI, Melton DA (2004) Adult pancreatic beta-cells are formed by self-duplication rather than stem-cell differentiation. Nature 429:41–46

Erlandsen SL, Hegre OD, Parsons JA, McEvoy RC, Elde RP (1976) Pancreatic islet cell hormones distribution of cell types in the islet and evidence for the presence of somatostatin and gastrin within the D cell. J Histochem Cytochem 24:883–897

Finegood DT, Scaglia L, Bonner-Weir S (1995) Dynamics of beta-cell mass in the growing rat pancreas. Estimation with a simple mathematical model. Diabetes 44:249–256

Grube D, Deckert I, Speck PT, Wagner HJ (1983) Immunohistochemistry and microanatomy of the islets of Langerhans. Biomed Res 4:25–36

Hastings HM, Schneider BS, Schreiber MA, Gorray K, Maytal G, Maimon J (1992) Statistical geometry of pancreatic islets. Proc R Soc Lond B Biol Sci 250:257–261

Hoang DT, Matsunari H, Nagaya M, Nagashima H, Mills JM, Witkowski P, Pewiwal V, Hara M, Jo J (2014). A conserved rule for pancreatic islet organization. PloS ONE 9:e110384

Jo J, Hara M, Ahlgren U, Sorenson R, Periwal V (2012) Mathematical models of pancreatic islet size distributions. Islets 4:10–19

Jo J, Hornblad A, Kilimnik G, Hara M, Ahlgren U, Periwal V (2013) The fractal spatial distribution of pancreatic islets in three dimensions: a self-avoiding growth model. Phys Biol 10:036009

Jo J, Kilimnik G, Kim A, Guo C, Periwal V, Hara M (2011) Formation of pancreatic islets involves coordinated expansion of small islets and fission of large interconnected islet-like structures. Biophys J 101:565–574

Kilimnik G, Kim A, Jo J, Miller K, Hara M (2009) Quantification of pancreatic islet distribution in situ in mice. Am J Physiol Endocrinol Metab 297:E1331–E1338

Kilimnik G, Zhao B, Jo J, Periwal V, Witkowski P, Misawa R, Hara M (2011) Altered islet composition and disproportionate loss of large islets in patients with type 2 diabetes. PLoS ONE 6:e27445

Kim A, Kilimnik G, Guo C, Sung J, Jo J, Periwal V, Witkowski P, Dilorio P, Hara M (2011) Computer-assisted large-scale visualization and quantification of pancreatic islet mass, size distribution and architecture. J Vis Exp 49:e2471

Kim A, Miller K, Jo J, Kilimnik G, Wojcik P, Hara M (2009) Islet architecture: a comparative study. Islets 1:129–136

Koh DS, Cho JH, Chen L (2012) Paracrine interactions within islets of Langerhans. J Mol Neurosci 48:429–440

Lammert E, Cleaver O, Melton D (2001) Induction of pancreatic differentiation by signals from blood vessels. Science 294:564–567

Lee JH, Jo J, Hardikar AA, Periwal V, Rane SG (2010) Cdk4 regulates recruitment of quiescent beta-cells and ductal epithelial progenitors to reconstitute beta-cell mass. PLoS ONE 5:e8653

Lysy PA, Weir GC, Bonner-Weir S (2013) Making beta cells from adult cells within the pancreas. Curr Diabetes Rep 13:695–703

Magenheim J, Ilovich O, Lazarus A, Klochendler A, Ziv O, Werman R, Hija A, Cleaver O, Mishani E, Keshet E et al (2011) Blood vessels restrain pancreas branching, differentiation and growth. Development 138:4743–4752

Matveyenko AV, Butler PC (2008) Relationship between beta-cell mass and diabetes onset. Diabetes Obes Metab 10(Suppl 4):23–31

Meier JJ, Butler AE, Saisho Y, Monchamp T, Galasso R, Bhushan A, Rizza RA, Butler PC (2008) Beta-cell replication is the primary mechanism subserving the postnatal expansion of beta-cell mass in humans. Diabetes 57:1584–1594

Miller K, Kim A, Kilimnik G, Jo J, Moka U, Periwal V, Hara M (2009) Islet formation during the neonatal development in mice. PLoS ONE 4:e7739

Mogilner A, Wollman R, Marshall WF (2006) Quantitative modeling in cell biology: what is it good for? Dev Cell 11:279–287

Notkins AL, Lernmark A (2001) Autoimmune type 1 diabetes: resolved and unresolved issues. J Clin Investig 108:1247–1252

Orci L (1976) The microanatomy of the islets of Langerhans. Metab, Clin Exp 25:1303–1313

Poudel A, Savari O, Striegel DA, Periwal V, Taxy J, Millis JM, Witkowski P, Atkinson MA, Hara M (2015) Beta-cell destruction and preservation in childhood and adult onset type 1 diabetes. Endocrine

Prentki M, Nolan CJ (2006) Islet beta cell failure in type 2 diabetes. J Clin Investig 116:1802–1812

Rane SG, Dubus P, Mettus RV, Galbreath EJ, Boden G, Reddy EP, Barbacid M (1999) Loss of Cdk4 expression causes insulin-deficient diabetes and Cdk4 activation results in beta-islet cell hyperplasia. Nat Genet 22:44–52

Rhodes CJ (2005) Type 2 diabetes-a matter of beta-cell life and death? Science 307:380–384

Scaglia L, Cahill CJ, Finegood DT, Bonner-Weir S (1997) Apoptosis participates in the remodeling of the endocrine pancreas in the neonatal rat. Endocrinology 138:1736–1741

Schneider BS, Hastings HM, Maytal G (1996) The spatial distribution of pancreatic islets follows a universal power law. Proc Biol Sci 263:129–131

Sherman A, Rinzel J (1991) Model for synchronization of pancreatic beta-cells by gap junction coupling. Biophys J 59:547–559

Slack JM (1995) Developmental biology of the pancreas. Development 121:1569–1580

Soria B, Andreu E, Berna G, Fuentes E, Gil A, Leon-Quinto T, Martin F, Montanya E, Nadal A, Reig JA et al (2000) Engineering pancreatic islets. Pflug Arch 440:1–18

Steinberg MS (1963) Reconstruction of tissues by dissociated cells. Some morphogenetic tissue movements and the sorting out of embryonic cells may have a common explanation. Science 141:401–408

Teta M, Long SY, Wartschow LM, Rankin MM, Kushner JA (2005) Very slow turnover of beta-cells in aged adult mice. Diabetes 54:2557–2567

Teta M, Rankin MM, Long SY, Stein GM, Kushner JA (2007) Growth and regeneration of adult beta cells does not involve specialized progenitors. Dev Cell 12:817–826

Tomlin CJ, Axelrod JD (2007) Biology by numbers: mathematical modelling in developmental biology. Nat Rev Genet 8:331–340

Unger RH, Orci L (2010) Paracrinology of islets and the paracrinopathy of diabetes. Proc Natl Acad Sci USA 107:16009–16012

Chapter 4
Epigenetic Regulation of Islet Development and Regeneration

Satyajeet Khare, Prachi Katre and Sanjeev Galande

Introduction

Glucose is a monosaccharide with multiple metabolic paths. It is absorbed from the gut with the help of specialized transport proteins called glucose transporters (GLUTs). The absorbed glucose is either used immediately by cells as a source of energy, stored in liver and muscles as glycogen, or is converted to triglycerides by adipocytes. Glucose can also be synthesized in the liver by the process of gluconeogenesis. Glucose concentration in blood increases upon intake of food or gluconeogenesis by liver and decreases upon uptake of glucose into the cells of the body (discussed in Chap. 1). Cells of the nervous system have an absolute requirement for a continuous supply of glucose for their survival. Therefore, regulation of blood glucose levels and more precisely, its fine tuning is of utmost importance to the body. The homeostatic regulation of the concentration of extracellular glucose is achieved by controlling the rates of glycogen synthesis and uptake of glucose. These activities are primarily governed by the hormones secreted by the pancreas (Nussey and Whitehead 2001).

S. Khare · S. Galande
Indian Institute of Science Education and Research, Pune 411008, India

P. Katre
Diabetes Unit, KEM Hospital Research Centre, Pune 400012, India

S. Galande (✉)
Centre of Excellence in Epigenetics, Indian Institute of Science Education and Research, Dr Homi Bhabha Road, Pune 411008, India
e-mail: sanjeev@iiserpune.ac.in

A.A. Hardikar (ed.), *Pancreatic Islet Biology*, Stem Cell Biology and Regenerative Medicine, DOI 10.1007/978-3-319-45307-1_4

Pancreas Structure and Function

The pancreas is an exocrine as well as an endocrine organ. In vertebrates, it lies in the upper left abdomen at the junction of stomach and duodenum. The pancreas contains two different populations of cells that regulate two major physiological processes: digestion and glucose metabolism. The exocrine cells of the pancreas are arranged in acini and ducts and secrete digestive enzymes into the duodenum of the digestive tract (Pandol 2010). The endocrine cells of the pancreas are arranged in pockets known as the 'Islets of Langerhans' embedded within the acinar tissue and comprise approximately 1–1.5 % of the total organ mass. Pancreas of an adult human contains about a million islets distributed throughout the organ (Nussey and Whitehead 2001). Five different types of cells in islets secrete hormones directly into the bloodstream and are discussed in Chap. 1. The *INS* gene, which codes for pro-insulin, is under stringently controlled transcriptional regulation in β-cells and is a subject of intense scientific research (Andrali et al. 2008).

Diabetes Mellitus

Loss of β-cell mass and function is associated with diabetes mellitus. Type 1 diabetes mellitus (T1DM) is caused by autoimmune targeting of the insulin-producing β-cells leading to the loss of insulin-producing cells and increased blood (and urine) glucose levels. Epigenetic and/or environmental factors have been shown to play an important role in the etiology of T1DM (Dang et al. 2013). Altered epigenetic regulation could affect T1DM susceptibility in number of ways. The intrauterine environment, especially during early periods such as post-fertilization, is presumably the most important environment to affect T1DM-susceptibility, the others being pancreas development and immune system maturation (MacFarlane 2012).

The type 2 diabetes mellitus (T2DM) results from high insulin demand due to the insulin resistance in peripheral tissues triggering β-cell mass expansion and hyperinsulinemia. This, in turn, leads to gradual β-cell exhaustion and dysfunction, eventually instigating loss of β-cell mass by apoptosis. Obesity, reduced physical activity and aging increase the susceptibility to T2DM. While growing body of literature has suggested a role for epigenetic factors in the complex interplay between genes and the environment, molecular mechanisms linking environmental factors and T2DM are currently being elucidated (Nielsen et al. 2014; Olden et al. 2014).

The global prevalence of diabetes has been proposed to increase from 6.4 % in 2010 to 7.7 % in adults (aged 20–79 years) by 2030 (IDF 2014; Shaw et al. 2010).The increased incidence of diabetes calls for the development of curative therapies, which could eventually replace insulin injection and oral anti-diabetics treatment to patients. The Edmonton protocol for islet transplantation is limited by the lack of islet donor availability (Shapiro et al. 2006). Therefore, preserving a functional β-cell mass and finding novel sources of insulin-secreting β-cells are the topics of intense investigation. Replenishment of the β-cell mass could be achieved by triggering β-cell proliferation. However, this requires a thorough understanding of the signals and factors that influence β-cell degeneration and regeneration. Thus, detailed understanding of the islet development and β-cell regeneration is essential for the management of either type of diabetes.

It is important to describe in brief the transcription factor network in islet cell development and then discuss basic epigenetic mechanism in islet function. The focus will be mainly on the DNA methylation, histone modifications and small molecular epigenetic modulators in regulation of β-cell mass and function. We then briefly describe non-coding RNA expression in islet development which will be discussed in greater details elsewhere in this book.

Pancreatic Development and Regeneration

Embryonic Pancreatic Development

Embryonic pancreatic development in mice begins at E9.5 as two outpouchings of the endodermal lining of the duodenum just distal to the forming stomach. The dorsal and ventral outpouchings thus formed along the gut tube are termed as "pancreatic buds". Subsequent morphogenesis of the pancreas can be divided into two overlapping waves of development: the primary and the secondary transition. The "primary transition" between E9.5 and 12.5 involves organ determination and corresponds to period of active proliferation of pancreatic progenitors. At E11.5, the duodenum rotation brings the two buds into proximity for their fusion to form the definitive pancreas (Fig. 4.1). At E12.5, the evagination of the pancreatic epithelium into the surrounding mesenchyme and compartmentalization of this epithelium into "tip" and "trunc" domains begins (Pan and Wright 2011).

The morphogenic transformation of the pancreatic epithelium, termed as "secondary transition" corresponds to specification of multipotent precursor cells toward differential lineages. At E13.5, the pancreatic epithelium expands extensively. While pro-acini differentiate from distal "tip" domain synthesizing high levels of digestive enzymes, mature endocrine cells emerge from within the "trunc" domain of epithelium. Endocrine cells that leave the epithelium assemble into clustered endocrine islets and often stay in proximity to their parent ducts. Fully differentiated insulin-expressing β-cell and glucagon expressing α-cells emerge between E13 and E15. Within the next 24 h, the somatostatin expressing δ-cells appear and over the next few days (E14-18), endocrine cells coalesce into small aggregates. Shortly before birth, at E18, the γ(PP)-cells differentiate and endocrine cells begin to form well organized islets of Langerhans (Pan and Wright 2011). After birth, the predominant mechanism for islet cell mass maintenance is self-duplication which we will discuss a little later.

Transcription Factor Network in Pancreas Development

Several members of common signaling pathways, the Hedgehog system, the homeobox genes and Notch signaling are involved in the mechanisms governing early pancreas development. A specific program of transcription factor (TF) gene expression is subsequently activated that defines the fate of pancreatic endocrine progenitors (Fig. 4.1).

At E8-8.5 in the mouse, before the dorsal pancreatic bud first appears, early "pre-pancreatic" endoderm selectively expresses two homeodomain transcription factors, Mnx1 (Motor neuron and pancreas homeobox-1) and Pdx-1 (Pancreatic and duodenal homeobox-1). Pdx-1, also known as insulin promoter factor-1 (Ipf-1) in humans, is necessary for pancreatic development and β-cell maturation (Ahlgren et al. 1998). At later stages (E12.5), Pdx-1 expression decreases and it remains restricted in β-cells where it plays role in insulin gene expression. Mnx1,

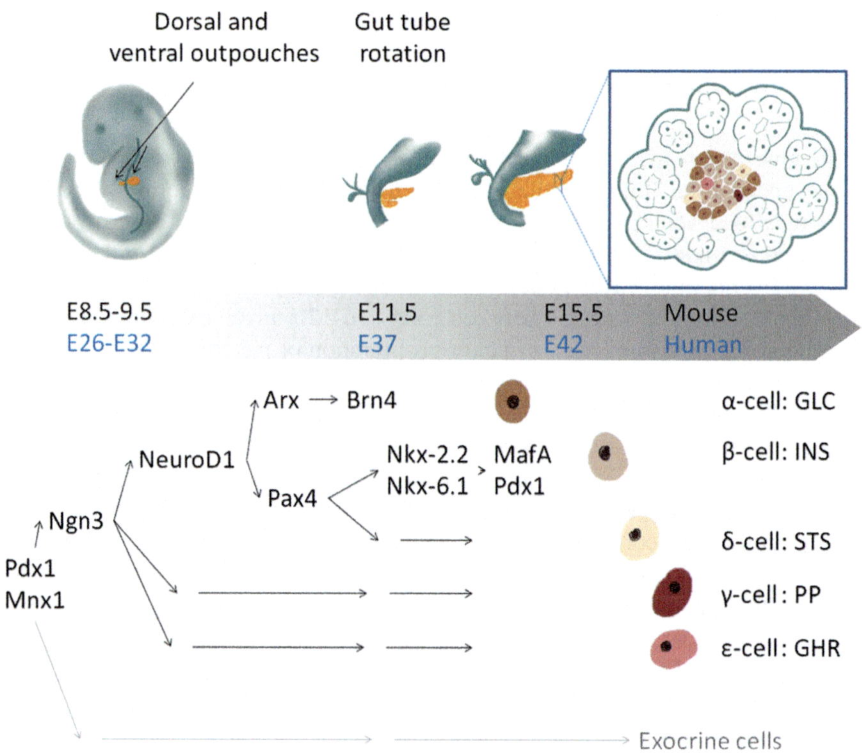

Fig. 4.1 Schematic representation of the transcription factor expression during pancreatic islet development. Pancreas development begins with the development of dorsal and ventral outpouches that are brought together by rotation of the gut tube for the formation of pancreas. Pdx1 and Mnx1 expression marks the pre-pancreatic endoderm, whereas Ngn3 specifies the endocrine lineage that forms the 'islets of Langerhans'. Pax4 expression defines the β/δ lineage and the late transcription factors specify specific functional cell-types (adapted from Naujok et al. 2011)

also known as HB9 or HLXB9, has also been found to be essential for normal pancreas development and function (Harrison et al. 1999). Mnx1 is down-regulated when the pancreatic buds form but it re-expresses later (E17.5) in the differentiated β-cells.

Specification of the endocrine cells in the developing pancreatic endoderm is determined by expression of bHLH transcription factor neurogenin-3 (Ngn-3) (Gradwohl et al. 2000). Ngn-3 expression is negatively regulated by Notch signaling (Apelqvist et al. 1999) and is detected from E8.5 to 15.5 following which it reduces to extremely low levels. Ngn3 activation leads to the expression of additional transcription factor, NeuroD1 (Anderson et al. 2009). NeuroD1 is expressed slightly later than Ngn-3 during pancreatic development; but, unlike Ngn-3, it persists in the mature islets, predominantly in β-cells. NeuroD1 seems to be necessary for islet development and insulin gene expression and mutations in the gene may lead to T2DM (Malecki et al. 1999).

Further specification into specific cell subtype seems to be determined before the expression of hormones begins. The early factors Arx, Pax4, Nkx-2.2, and Nkx-6.1 allow lineage specification of endocrine precursor cells toward mono-hormone-expressing cell type. Arx and Pax4 mutually inhibit each other's transcription and favor exclusive lineages. While expression of Arx promotes α- and γ(PP)-cell lineages, Pax4 favors the β- and δ-cell fates (Collombat et al. 2003, 2007). Nkx-2.2 and Nkx-6.1, the NKX-homeodomain factors, also play an important role in the cell fate specification toward the β-cell lineage and proper insulin secretion. Nkx-2.2 expression starts early at E9.5, where along with Ngn-3, it regulates NeuroD1 expression (Anderson et al. 2009). Nkx-2.2 deficiency can lead to Diabetes (Sussel et al. 1998). Nkx-6.1 also lies downstream of the Nkx-2.2 and seems to be regulated by Nkx-2.2 and Pdx-1. In the developing mouse pancreas, Nkx-6.1 protein can be detected at E10.5-12.5. With the start of the secondary transition, around E13-15.5, Nkx-6.1 becomes restricted to insulin-expressing cells and scattered ductal and peri-ductal cells (Sander et al. 2000; Watada et al. 2000). Nkx-6.1 is also required for the development of the β-cells and glucose-stimulated insulin secretion (Schisler et al. 2005).

Late factors, Pax6, Isl1, Brn4, MafA, and MafB function in the final steps of islet cell differentiation. Pax6 (also known as Aniridia Type II protein; AN2) and Isl1 are expressed in all islet cells and their loss causes defects in the generation of all endocrine cell subtypes. In addition, both factors activate islet hormone gene expression, with Pax6 implicated in glucagon, insulin and somatostatin expression (Sander et al. 1997). Isl1 also expresses at early stages and is necessary for the development of the dorsal exocrine pancreas (Ahlgren et al. 1997). In β-cell MafA, a potent regulator of the insulin gene and β-cell function is a direct target of Isl1 (Du et al. 2009). While MafA is β-cell specific in expression, MafB is required for both α-cell and β-cell differentiation/maturation. Expression of Brn4, another late factor, begins at E10 in pancreas just before Pax6 (Heller et al. 2004). Brn4 is activated in mature islet cells, but it is restricted to non-β-cells, predominantly α-cells, where it regulates glucagon gene expression (Ben-Othman et al. 2013; Wilson et al. 2003).

Regeneration of β*-Cells in Adults*

In humans, β-cell mass expands several folds from birth to the first few years of childhood. After this initial growth burst, the replication potential of β-cells gradually declines until adulthood. Interestingly, β-cell replication is the primary mechanism for maintaining postnatal β-cell mass (Dor et al. 2004). At molecular level, replication of β-cells is controlled by a network of finely tuned cell cycle regulators. For example, cyclin-dependent kinase inhibitor 2A (*CDKN2A*), a negative regulator of the proliferation of normal cells, is highly expressed in adult islets and is associated with the reduction in β-cell replication observed during aging. Various cyclins, on the other hand, have been shown to play a role in the positive regulation of β-cell proliferation in pancreas. The increase the β-cell mass in adulthood is regulated by different physiological processes to compensate for the increased metabolism (Vetere et al. 2014).

A marked increase in β-cell population is observed during pregnancy. Pregnancy-related hormones often interfere with insulin receptors in peripheral target tissues leading to insulin resistance. Whenever required, pancreas increases the number of insulin-producing β-cells to augment insulin secretion. However, failure of pancreas to respond in this manner can lead to gestational diabetes mellitus. Islet cells overexpress the receptors for hormones lactogen, prolactin and growth hormone during pregnancy which in turn regulate the β-cell expansion (Nielsen et al. 1999). Gut-derived hormones—incretins—stimulate a decrease in blood glucose levels by increasing the amount of insulin released from β-cells. The two main candidate Incretins are glucagon-like peptide-1 (GLP-1) and glucose-dependent insulin-tropic polypeptide (GIP). Upon food intake, duodenal L- and K-cells secrete GLP-1 and GIP, respectively. The incretin-receptor activation leads to insulin secretion, induction of β-cell mass expansion, and enhanced resistance to apoptosis creating a possibility of their therapeutic use in treatment of T2DM (Baggio and Drucker 2007).

Thus, the regulation of β-cell mass is orchestrated by systemic metabolic state, cross-talk between organs, and interactions between cells in the pancreas. Therefore, several ways might exist to trigger endogenous mechanisms of β-cell regeneration. The insights into the mechanisms that control the generation of pancreatic cells during development and adult body might eventually be applicable to the development of strategies for replenishing the β-cells in patients with diabetes.

Islet Plasticity

Cellular plasticity is defined by the capacity of a specialized cell type to differentiate into another cell type to compensate for the loss of function. Thus, the inter-conversion of pancreatic cells into β-cells might be harnessed for novel regenerative therapies. Several studies have focused their efforts on the generation

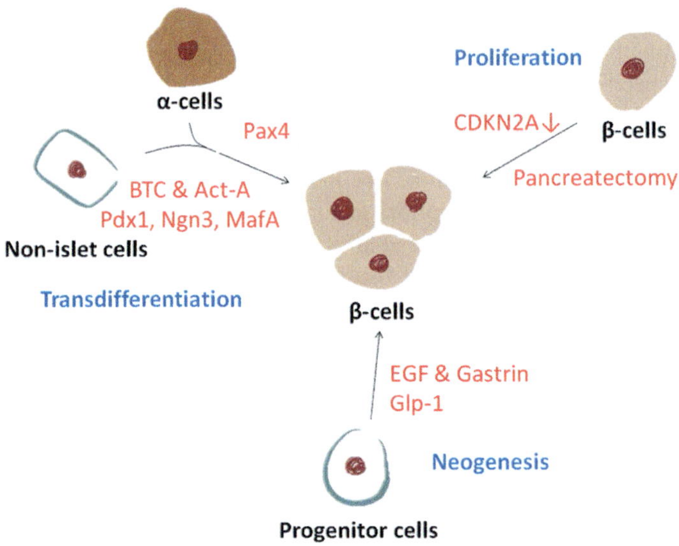

Fig. 4.2 Transcription factors involved in regeneration of β-cells of islets. Proliferation of existing mass of β-cells, neogenesis from progenitor cell population and reprogramming of non-β-cells into β-cell phenotype (adapted from Banerjee et al. 2005)

of β-cells from other pancreatic cells, while others have concentrated on proliferation of existing β-cell mass or differentiation of progenitor cells into β-cell phenotype (Fig. 4.2).

Betacellulin (BTC) is a member of the epidermal growth factor family originally isolated from the conditioned medium of insulin-producing β-cell carcinoma cells. BTC, in combination with Activin-A, has been shown to reprogram amylase-secreting rat exocrine AR42 J cells into insulin-secreting cells (Mashima et al. 1996). Another gut hormone, GLP-1, has also been shown to promote islet regeneration after partial pancreatectomy (De Leon et al. 2003). Epidermal growth factor (EGF) and gastrin (a hormone, normally expressed by fetal islets) have been shown to induce islet regeneration in vivo (Rooman and Bouwens 2004).

Transcription factors in β-cell development have been subject to immense focus due to their capability to reprogram one type of cell into another. Accordingly, expression of Pdx-1, which is central in regulating pancreatic development and islet cell function, activated endogenous insulin gene in mice liver tissues (Ferber et al. 2000). Three endocrine transcription factors (Ngn-3, Pdx-1 and MafA) were found to reprogram differentiated pancreatic exocrine cells into β-cell like phenotype via in vivo reprogramming in adult mice. Though these cells expressed β-cell-specific genes and reduced blood glucose levels in diabetic mice, they did not cluster to resemble islets, leaving some uncertainty about their maturation and functional state (Zhou et al. 2008). Several studies have also attempted inter-conversion of islet cells toward the β-cell fate. For example, Collombat et al. reported that ectopic expression of Pax4 in α-cells could drive their conversion to the β-cell fate (Collombat et al. 2009). As mentioned before, Pax4 is expressed in differentiating β-cell precursors and down-regulates Arx. Vice versa, expression of the α-cell-specific transcription factor Arx in β-cells leads to trans-differentiation into the α- or PP-cell lineage. This implies that the compensation of the lost or degenerated β-cells can be achieved from alternative endocrine cells by molecular reprogramming. BTC, mentioned earlier, has also been shown to act in combination with NeuroD transcription factors to reverse the diabetic phenotype in mice (Kojima et al. 2003). The role of hormones and transcription factors in islet plasticity and regeneration are also reviewed elsewhere for more information (Migliorini et al. 2014).

In the ongoing efforts to understand the reprogramming of cells toward the β-cell fate, accumulating evidence suggests that epigenetic memory of the source cell population can play a key role in its reprogramming (Bar-Nur et al. 2011). A better understanding of the heterogeneity and epigenetic status of endocrine cells therefore might open up a completely new field of research focusing on genetic and epigenetic manipulation to achieve reprogramming toward the β-cell fate. In the next part of the chapter, we will focus on epigenetic mechanisms involved in pancreatic islet development, function and their modulation by environmental factors. We will also discuss epigenetic dysregulation in onset of diabetes mellitus and therapeutic implications of the same.

Epigenetics

The term epigenetics was originally proposed by Conrad H. Waddington in the 1942 to define the heritable changes in gene expression that occur independent of a change in the DNA sequence (Waddington 2012). The most widely studied epigenetic mechanisms include DNA methylation, histone modifications, and expression of microRNAs and other non-coding RNAs.

DNA methylation was the first epigenetic mark to be discovered. Methylation of DNA occurs on Cytosine residue at the 5′ position (5mC) in "C" preceding "G" (CpG; denoting on the same strand and linked by a phosphate) dinucleotide context. Approximately 70 % of all CpG di-nucleotides are methylated, and most of the unmethylated CpGs are clustered in 'CpG islands' that are highly CG rich. CpG islands are often situated close to the promoter regions of the genes and can influence the affinity of transcription factors to the DNA binding sites (Bird 1986). DNA methylation is established and maintained by DNA methyltransferase family of enzymes (DNMT). There are three different DNMTs. The de novo methyltransferases DNMT3a and 3b methylate unmethylated CpG dinucleotides and therefore can establish new patterns of DNA methylation, whereas the maintenance methyltransferase DNMT1 predominantly methylates hemi-methylated DNA and enables faithful copy of the existing methylation pattern to the newly synthesized strand. Removal of methylation mark from DNA is facilitated by DNA demethylases, namely ten-eleven translocation (TET) and AID/APOBEC families of enzymes.

DNA methylation plays an active role in gene expression. DNA methylation at the promoters is commonly associated with gene silencing whereas that in the gene body is associated with expressed genes. DNA methylation also contributes to genomic imprinting, X-chromosome inactivation, suppression of transposable elements, and the establishment of cellular identities (Bhutani et al. 2011).

In eukaryotes, DNA and histone proteins assemble into nucleosomes to form chromatin. Amino acid residues present in the N-termini of histones undergo a variety of chemical modifications, namely: acetylation of lysines, methylation of lysine and arginine and phosphorylation of threonine and serine etc. These modifications, especially acetylation and methylation, have a regulatory role in transcription and contribute to cell fate determination. Histone acetylation is mediated by lysine acetyltransferases (KATs, also known as HATs) and reversed by lysine deacetylases (HDACs). Histone deacetylation is generally associated with increased DNA methylation and gene silencing. Histone methylation involves the transfer of a methyl group to either a lysine or an arginine residue by a process catalyzed by Lysine and Arginine methyltransferases. Lysine and arginine can undergo different extents of methylation (mono, di and trimethylation for lysine and mono and dimethylation for arginine). The opposite process, histone demethylation, depends upon Lysine and Arginine demethylases. Unlike acetylation, the function of histone methylation mark depends on the extent and exact location of the modification. For example, H3K4me3 (tri-methylation of lysine 4 on the histone H3) is associated with the activation of transcription, whereas the H3K27me3 is linked to repression of transcription. These epigenetic modifications are dynamically placed or removed in a lineage-specific manner during embryonic development and ultimately maintain cellular identities (Khare et al. 2012). Changes in histone modifications and DNA methylation can be studied by chromatin immunoprecipitation (Jayani et al. 2010), MedIP and bisulfite conversion-based techniques (Laird 2010) followed by high-throughput sequencing.

Although most of the mammalian genome is transcribed, only a small percentage encodes protein-coding genes (Mattick 2001). It is now evident that the non-protein coding portion of the transcriptome is of crucial functional importance in normal development and in disease etiology. In addition to DNA methylation and histone modifications, gene expression is also regulated by non-protein coding RNA the importance of which is particularly evident for a class of small non-coding RNAs (ncRNAs) called microRNAs (Esteller 2011). The miRNA modulates gene expression by binding to the untranslated regions of the target mRNA marking it for degradation. A single miRNA can target several genes depending on sequence homology (Kadamkode and Banerjee 2014). MicroRNAs have been shown to differentially express during pancreatic development and within the islets cells (Joglekar et al. 2009).

These distinct epigenetic mechanisms are often functionally connected to each other. For example, methylated DNA binding proteins can recruit both histone deacetylases and histone methyltransferases to nucleosomes, while methylation of histone H3 lysine 9 can in turn recruit DNA methyltransferases reinforcing chromatin compaction and silencing (Conerly and Grady 2010). These epigenetic

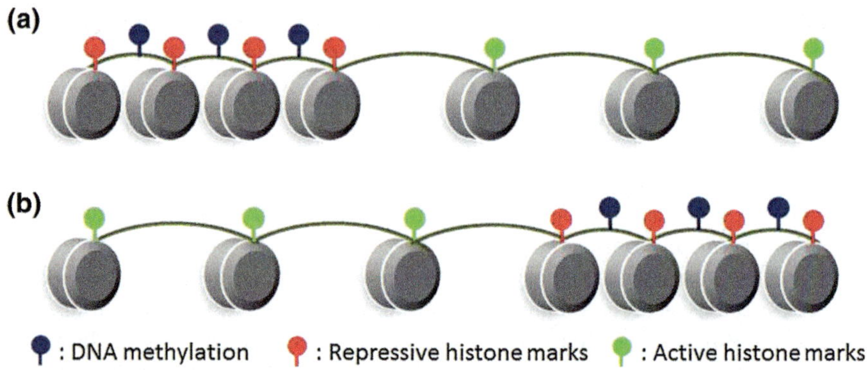

Fig. 4.3 DNA methylation and histone modifications as epigenetic modifications. **a** Chromatin conformation in islet cells from healthy individual; and **b** chromatin conformation in islet cells from Diabetic individual

mechanisms together affect various processes related to the islet cell development and in particular β-cell proliferation, mass and function and have been reviewed recently (Avrahami and Kaestner 2012; Dayeh and Ling 2015; De Jesus and Kulkarni 2014; Haumaitre 2013) (Fig. 4.3).

Epigenetic Regulation of β-Cell Identity, Mass and Function

Epigenetics of the β-Cell Identity

During pancreas development, transcription factors play a crucial role in exocrine and endocrine differentiation. Regulation of cell differentiation programs requires a complex transcriptional network. Gene regulation in such a scenario is regulated and maintained via epigenetic mechanisms that lead to context specific modifications of histones and/or DNA.

Nkx-2.2 is a transcription factor required for cell fate decisions in the pancreatic islets. Using Nkx-2.2 mutant mice, Nkx-2.2 was shown to be a part of a large repressor complex in β-cells. Nkx-2.2 was shown to preferentially recruit Grg3 and HDAC1 to the *Arx* promoter methylated by DNMT3a in β-cells. Interestingly, Nkx-2.2 is expressed in both α- and β-cells; therefore, study shows that Nkx-2.2 binding is epigenetically regulated during development and has differential effect on *Arx* expression in the two cell types. The differential DNA methylation appears to provide a platform for Nkx-2.2 binding and maintenance of repressed state of the α-cell-specific transcription factor Arx in β-cells, to assure cell identity maintenance and blocking of β- to α-cell conversion (Papizan et al. 2011).

The methylation of the *Arx* promoter in β-cells seems to be maintained by the maintenance methyltransferase DNMT1 postnatally. In an interesting observation, conditional deletion of DNMT1 in β-cells in mice, led to their selective cell fate

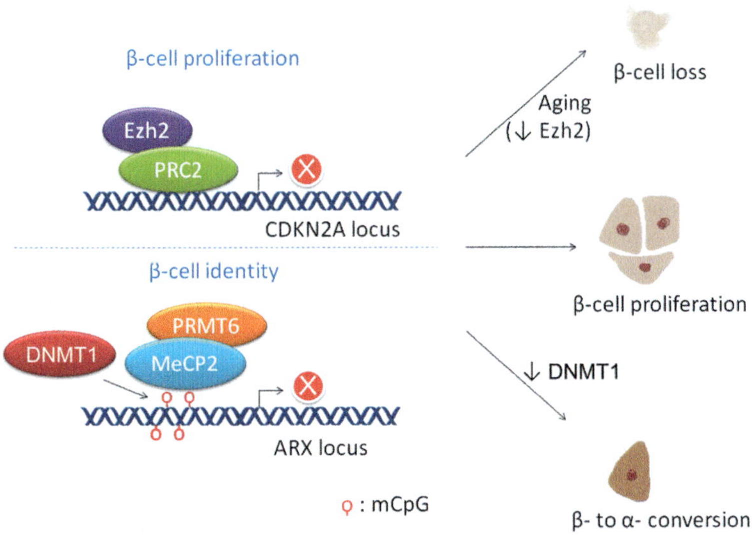

Fig. 4.4 Epigenetic regulation of β-cell development and proliferation. Proliferation of β-cells relies on repression of cell cycle inhibitor *CDKN2A* gene locus via PRC2 complex. Age associated reduction in Ezh2 levels results in de-repression of this locus which hampers proliferation potential of β-cells. The identity of β-cells is maintained by repression of α-cell specific transcription factor Arx. DNMT1 maintains repressed state of Arx post-natally by faithful transmission of methylation of *Arx* promoter during cell division. Accordingly, down-regulation of DNMT1 results in expression of *Arx* locus and conversion of β-cells to α-cell phenotype

transition into α-cell phenotype. *Arx* locus was found methylated and repressed in β-cells and hypomethylated and expressed in α-cells and DNMT1-deficient β-cells. The methylated *Arx* locus in β-cells was bound by methyl-binding protein MeCP2, which recruited PRMT6, an enzyme that methylates histone H3R2 resulting in repression of Arx (Fig. 4.4). This suggests that the propagation of the DNA methylation pattern forms a platform for transmitting histone modifications that regulate chromatin remodeling at the *Arx* locus in a repressive structure, which can be stably inherited through cell division, thus retaining the identity of β-cells (Dhawan et al. 2011).

Altogether, these findings suggest that DNMTs and MeCP2 together with Nkx-2.2, regulate DNA methylation of the *Arx* promoter specifically in β-cells, and recruit the histones modifiers HDAC1 and PRMT6 in order to silence Arx expression. These studies demonstrate that interactions between transcription factors and epigenetic regulators are crucial for the differentiation of islet cells.

Epigenetics of the β-Cell Proliferation

Proliferation of pancreatic islet β-cells is an important mechanism for self-renewal and adaptive islet expansion. As mentioned earlier, the replication of β-cells is negatively regulated by cell cycle regulators P16^{INK4A} and P19^{ARF} that are alternate reading frame products of the *cdkn2a* gene locus (also known as the *Ink4a/Arf* locus). The *cdkn2a* gene shows increased expression with age, and this locus has been linked with T2DM by genome-wide association studies (GWAS) (Scott et al. 2007). Interestingly, a recent study reported that polycomb complex protein Bmi-1-dependent modulation of the Ink4a/Arf expression levels is critical to regulate pancreatic β-cell proliferation during aging and regeneration. In aged islets, expression of the *Ink4a/Arf* locus was found to be associated with decreased Bmi-1 binding and loss of H2A ubiquitination. This epigenetic modification stimulated the recruitment of MLL1, a trithorax group (TrxG) protein, and increased H3K4me3 that augmented transcriptional activity from the *Ink4a/Arf* locus. During β-cell regeneration, increased Bmi-1 binding coincided with decreased H3K4me3, resulting in repression of the *Ink4a/Arf* locus and augmented β-cell proliferation (Dhawan et al. 2009).

The *Ink4a/Arf* locus also seems to be under control of another PcG protein Ezh2, a histone methyltransferase in Polycomb-Repressive complex 2 (PRC2) (Fig. 4.4). Decline in Ezh2 levels in aging islet β-cells leads to reduced histone H3 lysine 27 trimethylation (H3K27me3) at the *Ink4a/Arf* locus. Reduced repressive histone methylation marks lead to activation of *Ink4a/Arf* locus and increased expression of p16^{INK4a} and p19^{ARF}. Conditional deletion of Ezh2 in β-cells led to decreased H3K27me3 at the *Ink4a/Arf* locus, reduced β-cell proliferation, hypoglycemia and mild diabetes. Re-establishment of β-cell mass post-streptozotocin treatment was accompanied by increased Ezh2 expression (Chen et al. 2009). These findings suggest that PcG and TrxG proteins impart a combinatorial code of histone modifications on the *Ink4a/Arf* locus to control β-cell proliferation during aging and regeneration.

Epigenetic Regulation of β-Cell Function Through Non-Coding RNAs and MicroRNAs

In addition to DNA methylation and histone modifications, pancreatic islet cells may also be regulated by non-coding RNAs (ncRNAs). Conditional deletion of Dicer1 (an endoribonuclease essential in miRNA mediated post-transcriptional gene silencing) in the embryonic pancreas, results in lower β-cell mass and in impaired glucose tolerance (Mandelbaum et al. 2012). Indeed, overexpression of islet-specific miRNA, miR-375, was previously shown to suppress glucose-induced insulin secretion. Myotrophin, a protein involved in the final stages of insulin secretion, was identified as a target of miR-375 (Poy et al. 2004). Lipoapoptosis is an apoptotic program induced by free fatty acids. Lipoapoptosis of β-cells of islets is an important cause of β-cell dysfunction and loss of mass in diabetes. Interestingly, miR-375 was also shown to enhance lipoapoptosis in insulin-secreting NOD-derived β-cell line (Li et al. 2010) as well as an important microRNA expressed in human pancreas development as well (Joglekar et al. 2007, 2009).

Epigenetics of Stress Mediated β-Cell Death

Endoplasmic reticulum is the central organelle for folding and maturation of secretory and transmembrane proteins. Inability to meet folding demand can disrupt its function. The improperly folded proteins can lead to ER stress. Insulin-producing β-cells face ER-stress under excessive protein secretion requirement. As a response to ER-stress, cells initiate the unfolded protein response (UPR) and microRNAs play an important role in regulation of cellular response to ER stress. IRE-1 pathway is the most important UPR pathway. IRE-1α, the transmembrane ER-stress sensor (with RNase activity) induces thioredoxin-interacting protein (TXNIP) by reducing levels of a TXNIP destabilizing micro-RNA, miR-17. The IRE-1α pathway and anti-apoptotic microRNA scan be targets to develop effective treatments for cell degenerative diseases (Kadamkode and Banerjee 2014; Lerner et al. 2012).

Environmental Modifiers of Epigenetic Regulation

Intrauterine Environment, Nutrition and Diet

In addition to age, the intrauterine environment, diet and hormones also modulate epigenetic regulators of islet function. Though, these modifiers are described under separate headings, as can be seen, cross-talk exists between them which perhaps drives the genetic and epigenetic changes toward proper functioning of the pancreatic islet.

An earlier investigation in the Pima Indians showed that the children exposed to maternal diabetes in pregnancy were more likely to be obese and glucose intolerant as compared to the children born to pre-diabetic and non-diabetic mothers (Pettitt et al. 1993). In order to delineate the role of intrauterine environment from genetic factors, a subsequent study (Dabelea et al. 2000) investigated the risk of obesity and diabetes in the siblings who were discordant to intrauterine hyperglycemia (born before and after the development of maternal diabetes). This study showed that the risk of obesity and diabetes was higher in the siblings who were born after the diagnosis of maternal diabetes underlining the importance of intrauterine environment as an important factor in development of diabetes. Barker hypothesized that the environmental triggers in the crucial period of early life (e.g., during fetal life) can influence the risk of metabolic disease in later life (Barker 2004). The concept was later supported by number of studies, for example, children born during the Dutch hunger winter period (famine of 1944–1945) were at increased risk of non-communicable diseases (Heijmans et al. 2008). This is thought to be related to epigenetic changes induced by maternal undernutrition. Children conceived during the winter hunger had less DNA methylation of the imprinted insulin-like growth factor-II (*IGF2*) gene compared with their unexposed, same-sex siblings when studied six decades later. Several studies in animal models have begun to characterize epigenetic modifications that are influenced by the intrauterine environment (Lehnen et al. 2013). These studies elucidate the epigenetic route often followed by metabolic reprogramming.

Hnf4a is a transcription factor associated with regulation of insulin secretion and is down-regulated in T2DM. Hnf4a expression was shown to be epigenetically regulated by environmental factors such as maternal diet and aging in rat islets. Environmentally induced epigenetic changes at the Hnf4a enhancer altered its interaction with the P2 promoter. This altered interaction seems to affect Hnf4a transcription consequently increasing T2DM risk (Sandovici et al. 2011).

Another study suggested contribution of paternal diet in β-cell dysfunction in female offspring. The study used a rodent model to address the paternal metabolic reprogramming. High-fat diet (HFD) induced adiposity, glucose intolerance and insulin resistance in rats. When male rats on chronic HFD were crossed with control females, the female progeny exhibited an early onset of impaired insulin secretion and glucose intolerance that worsened with time. Altered expression of 642 genes was observed in pancreatic islets of female progeny. Most notable among these genes was *IL-13RA2*, a receptor for interleukin which is important in cell viability in pancreatic β-cells. IL-13RA2 expression exhibited the highest fold change, and the locus was hypomethylated in female offspring of HFD rats (Ng et al. 2010).

Circulating free fatty acids are often elevated in obese individuals and in patients with T2DM. When human islets were treated with Palmitate in vitro, insulin secretion was found to be impaired. Moreover, 1860 genes, including the candidates for T2DM and those involved in glucose metabolism, showed differential expression. Out of the differentially expressed genes, 290 genes exhibited differential methylation, while 37 genes showed differential expression in islets from T2DM patients suggesting role of DNA methylation in fat-induced dysregulation of genes in diabetes (Hall et al. 2014).

Exposure to a change in nutritional environment seems to set up a metabolic state in the parents that is best suited to that environment. This state is transmitted to the offspring through epigenetic marks in the parental gametes or developing fetus. If, however, nutrition status of the progeny differs from that of the parents, their metabolism will no longer be adapted to the environment. This kind of 'mismatch' could be a potential source of altered risk for developing complex metabolic disorders (Somer and Thummel 2014).

Hormones and Growth Factors

Platelet-derived growth factor receptor (Pdgfr) signaling controls age-dependent β-cell proliferation in pancreatic islets. With age, declining β-cell Pdgfr-α level was accompanied by reductions in Ezh2 levels and β-cell loss. Conditional inactivation of the *Pdgfra* gene in β-cells prevented mouse neonatal β-cell expansion and adult β-cell regeneration. Targeted human PDGFR-α activation in mouse β-cells stimulated Erk1/2 phosphorylation, leading to Ezh2-dependent expansion of adult β-cells. Adult human islets lack PDGF signaling competence, but exposure of juvenile human islets to PDGF-AA stimulated β-cell proliferation. Using complementary KO models, the authors concluded that Pdgfr-α controls β-cell proliferation by positively regulating Ezh2 expression via Erk and Rb/E2f signaling (Chen et al. 2011).

During pregnancy, maternal pancreatic islets grow to match dynamic physiological demands. Menin, an endocrine tumor suppressor protein negatively regulates islet growth in pregnant mice. Menin expression seems to be negatively regulated by pregnancy hormone prolactin via Bcl6 expression to stimulate β-cell proliferation (Karnik et al. 2007). Interestingly, menin is also an essential component of the MLL/SET1 H3K4 methyl-transferase complex. Menin-dependent histone methylation maintains the expression of cyclin-dependent kinase inhibitors $p27^{Kip1}$ and $p18^{INK4C}$ to prevent cell proliferation. Thus, reduction in Menin enhances proliferation of β-cells, providing an epigenetic mechanism for pregnancy-induced changes in islet expansion (Karnik et al. 2005).

The gut-derived hormones incretins (GLP-1 & GIP) regulate blood glucose levels. Insulin secretion is enhanced upon the binding of GLP-1 to its receptor (GLP-1-R) in pancreatic β-cells. Studies with $Glp1r^{-/-}$; Gipr-ipredKO) mice suggested that the incretins work via epigenetic repression of *Fxyd3*, an ion transport regulator, by promoter methylation. Accordingly, *Fxyd3* was up-regulated in dKO mice and diabetic islets and showed reduced promoter methylation. *Fxyd3* overexpression impaired glucose-stimulated insulin secretion linking epigenetic regulation of β-cell function to hormone action (Vallois et al. 2014).

Similar to the effect of nutrition, the effect of hormones can also be transgenerational leaving an epigenetic memory behind, predisposing the progeny to diabetes. Gestational diabetes mellitus (GDM) has been shown to be associated with high risk of diabetes in offspring. When F1 offspring of the GDM mice were mated with control mice, the F2 progeny exhibited higher impaired glucose tolerance (IGT). Insulin-like growth factor-II (*IGF2*) and non-coding RNA *H19* are two important genes involved in islet development and the pathogenesis of diabetes. While H19 gene is expressed only from maternally inherited chromosome, IGF2 is expressed only from the paternally inherited chromosome. In pancreatic islets of both F1 and F2 offspring of GDM parent, abnormal methylation status of the differentially methylated region (DMR) was found to down-regulate the expression of Igf2 and H19. Interestingly, altered *Igf2* and *H19* gene expression was found in sperm of adult F1-GDM mice, indicating that epigenetic changes in germ cells can contribute to the transgenerational transmission of diabetes (Ding et al. 2012).

Since epigenetic processes play a crucial role in β-cell maintenance, dysregulation of these processes can be expected to contribute to diabetes. Accordingly, many studies identify epigenetic alterations in clinical samples of both types of diabetes. Here, we discuss few of such studies.

Age and Other Factors

The replication potential of β-cells gradually declines with age and is associated with decreased insulin secretion. As mentioned earlier (Sects. "Hormones and

Growth Factors" and "Epigenetics of the β-cell proliferation"), Bmi1, a component of PRC1 complex, and Ezh2, a component of PRC2 complex, positively regulate islet cell growth by repressing *Ink4a/Arf* locus in islet β-cells and decline in expression during aging leading to reduction in the proliferation potential. Epigenetic regulation of key genes involved in islet proliferation and function has been studied extensively.

Pdx-1 is crucial for β-cell development and function and silencing of the *Pdx1* gene in pancreatic β-cells of mice causes diabetes. Studies on islets isolated from non-diabetic and T2DM patient donors showed decrease in the expression of *Pdx1* mRNA in the latter. Further, 10 CpG sites in the distal *Pdx1* promoter and enhancer regions were observed to be highly methylated in T2DM islets. Exposure to high glucose lead to reduced *Pdx1* expression, increased DNMT1 expression and Pdx1 methylation in clonal β-cells. Given the DNA methylation changes at *Pdx1* in T2DM and in β-cells exposed to hyperglycemia, epigenetic mechanisms seem to play a role in the development of T2DM through Pdx1 regulation (Yang et al. 2012).

Similarly, insulin gene expression is also regulated by its promoter methylation in a cell type-specific manner. The CpG sites in both mouse and human promoters are demethylated in insulin-producing pancreatic beta cells (Kuroda et al. 2009). When epigenetic regulation of insulin gene promoter was studied in islets from healthy individuals and T2DM patients, insulin content and glucose-stimulated insulin secretion was found to be reduced in the latter. The reduced expression correlated with higher levels of insulin promoter methylation (Yang et al. 2011). Insulin secretion is also dependent upon mitochondrial function. The Peroxisome proliferator-activated receptor gamma coactivator 1-α (PGC-1α or PPARGC-1α) is a master regulator of mitochondrial genes including those involved in oxidative phosphorylation. In the first study that linked epigenetic modification to diabetes, PGC-1α was demonstrated to be reduced up to 90 % in expression in islets from T2DM patients. Furthermore, the *Ppargc1a* gene promoter was shown to have a twofold increase in methylation in T2DM islets as compared with controls (Ling et al. 2008). Together these results indicate that insulin expression and secretion are subject to epigenetic regulation which is often disrupted in diabetes.

We previously discussed the role of incretins in regulation of glucose levels through epigenetic mechanisms (Sect. "Hormones and Growth Factors"). Interestingly, the expression of incretins itself is also regulated by epigenetic mechanisms. When DNA methylation levels of CpGs close to the transcription start site of Glp-1 receptor (GLP1R) were analyzed in pancreatic islets from non-diabetic and T2DM patients, one CpG unit showed a small increase in DNA methylation in islets from donors with T2DM as compared to healthy donors (Hall et al. 2013). These results suggest that epigenetic regulation has dual control over incretin-mediated insulin secretion and that dysregulation of incretin signaling and incretin function can be associated with diabetes.

Specific miRNAs are also implicated in the etiology of T2DM. A cluster of microRNAs was found in an imprinted locus on human chromosome 14q32 that was dramatically down-regulated in islets from T2DM organ donors. The disease-relevant targets of the chromosome 14q32 microRNAs were found to be involved in β-cell apoptosis. The down-regulation of this microRNA locus by hypermethylation of its promoter perhaps leads to activation of the apoptotic pathway in β-cells in diabetes patients (Kameswaran et al. 2014).

Genome-Wide Studies in diabetes

Genome-wide studies have also been performed for epigenetic analysis of diabetes patient samples. Volkmar and colleagues analyzed human islet DNA methylation profiles from T2DM and non-diabetic cadaveric organ donors and identified 276 differentially methylated CpG sites affiliated with 254 genes. Functional annotation of the aberrantly methylated genes highlighted pathways implicated in β-cell survival and function suggesting involvement of epigenetic dysregulation in diabetic islets (Volkmar et al. 2012).

Dayeh et al. performed DNA methylome and gene expression analyses in pancreatic islets from non-diabetic and T2DM donors. The study identified 1649 CpG sites and 853 genes, including diabetes susceptibility genes *TCF7L2*, and *KCNQ1*, with differential DNA methylation in T2DM islets. Increased expression and decreased methylation of CDKN1A (regulator of cell cycle progression) and PDE7B (regulator of cAMP signaling) were also observed in T2DM islets (Dayeh et al. 2014).

In another study, genome-wide methylation quantitative traits loci (mQTL) were analyzed in islets in order study the genetic and epigenetic association with function. The study identified 67,438 SNP-CpG pairs in cis and 2562 significant SNP-CpG pairs in trans including reported diabetes gene loci (e.g., *ADCY5*, *KCNJ11*, *HLA-DQA1*, *INS*, *PDX1* and *GRB10*). Follow-up analyses further identified mQTLs associated with gene expression and insulin secretion in human islets (Olsson et al. 2014).

More recently, Pasquali et al. (2014) identified genomic sequences targeted by islet transcription factors to drive islet-specific gene activity. The study showed that most target sequences resided in clusters of enhancers that form physical three-dimensional chromatin domains. Finally, sequence variants associated with T2DM and glycemic states that were identified from GWAS studies were found to disrupt DNA binding and islet enhancer activity. The study illustrated how islet transcription factors interact functionally with the epigenome and provided systematic evidence that the dysregulation of islet enhancers is relevant to the mechanisms underlying T2DM.

Taken together, these studies indicate that epigenetic regulation of the islet development plays a key role in T2DM etiology. The genetic makeup and the environmental factors modulate the epigenetic machinery either in a site-specific or genome-wide scale to favor disruption of sugar metabolism and onset of diabetes (Fig. 4.5). The environmental factors that affect islet development and their effects on β-cell regeneration are tabulated in Table 4.1.

T1DM, on the other hand, is characterized by destruction of islet β-cell by misregulated immune cells. Factors involved in immune response as well as those

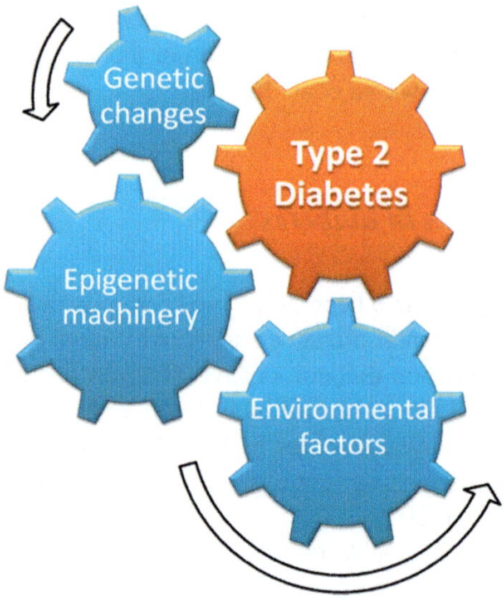

Fig. 4.5 Epigenetic basis of onset of the Type 2 diabetes mellitus. The environmental factors (obesity, aging etc.) as well as genetic factors influence onset of T2DM. Both environmental factors and genetic factors modulate the epigenetic machinery (DNA methylation, histone modifications and microRNA expression etc.) to effect a cascade of events that lead to diabetes

Table 4.1 The epigenetic connection: list of environmental factors that affect β-cell regeneration

Category	Factors	Epigenetic change	Effect
Age	Ezh2, Bmi1	Down-regulation	Decrease in β-cell population
	Ink4/Arf	Demethylation of the gene locus	Decrease in β-cell population
	Pdx-1	Promoter methylation	Decrease in β-cell population
Nutrition and Diet	High-fat diet	IL-13RA hypomethylation	β-cell dysfunction
	Palmitate intake	Global changes in DNA methylation	β-cell dysfunction
Hormones	PDGF	Ezh2 activation	Increase in β-cells population
	Pregnancy-related hormones	Menin-mediated repression of cell cycle inhibitors	Increase in β-cells population
	Incretins	Repression of Fxyd3	Increase in β-cells function

involved in β-cell development play a key role in onset and progression of T1DM. As mentioned earlier, the involvement of factors involved in islet development, β-cell differentiation is relatively less well established in T1DM. The insulin (INS) region is one of the most important loci associated with T1DM. When DNA methylation patterns of the 7 CpGs proximal to the transcription start site in the *INS* gene promoter were studied, they revealed altered methylation pattern. However, a causal relationship could not be established between changes in DNA methylation and T1DM (Fradin et al. 2012). Further studies are required to understand the functional role of epigenetic regulators in T1DM.

Epigenome Studies for Cell Type-Specific Differences in Islet Cells

While locus specific changes in epigenetic marks in β-cells (Sect. "Age and Other Factors") underscore their importance in islet-specific gene expression, altered epigenome profile in diabetes outlines its dysregulation under the disease condition. Considering genomic equivalence of all somatic cells of the body, it is clear that the epigenome profile is crucial in separating one cell type into another. Identification of the epigenetic differences between islet cell types can therefore be crucial in development of strategies to reprogramming one islet cell into another.

When chromatin from human islets was profiled using formaldehyde-assisted isolation of regulatory elements followed by high-throughput sequencing (FAIRE-Seq) analysis, 3300 physically linked clusters of islet-selective open chromatin sites were identified. Interestingly, analysis of risk alleles showed that rs7903146, an intronic variant of Tcf7L2, a master regulator of insulin production, was present in a FAIRE enriched site (Gaulton et al. 2010).

Investigation of epigenetic marks on β-, α-, and exocrine-cells at genome-wide level revealed that α-cells exhibited higher number of genes harboring bivalent chromatin marks, specifically the H3K4me3 (a mark of active chromatin) and H3K27me3 (a mark of repressed chromatin). Remarkably, in α-cells, the β-cell signature genes involved in transcriptional regulation were among the bivalently marked genes in α-cells. These results provided a possible explanation for the trans-differentiation potential of the α-cells and also suggested that epigenomic manipulation could enable cellular reprogramming of islet cells. Accordingly, treatment of human islets with a histone methyltransferase inhibitor caused a partial conversion of α-cells toward the β-cell fate, indicating that the epigenomic modulators might serve as means to reprogram endocrine cells to develop as β-cells (Bramswig et al. 2013; Bramswig and Kaestner 2014).

Human islet-derived precursor cells (hIPCs) are mesenchymal cells derived in vitro from adult pancreas. hIPCs do not express insulin but proliferate freely and can be differentiated to epithelial cells that express insulin. hIPCs have been studied with the goal of obtaining large quantities of insulin-producing cells suitable for transplantation into patients suffering from type 1 diabetes. When patterns of histone modifications were studied, hIPCs showed significant levels of active chromatin modifications over the insulin gene. This indicated that hIPCs, which do not express the insulin gene, uniquely exhibit epigenetic marks that could poise them for activation of insulin expression (Mutskov et al. 2007).

The genome-wide studies mentioned above confirm the epigenome differences between islet cell-types in turn indicating that alterations in epigenetic regulators can lead to their inter-conversion via intermediary transcription factors or directly. Below we describe some examples of epigenetic modifiers in in vivo and in vitro regeneration of β-cells.

Epigenetics of Reprogramming and Therapeutic Ramifications

Epigenetics of in Vivo regeneration

Intrauterine growth restriction (IUGR) refers to a condition in which an unborn baby is smaller than expected due to a slower rate of growth inside the womb. IUGR has been linked to the onset of diabetes in adulthood. When epigenetic regulation of *Pdx1* gene was studied following IUGR, repressive histone marks were observed at the *Pdx1* promoter in fetal and neonatal rodents. These repressive marks and the reduction in *Pdx1* expression could be reversed by HDAC inhibitors until neonatal stage; however; after the onset of diabetes in adults, DNA methylation mediated permanent silencing of *Pdx1* locus (Park et al. 2008). Exendin-4 is a hormone of reptile origin that increases *Pdx1* expression and prevents the development of diabetes in the IUGR rat. Administered of exendin-4 to the newborn offspring of IUGR rats increased association of upstream stimulatory factor-1 and histone acetyltransferase PCAF at the proximal promoter of *Pdx1*. Histone H3 acetylation and tri-methylation of H3K4 were permanently increased, whereas DNMT1 binding and subsequent DNA methylation were prevented at the proximal promoter of *Pdx1* in IUGR islets suggesting epigenetic mode of action of this hormone analog (Pinney et al. 2011).

HDAC Inhibitors and Other Small Molecules

Histone deacetylases (HDACs) have an important role in the developmental regulation of the pancreas. Based on sequence similarity, catalytic sites, and cofactor dependency, mammalian HDACs are grouped into the class I, IIa, IIb, III (sirtuins) and class IV. HDACs can repress gene expression and regulate cell proliferation, differentiation, and function in various tissues. Haumaitre and colleagues discovered that pancreatic treatment with different HDAC inhibitors (HDACi) modified the timing and determination of pancreatic cell fate. Importantly, HDACi treatment promoted the NGN3 positive pro-endocrine lineage. Treatment with trichostatin A (TSA) and sodium butyrate (NaB), two inhibitors of both class I and class II HDACs, enhanced the pool of β-cells (Haumaitre et al. 2008). Further, HDAC expression analysis in islet cells revealed restricted expression of class IIa HDACs (HDAC 4, 5, 7, 9) in β- and δ-cells of pancreas. Interestingly, rats lacking Hdac 5 and 9 had an increasing number of β-cell, and those lacking Hdac4 and 5 revealed a greater pool of δ-cells (Lenoir et al. 2011). Together, these observations highlight the roles of HDACs at key points in exocrine and endocrine differentiation and their potential applications in cell replacement therapies in diabetes.

In another study, two potential anticancer agents, Romidepsin (Romi), an HDAC inhibitor and 5-Azacytidine (5-azaC), a chemical analog of Cytidine that cannot be methylated, were used in combination with Pdx1 to induce expression of insulin from primary human dermal fibroblasts (hDFs). The study showed that fully differentiated non-islet cells can be transdifferentiated into islet-like cells by combining epigenetic modulation with transcription factor modulation (Katz et al. 2013).

Recently, short treatment with I-BET151, a small-molecule inhibitor of a family of bromodomain-containing transcriptional regulators was shown to irreversibly suppress development of diabetes in non-obese diabetic (NOD) mice, an animal model for T1DM. The inhibitor induced pancreatic macrophages to adopt an anti-inflammatory phenotype (impacting the NF-κB pathway in particular) and elicited regeneration of islet β-cells. I-BET151 thus achieved a 'combination therapy' often advocated for T1DM (Fu et al. 2014).

Future Directions

In the past few decades, we have significantly improved our understanding of the mechanistic actions of the factors that shape islet function at cellular level. Various studies have clearly suggested that environmental factors can create cellular memory that pre-disposes cells to disease conditions later in life. These studies underscore the importance of epigenetic regulation in islet function.

While epigenetic differences between islet cell-types underscore the possibility of their inter-conversion, the small molecular inhibitors provide a tool for editing of their epigenomes. Considerable progress has been made in regeneration of β-cells from pancreatic and non-pancreatic cells; however, site-specific modulation of epigenome might represent the next step in β-cell regeneration. Such studies will not only require more detailed understanding of the β-cell reprogramming, but will also necessitate strategies for better targeting and efficient delivery of drugs to the target organ. Additionally, along with insulin expression, strategies to improve insulin release need to take the center stage.

The combinatorial action of small molecular epigenetic regulators on β-cells and the target cells of the body (in T2DM) or the cells of the immune system (in T1DM) needs to be used to its fullest potential. However, while testing these molecules on animal systems such as rodents, the inherent differences between the rodents and the human system also need to be taken into consideration. While developing the treatment modalities for diabetes, epigenetic studies can also be used for diabetes prevention. Studies addressing reversal of the adverse epigenetic memory left by the environmental factors on islet cells are required to be undertaken more vigorously. With these research directions along with some lifestyle intervention efforts, significant progress can be expected to be made in coming years with a hope to eradicate better manage this complex metabolic disorder.

Acknowledgments SPK would like to thank IISER-Pune for Postdoctoral research fellowship. The authors would like to thank Dr. Puli Chandramouli Reddy, Dr. Ashwin Kelkar and members of the SG laboratory for discussion and suggestions. PK would like to thank Dr. CS Yajnik, Diabetes Unit, KEM Hospital Research Centre, Pune, for support. Work in SG laboratory is supported by grants from the Centre of Excellence (CoE) in Epigenetics program and Centre of Excellence in Biotechnology (CEIB) program from the Department of Biotechnology.

References

Ahlgren U, Jonsson J, Jonsson L, Simu K, Edlund H (1998) Beta-cell-specific inactivation of the mouse Ipf1/Pdx1 gene results in loss of the beta-cell phenotype and maturity onset diabetes. Genes Dev 12:1763–1768

Ahlgren U, Pfaff SL, Jessell TM, Edlund T, Edlund H (1997) Independent requirement for ISL1 in formation of pancreatic mesenchyme and islet cells. Nature 385:257–260

Anderson KR, Torres CA, Solomon K, Becker TC, Newgard CB, Wright CV, Hagman J, Sussel L (2009) Cooperative transcriptional regulation of the essential pancreatic islet gene NeuroD1 (beta2) by Nkx2.2 and neurogenin 3. J Biol Chem 284:31236–31248

Andrali SS, Sampley ML, Vanderford NL, Ozcan S (2008) Glucose regulation of insulin gene expression in pancreatic beta-cells. Biochem J 415:1–10

Apelqvist A, Li H, Sommer L, Beatus P, Anderson DJ, Honjo T, Hrabe de Angelis M, Lendahl U, Edlund H (1999) Notch signalling controls pancreatic cell differentiation. Nature 400:877–881

Avrahami D, Kaestner KH (2012) Epigenetic regulation of pancreas development and function. Semin Cell Dev Biol 23:693–700

Baggio LL, Drucker DJ (2007) Biology of incretins: GLP-1 and GIP. Gastroenterology 132:2131–2157

Banerjee M, Kanitkar M, Bhonde RR (2005) Approaches towards endogenous pancreatic regeneration. Rev Diabet Stud 2:165–176

Bar-Nur O, Russ HA, Efrat S, Benvenisty N (2011) Epigenetic memory and preferential lineage-specific differentiation in induced pluripotent stem cells derived from human pancreatic islet beta cells. Cell Stem Cell 9:17–23

Barker DJ (2004) The developmental origins of well-being. Philos Trans R Soc Lond B Biol Sci 359:1359–1366

Ben-Othman N, Courtney M, Vieira A, Pfeifer A, Druelle N, Gjernes E, Faurite B, Avolio F, Collombat P (2013) From pancreatic islet formation to beta-cell regeneration. Diabetes Res Clin Pract 101:1–9

Bhutani N, Burns DM, Blau HM (2011) DNA demethylation dynamics. Cell 146:866–872

Bird AP (1986) CpG-rich islands and the function of DNA methylation. Nature 321:209–213

Bramswig NC, Everett LJ, Schug J, Dorrell C, Liu C, Luo Y, Streeter PR, Naji A, Grompe M, Kaestner KH (2013) Epigenomic plasticity enables human pancreatic alpha to beta cell reprogramming. J Clin Investig 123:1275–1284

Bramswig NC, Kaestner KH (2014) Transcriptional and epigenetic regulation in human islets. Diabetologia 57:451–454

Chen H, Gu X, Liu Y, Wang J, Wirt SE, Bottino R, Schorle H, Sage J, Kim SK (2011) PDGF signalling controls age-dependent proliferation in pancreatic beta-cells. Nature 478:349–355

Chen H, Gu X, Su IH, Bottino R, Contreras JL, Tarakhovsky A, Kim SK (2009) Polycomb protein Ezh2 regulates pancreatic beta-cell Ink4a/Arf expression and regeneration in diabetes mellitus. Genes Dev 23:975–985

Collombat P, Hecksher-Sorensen J, Krull J, Berger J, Riedel D, Herrera PL, Serup P, Mansouri A (2007) Embryonic endocrine pancreas and mature beta cells acquire alpha and PP cell phenotypes upon Arx misexpression. J Clin Investig 117:961–970

Collombat P, Mansouri A, Hecksher-Sorensen J, Serup P, Krull J, Gradwohl G, Gruss P (2003) Opposing actions of Arx and Pax4 in endocrine pancreas development. Genes Dev 17:2591–2603

Collombat P, Xu X, Ravassard P, Sosa-Pineda B, Dussaud S, Billestrup N, Madsen OD, Serup P, Heimberg H, Mansouri A (2009) The ectopic expression of Pax4 in the mouse pancreas converts progenitor cells into alpha and subsequently beta cells. Cell 138:449–462

Conerly M, Grady WM (2010) Insights into the role of DNA methylation in disease through the use of mouse models. Dis Model Mech 3:290–297

Dabelea D, Hanson RL, Lindsay RS, Pettitt DJ, Imperatore G, Gabir MM, Roumain J, Bennett PH, Knowler WC (2000) Intrauterine exposure to diabetes conveys risks for type 2 diabetes and obesity: a study of discordant sibships. Diabetes 49:2208–2211

Dang MN, Buzzetti R, Pozzilli P (2013) Epigenetics in autoimmune diseases with focus on type 1 diabetes. Diabetes Metab Res Rev 29:8–18

Dayeh T, Ling C (2015) Does epigenetic dysregulation of pancreatic islets contribute to impaired insulin secretion and type 2 diabetes? Biochem Cell Biol 4.1–11

Dayeh T, Volkov P, Salo S, Hall E, Nilsson E, Olsson AH, Kirkpatrick CL, Wollheim CB, Eliasson L, Ronn T et al (2014) Genome-wide DNA methylation analysis of human pancreatic islets from type 2 diabetic and non-diabetic donors identifies candidate genes that influence insulin secretion. PLoS Genet 10:e1004160

De Jesus DF, Kulkarni RN (2014) Epigenetic modifiers of islet function and mass. Trends Endocrinol Metab 25:628–636

De Leon DD, Deng S, Madani R, Ahima RS, Drucker DJ, Stoffers DA (2003) Role of endogenous glucagon-like peptide-1 in islet regeneration after partial pancreatectomy. Diabetes 52:365–371

Dhawan S, Georgia S, Tschen SI, Fan G, Bhushan A (2011) Pancreatic beta cell identity is maintained by DNA methylation-mediated repression of Arx. Dev Cell 20:419–429

Dhawan S, Tschen SI, Bhushan A (2009) Bmi-1 regulates the Ink4a/Arf locus to control pancreatic beta-cell proliferation. Genes Dev 23:906–911

Ding GL, Wang FF, Shu J, Tian S, Jiang Y, Zhang D, Wang N, Luo Q, Zhang Y, Jin F et al (2012) Transgenerational glucose intolerance with Igf2/H19 epigenetic alterations in mouse islet induced by intrauterine hyperglycemia. Diabetes 61:1133–1142

Dor Y, Brown J, Martinez OI, Melton DA (2004) Adult pancreatic beta-cells are formed by self-duplication rather than stem-cell differentiation. Nature 429:41–46

Du A, Hunter CS, Murray J, Noble D, Cai CL, Evans SM, Stein R, May CL (2009) Islet-1 is required for the maturation, proliferation, and survival of the endocrine pancreas. Diabetes 58:2059–2069

Esteller M (2011) Non-coding RNAs in human disease. Nat Rev Genet 12:861–874

Ferber S, Halkin A, Cohen H, Ber I, Einav Y, Goldberg I, Barshack I, Seijffers R, Kopolovic J, Kaiser N et al (2000) Pancreatic and duodenal homeobox gene 1 induces expression of insulin genes in liver and ameliorates streptozotocin-induced hyperglycemia. Nat Med 6:568–572

Fradin D, Le Fur S, Mille C, Naoui N, Groves C, Zelenika D, McCarthy MI, Lathrop M, Bougneres P (2012) Association of the CpG methylation pattern of the proximal insulin gene promoter with type 1 diabetes. PLoS ONE 7:e36278

Fu W, Farache J, Clardy SM, Hattori K, Mander P, Lee K, Rioja I, Weissleder R, Prinjha RK, Benoist C et al (2014) Epigenetic modulation of type-1 diabetes via a dual effect on pancreatic macrophages and beta cells. eLife 3:e04631

Gaulton KJ, Nammo T, Pasquali L, Simon JM, Giresi PG, Fogarty MP, Panhuis TM, Mieczkowski P, Secchi A, Bosco D et al (2010) A map of open chromatin in human pancreatic islets. Nat Genet 42:255–259

Gradwohl G, Dierich A, LeMeur M, Guillemot F (2000) neurogenin3 is required for the development of the four endocrine cell lineages of the pancreas. Proc Natl Acad Sci USA 97:1607–1611

Hall E, Dayeh T, Kirkpatrick CL, Wollheim CB, Dekker Nitert M, Ling C (2013) DNA methylation of the glucagon-like peptide 1 receptor (GLP1R) in human pancreatic islets. BMC Med Genet 14:76

Hall E, Volkov P, Dayeh T, Bacos K, Ronn T, Nitert MD, Ling C (2014) Effects of palmitate on genome-wide mRNA expression and DNA methylation patterns in human pancreatic islets. BMC Med 12:103

Harrison KA, Thaler J, Pfaff SL, Gu H, Kehrl JH (1999) Pancreas dorsal lobe agenesis and abnormal islets of Langerhans in Hlxb9-deficient mice. Nat Genet 23:71–75

Haumaitre C (2013) Epigenetic regulation of pancreatic islets. Curr DiabRep 13:624–632

Haumaitre C, Lenoir O, Scharfmann R (2008) Histone deacetylase inhibitors modify pancreatic cell fate determination and amplify endocrine progenitors. Mol Cell Biol 28:6373–6383

Heijmans BT, Tobi EW, Stein AD, Putter H, Blauw GJ, Susser ES, Slagboom PE, Lumey LH (2008) Persistent epigenetic differences associated with prenatal exposure to famine in humans. Proc Natl Acad Sci USA 105:17046–17049

Heller RS, Stoffers DA, Liu A, Schedl A, Crenshaw EB 3rd, Madsen OD, Serup P (2004) The role of Brn4/Pou3f4 and Pax6 in forming the pancreatic glucagon cell identity. Dev Biol 268:123–134

IDF (2014) IDF Diabetes Atlas update poster. International Diabetes Federation, Brussels

Jayani RS, Ramanujam PL, Galande S (2010) Studying histone modifications and their genomic functions by employing chromatin immunoprecipitation and immunoblotting. Methods Cell Biol 98:35–56

Joglekar MV, Joglekar VM, Hardikar AA (2009) Expression of islet-specific microRNAs during human pancreatic development. Gene Expr Patterns 9:109–113

Joglekar MV, Parekh VS, Hardikar AA (2007) New pancreas from old: microregulators of pancreas regeneration. Trends Endocrinol Metab 18:393–400

Kadamkode V, Banerjee G (2014) Micro RNA: an epigenetic regulator of type 2 diabetes. MicroRNA 3:86–97

Kameswaran V, Bramswig NC, McKenna LB, Penn M, Schug J, Hand NJ, Chen Y, Choi I, Vourekas A, Won KJ et al (2014) Epigenetic regulation of the DLK1-MEG3 microRNA cluster in human type 2 diabetic islets. Cell Metab 19:135–145

Karnik SK, Chen H, McLean GW, Heit JJ, Gu X, Zhang AY, Fontaine M, Yen MH, Kim SK (2007) Menin controls growth of pancreatic beta-cells in pregnant mice and promotes gestational diabetes mellitus. Science 318:806–809

Karnik SK, Hughes CM, Gu X, Rozenblatt-Rosen O, McLean GW, Xiong Y, Meyerson M, Kim SK (2005) Menin regulates pancreatic islet growth by promoting histone methylation and expression of genes encoding p27Kip1 and p18INK4c. Proc Natl Acad Sci USA 102:14659–14664

Katz LS, Geras-Raaka E, Gershengorn MC (2013) Reprogramming adult human dermal fibroblasts to islet-like cells by epigenetic modification coupled to transcription factor modulation. Stem Cells Dev 22:2551–2560

Khare SP, Habib F, Sharma R, Gadewal N, Gupta S, Galande S (2012) HIstome—a relational knowledgebase of human histone proteins and histone modifying enzymes. Nucleic Acids Res 40:D337–D342

Kojima H, Fujimiya M, Matsumura K, Younan P, Imaeda H, Maeda M, Chan L (2003) NeuroD-betacellulin gene therapy induces islet neogenesis in the liver and reverses diabetes in mice. Nat Med 9:596–603

Kuroda A, Rauch TA, Todorov I, Ku HT, Al-Abdullah IH, Kandeel F, Mullen Y, Pfeifer GP, Ferreri K (2009) Insulin gene expression is regulated by DNA methylation. PLoS ONE 4:e6953

Laird PW (2010) Principles and challenges of genomewide DNA methylation analysis. Nat Rev Genet 11:191–203

Lehnen H, Zechner U, Haaf T (2013) Epigenetics of gestational diabetes mellitus and offspring health: the time for action is in early stages of life. Mol Hum Reprod 19:415–422

Lenoir O, Flosseau K, Ma FX, Blondeau B, Mai A, Bassel-Duby R, Ravassard P, Olson EN, Haumaitre C, Scharfmann R (2011) Specific control of pancreatic endocrine beta- and delta-cell mass by class IIa histone deacetylases HDAC4, HDAC5, and HDAC9. Diabetes 60:2861–2871

Lerner AG, Upton JP, Praveen PV, Ghosh R, Nakagawa Y, Igbaria A, Shen S, Nguyen V, Backes BJ, Heiman M et al (2012) IRE1alpha induces thioredoxin-interacting protein to activate the NLRP3 inflammasome and promote programmed cell death under irremediable ER stress. Cell Metab 16:250–264

Li Y, Xu X, Liang Y, Liu S, Xiao H, Li F, Cheng H, Fu Z (2010) miR-375 enhances palmitate-induced lipoapoptosis in insulin-secreting NIT-1 cells by repressing myotrophin (V1) protein expression. Int J Clin Exp Pathol 3:254–264

Ling C, Del Guerra S, Lupi R, Ronn T, Granhall C, Luthman H, Masiello P, Marchetti P, Groop L, Del Prato S (2008) Epigenetic regulation of PPARGC1A in human type 2 diabetic islets and effect on insulin secretion. Diabetologia 51:615–622

MacFarlane AJ (2012) Epigenetic epidemiology of type 1 diabetes. In: Michels KB (ed) Epigenetic epidemiology. Springer, Dordrecht, pp 377–400

Malecki MT, Jhala US, Antonellis A, Fields L, Doria A, Orban T, Saad M, Warram JH, Montminy M, Krolewski AS (1999) Mutations in NEUROD1 are associated with the development of type 2 diabetes mellitus. Nat Genet 23:323–328

Mandelbaum AD, Melkman-Zehavi T, Oren R, Kredo-Russo S, Nir T, Dor Y, Hornstein E (2012) Dysregulation of Dicer1 in beta cells impairs islet architecture and glucose metabolism. Exp Diabetes Res 2012:470302

Mashima H, Ohnishi H, Wakabayashi K, Mine T, Miyagawa J, Hanafusa T, Seno M, Yamada H, Kojima I (1996) Betacellulin and activin A coordinately convert amylase-secreting pancreatic AR42 J cells into insulin-secreting cells. J Clin Investig 97:1647–1654

Mattick JS (2001) Non-coding RNAs: the architects of eukaryotic complexity. EMBO Rep 2:986–991

Migliorini A, Bader E, Lickert H (2014) Islet cell plasticity and regeneration. Mol Metab 3:268–274

Mutskov V, Raaka BM, Felsenfeld G, Gershengorn MC (2007) The human insulin gene displays transcriptionally active epigenetic marks in islet-derived mesenchymal precursor cells in the absence of insulin expression. Stem Cells 25:3223–3233

Naujok O, Burns C, Jones PM, Lenzen S (2011) Insulin-producing surrogate beta-cells from embryonic stem cells: are we there yet? Mol Ther 19:1759–1768

Ng SF, Lin RC, Laybutt DR, Barres R, Owens JA, Morris MJ (2010) Chronic high-fat diet in fathers programs beta-cell dysfunction in female rat offspring. Nature 467:963–966

Nielsen JH, Haase TN, Jaksch C, Nalla A, Sostrup B, Nalla AA, Larsen L, Rasmussen M, Dalgaard LT, Gaarn LW et al (2014) Impact of fetal and neonatal environment on beta cell function and development of diabetes. Acta Obstet Gynecol Scand 93:1109–1122

Nielsen JH, Svensson C, Galsgaard ED, Moldrup A, Billestrup N (1999) Beta cell proliferation and growth factors. J Mol Med 77:62–66

Nussey SS, Whitehead SA (2001) Endocrinology: an integrated approach. BIOS Scientific Publishers, Oxford

Olden K, Lin YS, Gruber D, Sonawane B (2014) Epigenome: biosensor of cumulative exposure to chemical and nonchemical stressors related to environmental justice. Am J Public Health 104:1816–1821

Olsson AH, Volkov P, Bacos K, Dayeh T, Hall E, Nilsson EA, Ladenvall C, Ronn T, Ling C (2014) Genome-wide associations between genetic and epigenetic variation influence mRNA expression and insulin secretion in human pancreatic islets. PLoS Genet 10:e1004735

Pan FC, Wright C (2011) Pancreas organogenesis: from bud to plexus to gland. Dev Dyn 240:530–565

Pandol SJ (2010) The exocrine pancreas. Morgan & Claypool Life Sciences, San Rafael

Papizan JB, Singer RA, Tschen SI, Dhawan S, Friel JM, Hipkens SB, Magnuson MA, Bhushan A, Sussel L (2011) Nkx2.2 repressor complex regulates islet beta-cell specification and prevents beta-to-alpha-cell reprogramming. Genes Dev 25:2291–2305

Park JH, Stoffers DA, Nicholls RD, Simmons RA (2008) Development of type 2 diabetes following intrauterine growth retardation in rats is associated with progressive epigenetic silencing of Pdx1. J Clin Investig 118:2316–2324

Pasquali L, Gaulton KJ, Rodriguez-Segui SA, Mularoni L, Miguel-Escalada I, Akerman I, Tena JJ, Moran I, Gomez-Marin C, van de Bunt M et al (2014) Pancreatic islet enhancer clusters enriched in type 2 diabetes risk-associated variants. Nat Genet 46:136–143

Pettitt DJ, Nelson RG, Saad MF, Bennett PH, Knowler WC (1993) Diabetes and obesity in the offspring of Pima Indian women with diabetes during pregnancy. Diabetes Care 16:310–314

Pinney SE, Jaeckle Santos LJ, Han Y, Stoffers DA, Simmons RA (2011) Exendin-4 increases histone acetylase activity and reverses epigenetic modifications that silence Pdx1 in the intrauterine growth retarded rat. Diabetologia 54:2606–2614

Poy MN, Eliasson L, Krutzfeldt J, Kuwajima S, Ma X, Macdonald PE, Pfeffer S, Tuschl T, Rajewsky N, Rorsman P et al (2004) A pancreatic islet-specific microRNA regulates insulin secretion. Nature 432:226–230

Rooman I, Bouwens L (2004) Combined gastrin and epidermal growth factor treatment induces islet regeneration and restores normoglycaemia in C57Bl6/J mice treated with alloxan. Diabetologia 47:259–265

Sander M, Neubuser A, Kalamaras J, Ee HC, Martin GR, German MS (1997) Genetic analysis reveals that PAX6 is required for normal transcription of pancreatic hormone genes and islet development. Genes Dev 11:1662–1673

Sander M, Sussel L, Conners J, Scheel D, Kalamaras J, Dela Cruz F, Schwitzgebel V, Hayes-Jordan A, German M (2000) Homeobox gene Nkx6.1 lies downstream of Nkx2.2 in the major pathway of beta-cell formation in the pancreas. Development 127:5533–5540

Sandovici I, Smith NH, Nitert MD, Ackers-Johnson M, Uribe-Lewis S, Ito Y, Jones RH, Marquez VE, Cairns W, Tadayyon M et al (2011) Maternal diet and aging alter the epigenetic control of a promoter-enhancer interaction at the Hnf4a gene in rat pancreatic islets. Proc Natl Acad Sci USA 108:5449–5454

Schisler JC, Jensen PB, Taylor DG, Becker TC, Knop FK, Takekawa S, German M, Weir GC, Lu D, Mirmira RG et al (2005) The Nkx6.1 homeodomain transcription factor suppresses glucagon expression and regulates glucose-stimulated insulin secretion in islet beta cells. Proc Natl Acad Sci USA 102:7297–7302

Scott LJ, Mohlke KL, Bonnycastle LL, Willer CJ, Li Y, Duren WL, Erdos MR, Stringham HM, Chines PS, Jackson AU et al (2007) A genome-wide association study of type 2 diabetes in Finns detects multiple susceptibility variants. Science 316:1341–1345

Shapiro AM, Ricordi C, Hering BJ, Auchincloss H, Lindblad R, Robertson RP, Secchi A, Brendel MD, Berney T, Brennan DC et al (2006) International trial of the Edmonton protocol for islet transplantation. N Engl J Med 355:1318–1330

Shaw JE, Sicree RA, Zimmet PZ (2010) Global estimates of the prevalence of diabetes for 2010 and 2030. Diabetes Res Clin Pract 87:4–14

Somer RA, Thummel CS (2014) Epigenetic inheritance of metabolic state. Curr Opin Genet Dev 27:43–47

Sussel L, Kalamaras J, Hartigan-O'Connor DJ, Meneses JJ, Pedersen RA, Rubenstein JL, German MS (1998) Mice lacking the homeodomain transcription factor Nkx2.2 have diabetes due to arrested differentiation of pancreatic beta cells. Development 125:2213–2221

Vallois D, Niederhauser G, Ibberson M, Nagaray V, Marselli L, Marchetti P, Chatton JY, Thorens B (2014) Gluco-incretins regulate beta-cell glucose competence by epigenetic silencing of Fxyd3 expression. PLoS ONE 9:e103277

Vetere A, Choudhary A, Burns SM, Wagner BK (2014) Targeting the pancreatic beta-cell to treat diabetes. Nat Rev Drug Discovery 13:278–289

Volkmar M, Dedeurwaerder S, Cunha DA, Ndlovu MN, Defrance M, Deplus R, Calonne E, Volkmar U, Igoillo-Esteve M, Naamane N et al (2012) DNA methylation profiling identifies epigenetic dysregulation in pancreatic islets from type 2 diabetic patients. EMBO J 31:1405–1426

Waddington CH (2012) The epigenotype. 1942. Int J Epidemiol 41:10–13

Watada H, Mirmira RG, Leung J, German MS (2000) Transcriptional and translational regulation of beta-cell differentiation factor Nkx6.1. J Biol Chem 275:34224–34230

Wilson ME, Scheel D, German MS (2003) Gene expression cascades in pancreatic development. Mech Dev 120:65–80

Yang BT, Dayeh TA, Kirkpatrick CL, Taneera J, Kumar R, Groop L, Wollheim CB, Nitert MD, Ling C (2011) Insulin promoter DNA methylation correlates negatively with insulin gene expression and positively with HbA(1c) levels in human pancreatic islets. Diabetologia 54:360–367

Yang BT, Dayeh TA, Volkov PA, Kirkpatrick CL, Malmgren S, Jing X, Renstrom E, Wollheim CB, Nitert MD, Ling C (2012) Increased DNA methylation and decreased expression of PDX-1 in pancreatic islets from patients with type 2 diabetes. Mol Endocrinol 26:1203–1212

Zhou Q, Brown J, Kanarek A, Rajagopal J, Melton DA (2008) In vivo reprogramming of adult pancreatic exocrine cells to beta-cells. Nature 455:627–632

Chapter 5
Pancreatic Regeneration After Partial Pancreatectomy in Rodents

Wan-Chun Li, Chang-Yi Chen, Hung-Yu Chien and Susan Bonner-Weir

Introduction

Medical interest in regeneration is often focused on the repair of damaged tissues (Chen et al. 2014; Lee et al. 2014). In nature, an endogenous regeneration program to generate new target cells/tissues/organs in response to tissue loss has been reported in a wide range of animals, from *Cnidarians* to mammals where it has been accepted that lower vertebrates exhibit greater regeneration potential compared to higher vertebrates (Slack 2003). Pancreatic regeneration is of particular interest since diabetes results from an in adequate mass of functional insulin-producing beta cells.

While tissues with high turnover rate such as skin epithelium (Fuchs and Nowak 2008), hepatocytes (Michalopoulos 2013) and intestinal cells (Barker 2014) have dynamic cell regrowth, the regeneration potential of adult pancreatic cells, both for exocrine and for endocrine component, had been regarded as minimal (Lehv and Fitzgerald 1968). Yet, more recently, several models of pancreatic regeneration, including partial duct ligation (Wang et al. 1995; Xu et al. 2008),

W.-C. Li · C.-Y. Chen
Institute of Oral Biology, School of Dentistry,
National Yang-Ming University, Taipei, Taiwan

H.-Y. Chien
Department of Endocrinology and Metabolism,
Taipei City Hospital, Ren-Ai Branch, Taipei, Taiwan

S. Bonner-Weir (✉)
Section on Islet Cell and Regenerative Biology, Research Division,
Joslin Diabetes Center, Harvard Medical School, Boston, MA, USA
e-mail: susan.bonner-weir@harvard.edu

A.A. Hardikar (ed.), *Pancreatic Islet Biology*, Stem Cell Biology
and Regenerative Medicine, DOI 10.1007/978-3-319-45307-1_5

caerulein treatment (Jensen et al. 2005), diptheria toxin ablation of specific cell types (Thorel et al. 2011; Criscimanna et al. 2011) and partial pancreatectomy (Bonner-Weir et al. 1983), have supported a more robust regeneration of both exocrine and endocrine pancreatic components.

The partial pancreatectomy model in rodents has been an interest of ours, first as a model of mechanical reduction of beta cell mass resulting by 4 weeks after surgery in chronic moderate hyperglycemia reflecting that of type 2 diabetes, and second as a regeneration model in the first week or two after surgery. This model has been used for decades in many classic studies (Allen 1922; Lukens and Dohan 1942; Foglia 1944; Houssay et al. 1954; Martin and Lacy 1963) for the induction of diabetes through reduction of beta cell mass. We initially used it to study the effects of reduced beta cell mass on islet function (Bonner-Weir et al. 1983) and have since used it in a number of studies (Leahy et al. 1988; Jonas et al. 1999; Laybutt et al. 2002, 2003) but here will focus on it as a regeneration model. The model has been varied to be only 50–60 % reduction in rats without hyperglycemia (Leahy et al. 1988; Liu et al. 2000) and has been transferred to mice (Hardikar et al. 1999; Peshavaria et al. 2006; Ackermann Misfeldt et al. 2008).

The regenerative aspects of the model include all the mechanisms of beta cell growth that are known: self-duplication/replication of preexisting beta cells, neogenesis or differentiation of new beta cells from non-endocrine cells and hypertrophy of individual beta cells (Bonner-Weir et al. 1983, 1993). This model also provided the basis for our working hypothesis that the mature duct cells serve as a pool of progenitors and when stimulated regress to the progenitor phenotype, expand and then differentiate to new islets, acinar and duct cells (Bonner-Weir et al. 2004, 2010). The well-defined progression of the regeneration provides an excellent means to define the dynamic cellular and molecular mechanisms during regeneration.

Physiological Changes in Response to Pancreatectomy

The surgical approach we have used (Bonner-Weir et al. 1983) has direct lineage to that of Houssay. After anesthetization, the pancreatic tissue close to spleen and stomach are removed by gentle abrasion with cotton applicators. By using abrasion instead of resection, the larger blood vessels remain intact and the neighboring organs are not compromised. The remnant (residual pancreas) is anatomically well defined and contains the islet mass indicated by the percentage of total pancreatic tissue. For 85–90 % Px in rats, remnant comprised the tissue within 2 mm of the common bile duct and extending from the duct to the first part of the duodenum representing the upper portion of the head of pancreas considered to be embryologically of dorsal anlagen origin, while pancreatic tissues attached to distal part of duodenum was preserved in 60–70 % Px mice (Hardikar et al. 1999). The percentage of dissected pancreatic tissue mass positively correlated with regeneration capacity and possibly determines the differential molecular mechanisms during regeneration (Pearson et al. 1977; Bonner-Weir et al. 1988; Lee et al. 2006; Li et al. 2010). Interestingly, the effect of aging was also found to negatively correlate with replicatory capacity of acinar and beta cells, while duct-to-islet neogenesis activity seems unchanged between young and aged animals after Px (Tanigawa et al. 1997; Watanabe et al. 2005). Additionally, in the Px model, the regulation of the beta cell expansion by neogenesis was shown to differ from that of replication of the preexisting beta cells in pancreas with FoxM1 genetically ablated animals (Ackermann Misfeldt et al. 2008).

After 90 % Px surgery in rats, there is moderate hyperglycemia and insignificant changes for body weight and plasma insulin level compared to sham-operated animals during 12-week follow-up period (Bonner-Weir et al. 1983). The physiological parameters are stable up to one year after Px surgery. Hyperglycemia was not seen until 3–4 days after surgery probably because incomplete digestion of the food due to insufficient acinar enzymes. In fact, at the 1st week after Px, the exocrine tissue remnant secreted relatively less fluid, bicarbonate and digestive enzymes compared with control animals, but by 6 weeks post-Px the pancreatic exocrine function recovered (Sommer 1987). Growth of the exocrine tissue and the endocrine beta cells was discordant within the first weeks after Px, as indicated by the mitotic index of the different cell types (Brockenbrough et al. 1988). By 4 weeks after 90 % Px, the remnant that had been 10 % of the pancreas initially had grown to 27 % of the weight of the sham-operated pancreas with 40–45 % of the beta cell mass in sham-operated animals but with still moderately elevated morning fed glucose levels (Bonner-Weir et al. 1983; Xu et al. 1999). Since the beta cell mass doubles in the sham over this period, there is essentially an eight-fold increase in beta cell mass in the 4 weeks. Interestingly, the islet blood flow values of mice that underwent 60 % Px were significantly increased at the 2nd, 4th and 16th weeks after surgery (Jansson and Sandler 1989).

Dynamic Regeneration After Px Surgery

Within the first few days after surgery in the 90 % partially pancreatectomized rat, there is increased replication of the preexisting islet and exocrine (acinar and duct) tissues (Brockenbrough et al. 1988) and formation of both new islet and exocrine tissues as new lobes (Bonner-Weir et al. 1993; Li et al. 2010). By 24 h, there is massive replication within the pancreatic ductal tree as assessed by 5-bromo-2′-deoxyuridine (BrdU) incorporation or expression of proliferating marker proliferating cell nuclear antigen (PCNA) and Ki67 (Bonner-Weir et al. 1993; Sharma et al. 1999; Min et al. 2003; Li et al. 2010). Replication was detected as a wave (i.e., transient increase) first in the intrapancreatic common bile ducts (CBDs) then main ducts and eventually to smaller ductules indicating that ductal cells served as stem-like progenitors for compensation of tissue loss after Px. Indeed, we found that by 16 h post-PX, a time preceding that of BrdU incorporation, the duct epithelial cells of the CBD de-differentiated and had markedly reduced expression of ductal markers Hepatocyte Nuclear Factor 6 (HNF6), Sox9 and Hnf1α mRNA and protein as well as down-regulated duct epithelial markers cystic fibrosis transmembrane conductance regulator (Cftr) and E-cadherin transcripts. These data suggest that, by sensing tissue loss, mature ducts regress back to progenitor-like state to amplify cell numbers for tissue restoration (Li et al. 2010). Thus, tissue removal by Px-triggered regeneration via early formation of embryonic-like duct progenitors that could be later directed to become mature insulin-producing beta cells. The activation of adult stem/progenitor cells and subsequent cellular events postoperatively in Px model is illustrated in Fig. 5.1.

The regeneration process is highly dynamic. Less than 48 h after Px surgery, multiple specialized areas of duct-rich tissues, termed "foci of regeneration" (or for short, focal areas), can be observed on each histological section of pancreas. These foci consist of abundant stromal cells, increasingly branched ductules and differentiating acini and islets (Bonner-Weir et al. 1993; Li et al. 2010). They rapidly appeared and then disappeared over the week after Px (Bonner-Weir et al. 1993; Li et al. 2010). While lineage tracing with duct marker carbonic anhydrase II (CAII) showed ducts could postnatally give rise to both newly formed exocrine and endocrine cells (Inada et al. 2008), the lack of the genetic lineage tracing in

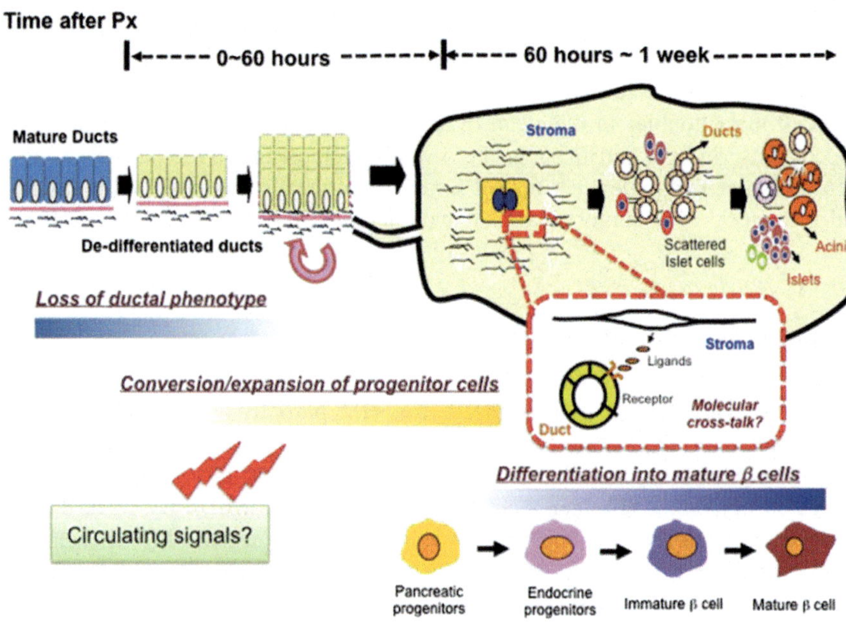

Fig. 5.1 Dynamic pancreatic regeneration in response to Px within the first week. Temporal cellular events in regenerating pancreas after Px surgery were illustrated. Initial de-differentiation or regression of mature duct epithelial cells to less-differentiated progenitors occurs within 24 h after Px, followed by expansion and redifferentiation into endocrine and exocrine cells following a program similar to that of the embryonic development. This mechanism of regeneration relies on the plasticity of the differentiation of cells within an organ

rats led us to examine the continuity of regenerating foci with the intrapancreatic common bile duct (also called the common pancreatic duct). The ducts in regenerating foci were continuous with the epithelium of CBDs and thus probably originated by expansion of the ductal epithelium (Li et al. 2010).

Within a single remnant pancreas after 90 % Px in rats, focal areas of various tissue compositions were detected: Some were predominantly mesenchymal cells with some ductal cross-section profiles (young foci), whereas in others, acinar cells are predominant (mature foci). Within 1 week, some lobes are fully mature and difficult to distinguish from preexisting lobes. The multiple different staged focal areas within the same pancreas provided ideal material to analyze dynamic regeneration progression both temporally and spatially without intra-animal variation (Li et al. 2010). As shown in Fig. 5.2, with temporal expression of developmentally key proteins in the advancing stages of focal areas, it could be concluded that ducts located in young regenerating foci expressed progenitor-like molecular pattern and the mature foci expressed proteins normally detected in postnatal or mature pancreatic tissues (Sharma et al. 1999; Li et al. 2010). For example, the frequency of MafA expressing ins+beta cells in young foci resembled more that of embryonic pancreas with a gradual increase in MafA+ins+in the maturing foci, even so the remnant pancreas of same animals had even greater proportion of mafA+insulin positive beta cells (Li et al. 2010).

Potential Molecular Players for Px-Mediated Regeneration

The factors important for pancreatic development could also be essential during regeneration. Much effort has been put in elucidating the expression and function of transcription factors or soluble molecules in regulating organogenesis and differentiation in embryonic pancreas (Collombat et al. 2006; Grapin-Botton et al. 2001; Murtaugh 2007). In addition to cell autonomous transcriptional regulators, external growth factors and downstream signaling molecules play an essential role in controlling pancreas development and could thus be potential candidates to regulate regeneration (Sumi and Tamura 2000). During pancreatic organogenesis, the relative anatomical position of embryonic pancreatic endoderm to adjacent mesodermal tissues suggested that molecular crosstalk of pancreatic progenitors and surrounding tissues may be important to trigger cellular commitment, cell proliferation and differentiation (Cleaver and Krieg 2001; Gittes 2009). The signals released from surrounding structures including notochord, aorta, septum transversum connective tissues and cardiac mesoderm are thought to be essential for the patterning of pancreatic endoderm (Wells and Melton 1999). For example, inhibition of Sonic hedgehog (Shh) signaling is required for formation of dorsal pancreatic endoderm. This idea is supported by the detection of smooth muscle-like mesenchyme and interstitial cells of Cajal, characteristic of the intestine, rather than pancreatic mesenchyme in the pPdx1-*Shh* transgenic mice (Shh signals driven by Pdx1 promoter) (Apelqvist et al. 1997; Hebrok 2003). Moreover,

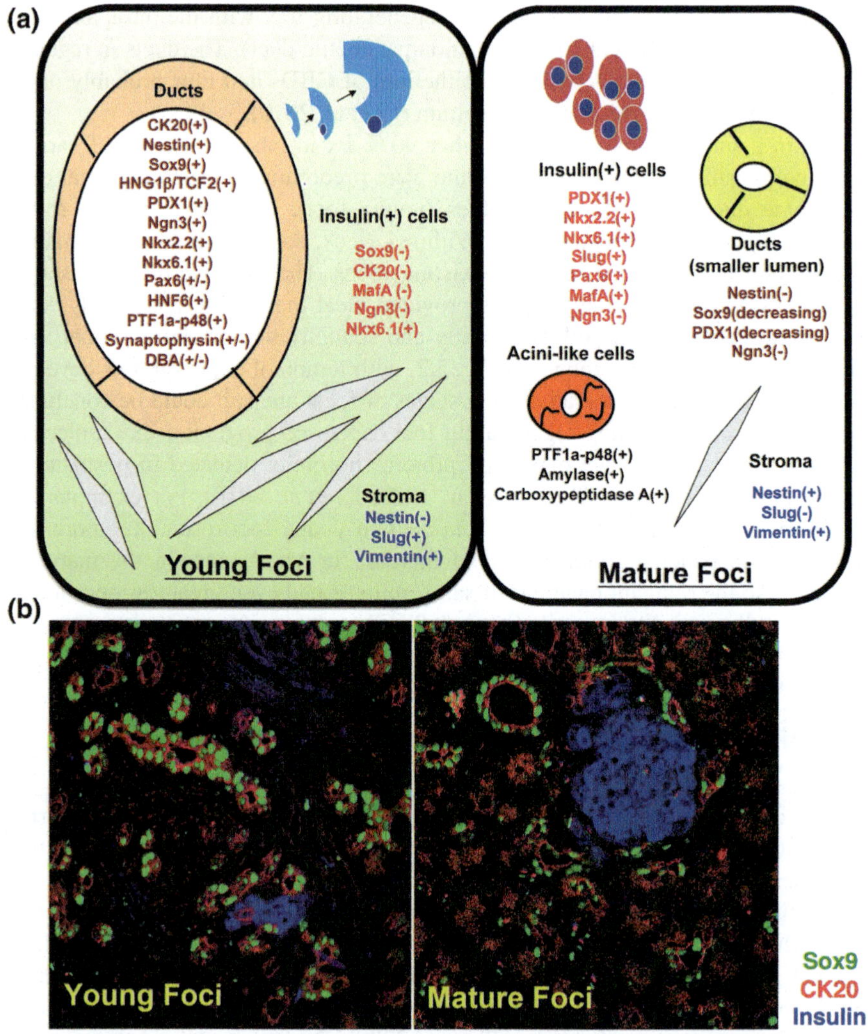

Fig. 5.2 Differential molecular profiles and birth of new beta cells during regeneration after Px. **a** Molecular expression profiles of young and mature regenerating foci comprising various cell compartments were summarized. Molecular expression patterns of young foci comprising of enriched open-lumened ductal progenitors, fibroblast-like stroma and scattered insulin+cells and mature foci containing less stroma and ductal progenitors but greater amount of acini-like cells was shown. **b** Immunofluorescent staining for ductal marker Sox9 (*green*), rat ductal epithelial marker Cytokeratin 20 (CK20, *red*) and insulin (*blue*) in young and mature foci was analyzed. New beta cells, both in young and mature foci, were seen budding from Sox9+duct progenitors suggesting that activated progenitors after Px could give rise to beta cells

it has further shown that activin-βB, a member of transforming growth factor beta (TGF-β) family, and fibroblast growth factor 2 (Fgf2) secreted from notochord are able to repress the expression of *Shh* therefore activating the pancreatic program in dorsal pancreatic endoderm (Hebrok et al. 1998; Kim et al. 1997). Other signals, including vascular endothelial growth factor (VEGF), epidermal growth factor (EGF) and insulin-like growth factor (IGF), play fundamental roles on pancreatic differentiation (Cras-Méneur et al. 2001; Lammert et al. 2001; Calderari et al. 2007). A list of potential molecular players important for Px-mediated regeneration is summarized in Table 5.1.

In fact, factors involving in pancreatic organogenesis have been shown to be important regulators for pancreas regeneration in response to Px. For example, insulin-like growth factor-1 (IGF-1) mRNA is increased in selected stromal cells as well as the proliferating ducts in Px rats (Smith et al. 1991). The vascular endothelial growth factor (VEGF), a potent angiogenic factor, was enriched at the transcriptional level after Px (Oberg-Welsh et al. 1997). Other novel molecules, including Reg/Pancreatic stone proteins (Terazono et al. 1988, 1990; Watanabe et al. 1994), gastrin (Xu et al. 1996; Kim et al. 2001; Téllez et al. 2011; Tellez and Montanya 2014), clusterin (Min et al. 2003; Lee et al. 2011) and IGF-downstream effectors Insulin receptor substrate (IRS)/Protein kinase B (PKB)/Akt signaling (Jetton et al. 2001; Peshavaria et al. 2006; Watanabe et al. 2008a, b) all have been reported as regulating pancreas regeneration after Px; however, some of these have been discounted by other studies (Smith et al. 1994; Togashi et al. 2014). Studies using state-of-art techniques such as laser capture microdissection (LCM), gene expression profiling and proteomics identified more Px-mediated regulators that should be characterized for their role during regeneration (Lim et al. 2002; Shin et al. 2005; Katsuta et al. 2005; Yang et al. 2006).

Future Scope and Discussion

The ducts in regenerating foci in Px rodents exhibit embryonic phenotype suggesting that normal developmental program is reactivated during regeneration (Sharma et al. 1999; Li et al. 2010). Better understanding of molecular basis to define injury-free and injury-activated pancreatic progenitors would pave the way to efficiently modulate the proliferation status and plasticity of pancreatic precursors.

Using Px as a regeneration model, the cellular and molecular basis for enhanced replication of ductal progenitors and age-independent neogenesis are being studied. However, the focus has mainly been on examination of the expression of intrinsic developmental-related transcription factors, while the temporal and spatial activities of stroma-mediated external effectors from the surrounding local environment or soluble molecules in systemic circulation are still not well defined. In addition, the key factors responsible in activating pancreatic progenitors and in promoting their differentiation during regeneration remained unclear. To address these issues, future effort could be put for the following studies: (1)

Table 5.1 Molecular changes in response to partial pancreatectomy surgery

Molecule	Effect	References
Cell cycle regulators (Cyclins/Cdks)	Overexpressed cyclins D1/E & reduced p15 and p27 cyclin inhibitors at Px+48 h in rat	Morisset et al. (1999)
Cholecystokinin-A receptor (CCK-Ar)	Protein/DNA content was significantly lower in CCK-Ar null OLETF rat than wild-type LETO Px rats	Miyasaka et al. (1997), Morisset et al. (1999)
Clusterin	Promote both exocrine and endocrine regeneration, both for neogenesis and replication	Min et al. (2003), Lee et al. (2011)
C-myc/H-ras	Activation of C-myc/H-ras in rat pancreatic remnant after Px	Calvo et al. (1991)
Epidermal growth factor (receptor)	EGF-EGFR binding activity regulates acinar regeneration in Px rat	Brockenbrough et al. (1988)
FoxM1	Required for increased beta cell proliferation, but not for beta cell neogenesis in Px mice	Ackermann Misfeldt et al. (2008)
Gastrin	Increased β-cell survival/mass and improved glucose tolerance	Xu et al. (1996), Kim et al. (2001), Téllez et al. (2011), Tellez and Montanya (2014)
GSK3beta	Intrapancreatic GSK3β knockdown leads to increased beta cell replication and differentiation and acinar growth in Px rat	Figeac et al. (2012)
Insulin-like growth factor	Enhanced IGF-1 in stroma of regenerating pancreas in Px rat; Upregulate IGF-1 at day 1 after Px in dog	Smith et al. (1991), Calvo et al. (1997)
Insulin receptor substrate-2/PKB/Akt	Up-regulated IRS2-PKB/Akt signaling activity → Increased Pdx1 expression	Jetton et al. (2001), Peshavaria et al. (2006), Watanabe et al. (2008a, b)
p42/p44 Mitogen activated protein kinase	Sustained activation of p42/p44 MAPKs at Px+48 h in rat	Morisset et al. (1999)
Reg/PSP	Enhanced beta cell replication amelioration of surgical diabetes	Terazono et al. (1988), Terazono et al. (1990), Watanabe et al. (1994)
Transforming growth factor beta/Interleukin 6	Increased serum TGFβ and IL6 on Day 7 and 14 after Px in rat	Farkas et al. (1997)
Vimentin	Co-expressed in embryonic and regenerating ductal epithelium	Ko et al. (2004)
Larger-scale screening	Candidate Px-associated molecules	
Differential display	WISP-1, Rap1B, VCAM-1, HIP	Lim et al. (2002)
Proteomic analysis	Reg III	Shin et al. (2005)
Proteomic analysis	Vimentin, CK8, L-plastin, hnRNP A2/B1, and AGAT	Yang et al. (2006)

to explore circulating molecules triggers for initiation of regenerating foci and (2) to uncover circulating and local regulators for maturation of regenerating foci. The clues could possibly come from histological observation of regenerating foci, while ducts in young regenerating foci are surrounded by mesenchymal stroma that becomes less prominent as regeneration proceeds (in intermediate or mature foci) suggesting a potential reciprocal interaction between ductal progenitors and stromal mesenchyme. To test this hypothesis, a more sophisticated assay, such as LCM, in combination with high throughput screening could be applied to systemically determine the expression of candidate molecules and their corresponding receptors among individual cell population, and thereby the local regeneration triggers could be defined. As for regeneration initiators, circulating factors (such as inflammatory molecules) differentially expressed in response to Px surgery could be the candidates for induction of regeneration. Indeed, previous studies have found increased serum levels of TGF-beta 1 and IL-6 (Farkas et al. 1997) and Fn14 which works through its receptor TWEAK (TNF-like weak inducer of apoptosis) (Wu et al. 2013) after pancreas resection indicating that these 3 factors might serve as regeneration triggers after Px.

In summary, even though molecular mechanisms during pancreas regeneration after pancreas resection have been widely elucidated, analysis for regeneration-associated molecules using injury-free model and the regeneration activity in human would be important to more feasibly apply the knowledge for treatment of diabetic patient clinically.

Acknowledgments The authors appreciate al the discussions with past laboratory members that contributed to the studies discussed here. Much of the work was originally supported by grants from the US National Institutes of Health R01 DK 066056 (S.B.-W.), P30 DK036836 as well as the Diabetes Research and Wellness Foundation and an important group of private donors. W.-C.L. was supported by a JDRF postdoctoral fellowship (3-2008-72).

References

Ackermann Misfeldt A, Costa RH, Gannon M (2008) Beta-cell proliferation, but not neogenesis, following 60 % partial pancreatectomy is impaired in the absence of FoxM1. Diabetes 57:3069–3077

Allen FM (1922) Experimental studies in diabetes. Series III. The pathology of diabetes 1. Hydropic degeneration of islands of Langerhans after partial pancreatectomy. J Metab Res 1:5–41

Apelqvist A, Ahlgren U, Edlund H (1997) Sonic hedgehog directs specialised mesoderm differentiation in the intestine and pancreas. Curr Biol 7:801–804

Barker N (2014) Adult intestinal stem cells: critical drivers of epithelial homeostasis and regeneration. Nat Rev Mol Cell Biol 15:19–33

Bonner-Weir S, Trent DF, Weir GC (1983) Partial pancreatectomy in the rat and subsequent defect in glucose-induced insulin release. J Clin Invest 71:1544–1553

Bonner-Weir S, Baxter LA, Schuppin GT et al (1993) A second pathway for regeneration of adult exocrine and endocrine pancreas. A possible recapitulation of embryonic development. Diabetes 42:1715–1720

Bonner-Weir S, Leahy JL, Weir GC (1988) Induced rat models of non-insulin-dependent diabetes mellitus. In: Shapiro E, Renold AE (eds) Lessons from animals diabetes II. John Libbey Publishers, London, pp 295–300

Bonner-Weir S, Toschi E, Inada A et al (2004) The pancreatic ductal epithelium serves as a potential pool of progenitor cells. Pediatr Diabetes 5(Suppl2):16–22

Bonner-Weir S, Li WC, Ouziel-Yahalom L et al (2010) Beta-cell growth and regeneration: replication is only part of the story. Diabetes 59:2340–2348

Brockenbrough JS, Weir GC, Bonner-Weir S (1988) Discordance of exocrine and endocrine growth after 90 % pancreatectomy in rats. Diabetes 37:232–236

Calderari S, Gangnerau MN, Thibault M et al (2007) Defective IGF2 and IGF1R protein production in embryonic pancreas precedes beta cell mass anomaly in the Goto-Kakizaki rat model of type 2 diabetes. Diabetologia 50:1463–1471

Calvo EL, Dusetti NJ, Cadenas MB et al (1991) Changes in gene expression during pancreatic regeneration: activation of c-myc and H-ras oncogenes in the rat pancreas. Pancreas 6:150–156

Calvo EL, Bernatchez G, Pelletier G et al (1997) Downregulation of IGF-I mRNA expression during postnatal pancreatic development and overexpression after subtotal pancreatectomy and acute pancreatitis in the rat pancreas. J Mol Endocrinol 18:233–242

Chen Y, Carter RL, Cho IK et al (2014) Cell-based therapies for Huntington's diseases. Drug Discov Today 19:980–984

Cleaver O, Krieg PA (2001) Notochord patterning of the endoderm. Dev Biol 234:1–12

Collombat P, Hecksher-Sorensen J, Serup P et al (2006) Specifying pancreatic endocrine cell fates. Mech Dev 123:501–512

Cras-Méneur C, Elghazi L, Czernichow P (2001) Epidermal growth factor increases undifferentiated pancreatic embryonic cells in vitro: a balance between proliferation and differentiation. Diabetes 50:1571–1579

Criscimanna A, Speicher JA, Houshmand G, Shiota C, Prasadan K, Ji B, Logsdon CD, Gittes GK, Esni F (2011) Duct cells contribute to regeneration of endocrine and acinar cells following pancreatic damage in adult mice. Gastroenterology 141:1451–1462

Farkas G Jr, Takács T, Mándi Y et al (1997) TGF-beta 1 and IL-6–new aspects in pancreas regeneration? Acta Chir Hung 36:89–91

Figeac F, Ilias A, Bailbe D et al (2012) Local in vivo GSK3β knockdown promotes pancreatic β cell and acinar cell regeneration in 90 % pancreatectomized rat. Mol Ther 20:1944–1952

Foglia VG (1944) Caracteristicas de la diabetes en la rata. Rev Soc Argent Biol 20:21–37

Fuchs E, Nowak JA (2008) Building epithelial tissues from skin stem cells. Cold Spring Harb Symp Quant Biol 73:333–350

Gittes GK (2009) Developmental biology of the pancreas: a comprehensive review. Dev Biol 326:4–35

Grapin-Botton A, Majithia AR, Melton DA (2001) Key events of pancreas formation are triggered in gut endoderm by ectopic expression of pancreatic regulatory genes. Genes Dev 15:444–454

Hardikar AA, Karandikar MS, Bhonde RR (1999) Effect of partial pancreatectomy on diabetic status in BALB/c mice. J Endocrinol 162:189–195

Hebrok M, Kim SK, Melton DA (1998) Notochord repression of endodermal Sonic hedgehog permits pancreas development. Genes Dev 12:1705–1713

Hebrok M (2003) Hedgehog signaling in pancreas development. Mech Dev 120:45–57

Houssay BA, Rodriquez RR, Cardeza AF (1954) Prevention of experimental diabetes with adrenal steroids. Endocrinology 54:550–552

Inada A, Nienaber C, Katsuta H et al (2008) Carbonic anhydrase II-positive pancreatic cells are progenitors for both endocrine and exocrine pancreas after birth. Proc Natl Acad Sci USA 105:19915–19919

Jansson L, Sandler S (1989) Pancreatic and islet blood flow in the regenerating pancreas after a partial pancreatectomy in adult rats. Surgery 106:861–866

Jensen JN, Cameron E, Garay MV et al (2005) Recapitulation of elements of embryonic development in adult mouse pancreatic regeneration. Gastroenterology 128:728–741

Jetton TL, Liu YQ, Trotman WE et al (2001) Enhanced expression of insulin receptor substrate-2 and activation of protein kinase B/Akt in regenerating pancreatic duct epithelium of 60 %-partial pancreatectomy rats. Diabetologia 44:2056–2065

Jonas J-C, Sharma A, Hasenkamp W, Ilkova H, Patane G, Laybutt R, Bonner-Weir S, Weir GC (1999) Chronic hyperglycemia triggers loss of pancreatic beta cell differentiation in an animal model of diabetes. J Biol Chem 274:14112–14121

Katsuta H, Koyanagi-Katsuta R, Shiiba M et al (2005) cDNA microarray analysis after laser microdissection in proliferating islets of partially pancreatectomized mice. Med Mol Morphol 38:30–35

Kim SK, Hebrok M, Melton DA (1997) Notochord to endoderm signaling is required for pancreas development. Development 124:4243–4252

Kim SW, Kim KH, Park SJ et al (2001) Endogenous gastrin stimulates regeneration of remnant pancreas after partial pancreatectomy. Dig Dis Sci 46:2134–2139

Ko SH, Suh SH, Kim BJ et al (2004) Expression of the intermediate filament vimentin in proliferating duct cells as a marker of pancreatic precursor cells. Pancreas 28:121–128

Lammert E, Cleaver O, Melton DA (2001) Induction of pancreatic differentiation by signals from blood vessels. Science 294:564–567

Laybutt DR, Sharma A, Sgroi DC, Gaudet J, Bonner-Weir S, Weir GC (2002) Genetic regulation of metabolic pathways in β-cells disrupted by hyperglycemia. J Biol Chem 277:10912–10921

Laybutt DR, Glandt M, Xu G, Ahn YB, Trivedi N, Bonner-Weir S, Weir GC (2003) Critical reduction in β-cell mass results in two distinct outcomes over time: adaptation with impaired glucose tolerance or decompensated diabetes. J Biol Chem 278:2997–3005

Leahy JL, Bonner-Weir S, Weir GC (1988) Minimal chronic hyperglycemia is a critical determinant of impaired insulin secretion following incomplete pancreatectomy in the rat. J Clin Investig 81:1407–1414

Lee AY, Mahler N, Best C et al (2014) Regenerative implants for cardiovascular tissue engineering. Transl Res 163:321–341

Lee CS, De León DD, Kaestner KH et al (2006) Regeneration of pancreatic islets after partial pancreatectomy in mice does not involve the reactivation of neurogenin-3. Diabetes 55:269–272

Lee SH, Hao E, Levine F (2011) β-Cell replication and islet neogenesis following partial pancreatectomy. Islets 3:188–195

Lehv M, Fitzgerald PJ (1968) Pancreatic acinar cell regeneration. IV. Regeneration after resection. Am J Pathol 53:513–535

Li WC, Rukstalis JM, Nishimura W et al (2010) Activation of pancreatic-duct-derived progenitor cells during pancreas regeneration in adult rats. J Cell Sci 123:2792–2802

Lim HW, Lee JE, Shin SJ et al (2002) Identification of differentially expressed mRNA during pancreas regeneration of rat by mRNA differential display. Biochem Biophys Res Commun 299:806–812

Liu YQ, Nevin PW, Leahy JL (2000) Beta-cell adaptation in 60 % pancreatectomy rats that preserves normoinsulinemia and normoglycemia. Am J Physiol Endocrinol Metab 279:E68–E73

Lukens FDW, Dohan FC (1942) Pituitarydiabetes in the cat; recovery following insulin or dietary treatment. Endocrinology 30:175–202

Martin JM, Lacy PE (1963) The prediabetic period in partially pancreatectomized rats. Diabetes 12:238–242

Michalopoulos GK (2013) Principles of liver regeneration and growth homeostasis. Compr Physiol 3:485–513

Min BH, Kim BM, Lee SH et al (2003) Clusterin expression in the early process of pancreas regeneration in the pancreatectomized rat. J Histochem Cytochem 51:1355–1365

Miyasaka K, Ohta M, Masuda M et al (1997) Retardation of pancreatic regeneration after partial pancreatectomy in a strain of rats without CCK-A receptor gene expression. Pancreas 14:391–399

Morisset J, Aliaga JC, Calvo EL et al (1999) Expression and modulation of p42/p44 MAPKs and cell cycle regulatory proteins in rat pancreas regeneration. Am J Physiol 277:G953–G959

Murtaugh LC (2007) Pancreas and beta-cell development: from the actual to the possible. Development 134:427–438

Oberg-Welsh C, Sandler S, Andersson A, Welsh M (1997) Effects of vascular endothelial growth factor on pancreatic duct cell replication and the insulin production of fetal islet-like cell clusters in vitro. Mol Cell Endocrinol 126:125–132

Pearson KW, Scott D, Torrance B (1977) Effects of partial surgical pancreatectomy in rats. I. Pancreatic regeneration. Gastroenterology 72:469–473

Peshavaria M, Larmie BL, Lausier J et al (2006) Regulation of pancreatic beta-cell regeneration in the normoglycemic 60 % partial-pancreatectomy mouse. Diabetes 55:3289–3298

Sharma A, Zangen DH, Reitz P (1999) The homeodomain protein IDX-1 increases after an early burst of proliferation during pancreatic regeneration. Diabetes 48:507–513

Shin JS, Lee JJ, Lee EJ et al (2005) Proteome analysis of rat pancreas induced by pancreatectomy. Biochim Biophys Acta 1749:23–32

Slack JM (2003) Regeneration research today. Dev Dyn 226:162–166

Smith FE, Rosen KM, Villa-Komaroff L et al (1991) Enhanced insulin-like growth factor I gene expression in regenerating rat pancreas. Proc Natl Acad Sci USA 88:6152–6156

Smith FE, Bonner-Weir S, Leahy JL et al (1994) Pancreatic Reg/pancreatic stone protein (PSP) gene expression does not correlate with beta-cell growth and regeneration in rats. Diabetologia 37:994–999

Sommer H (1987) Functional recovery of the exocrine pancreas in rats after partial resection. Eur Surg Res 19:318–322

Sumi S, Tamura K (2000) Frontiers of pancreas regeneration. J Hepatobiliary Pancreat Surg 7:286–294

Tanigawa K, Nakamura S, Kawaguchi M et al (1997) Effect of aging on B-cell function and replication in rat pancreas after 90 % pancreatectomy. Pancreas 15:53–59

Téllez N, Joanny G, Escoriza J et al (2011) Gastrin treatment stimulates β-cell regeneration and improves glucose tolerance in 95 % pancreatectomized rats. Endocrinology 152:2580–2588

Tellez N, Montanya E (2014) Gastrin induces ductal cell dedifferentiation and β-cell neogenesis after 90 % pancreatectomy. J Endocrinol 223:67–78

Terazono K, Yamamoto H, Takasawa S et al (1988) A novel gene activated in regenerating islets. J Biol Chem 263:2111–2114

Terazono K, Uchiyama Y, Ide M et al (1990) Expression of reg protein in rat regenerating islets and its co-localization with insulin in the beta cell secretory granules. Diabetologia 33:250–252

Thorel F, Damond N, Chera S, Wiederkehr A, Thorens B, Meda P, Wollheim CB, Herrera PL (2011) Normal glucagon signaling and β-cell function after near-total α-cell ablation in adult mice. Diabetes 60:2872–2882

Togashi Y, Shirakawa J, Orime K et al (2014) β-Cell proliferation after a partial pancreatectomy is independent of IRS-2 in mice. Endocrinology 155:1643–1652

Wang RN, Kloppel G, Bouwens L (1995) Duct- to islet-cell differentiation and islet growth in the pancreas of duct-ligated adult rats. Diabetologia 38:1405–1411

Watanabe T, Yonemura Y, Yonekura H et al (1994) Pancreatic beta-cell replication and amelioration of surgical diabetes by Reg protein. Proc Natl Acad Sci USA 91:3589–3592

Watanabe H, Saito H, Rychahou PG et al (2005) Aging is associated with decreased pancreatic acinar cell regeneration and phosphatidylinositol 3-kinase/Akt activation. Gastroenterology 128:1391–1404

Watanabe H, Saito H, Ueda J et al (2008a) Regulation of pancreatic duct cell differentiation by phosphatidylinositol-3 kinase. Biochem Biophys Res Commun 370:33–37

Watanabe H, Saito H, Nishimura H et al (2008b) Activation of phosphatidylinositol-3 kinase regulates pancreatic duodenal homeobox-1 in duct cells during pancreatic regeneration. Pancreas 36:153–159

Wells JM, Melton DA (1999) Vertebrate endoderm development. Annu Rev Cell Dev Biol 15:393–410

Wu F, Guo L, Jakubowski A et al (2013) TNF-like weak inducer of apoptosis (TWEAK) promotes beta cell neogenesis from pancreatic ductal epithelium in adult mice. PLoS ONE 8:e72132

Xu G, Stoffers DA, Habener JF, Bonner-Weir S (1999) Exendin-4 stimulates both β-cell replication and neogenesis resulting in increased β-cell mass and improved glucose tolerance in diabetic rats. Diabetes 48:2270–2276

Xu G, Sumi S, Koike M et al (1996) Role of endogenous hypergastrinemia in regenerating endocrine pancreas after partial pancreatectomy. Dig Dis Sci 41:2433–2439

Xu X, D'Hoker J, Stangé G, Bonné S, De Leu N, Xiao X, Van de Casteele M, Mellitzer G, Ling Z, Pipeleers D, Bouwens L, Scharfmann R, Gradwohl G, Heimberg H (2008) Beta cells can be generated from endogenous progenitors in injured adult mouse pancreas. Cell 132:197–207

Yang M, Liu W, Wang CY et al (2006) Proteomic analysis of differential protein expression in early process of pancreatic regeneration in pancreatectomized rats. Acta Pharmacol Sin 27:568–578

Chapter 6
Regulators of Beta-Cell Death and Regeneration

Shin Takasawa, Asako Itaya-Hironaka, Akiyo Yamauchi, Hiroyo Ota, Maiko Takeda, Sumiyo Sakuramoto-Tsuchida, Takanori Fujimura and Hiroki Tsujinaka

Introduction

Pancreatic β-cell deficiency underlies both type 1 and type 2 diabetes, and restoration or replacement of β-cell mass/function is therefore one of the logical solutions to therapy. While it has long been held that type 1 diabetes results from an irreversible loss of β-cells, and that type 2 diabetes is primarily caused by impaired insulin action, there is now increasing evidence linking both types of diabetes to defects in β-cell mass and insulin secretion. Pancreatic β-cells have traditionally been viewed as a quiescent cell population. However, several recent lines of evidence indicated that like most tissues the β-cell mass is dramatically regulated with ongoing β-cell regeneration throughout life to replenish lost or damaged β-cells. Based on our recent data concerning β-cell death, dysfunction, and regeneration, we discuss the regulation of pancreatic β-cell death, their function and regeneration during the lifetime.

β-Cell Death and its Regulators

Pancreatic β-cell failure resulting from β-cell death or dysfunction is a crucial event in the development of diabetes. Pancreatic β-cell death is the final and critical step in the development of type 1 diabetes. On the other hand, β-cell death as

S. Takasawa (✉) · A. Itaya-Hironaka · A. Yamauchi · H. Ota · M. Takeda · S. Sakuramoto-Tsuchida · T. Fujimura · H. Tsujinaka
Department of Biochemistry, Nara Medical University (NMU), Shijo-cho, Kashihara 634-8521, Japan
e-mail: shintksw@naramed-u.ac.jp

A.A. Hardikar (ed.), *Pancreatic Islet Biology*, Stem Cell Biology and Regenerative Medicine, DOI 10.1007/978-3-319-45307-1_6

well as β-cell dysfunction contributes to β-cell failure in type 2 diabetes. Recently, definition/nomenclature of cell death has been newly defined by a series of precise, measurable biochemical features (extrinsic apoptosis, caspase-dependent or caspase-independent intrinsic apoptosis, regulated necrosis, autophagic cell death, and mitotic catastrophe) by the Nomenclature Committee on Cell Death (Galluzzi et al. 2012). However, most of researchers in the field of pancreatic β-cell death and regeneration still use the terminology based on morphological criteria (apoptosis, autophagy, necrosis, and mitotic catastrophe) (Kroemer et al. 2005), and major field in pancreatic β-cell death research still focuses on necrosis and apoptosis. Thus, in this chapter, we focus the description and discussion on necrosis and apoptosis of β-cells and their regulators/modulators.

Necrosis of β-Cells and its Regulators

Diabetes can be studied at an experimental level using surgical techniques or chemical agents. Many of acute metabolic derangements of severe human insulinopenic diabetes can be reproduced by the removal of insulin-producing pancreatic β-cells of the islets of *Langerhans*. Von Mering and Minkowski first produced this artificial form of type 1 diabetes by removing the pancreases of dogs (von Mering and Minkowski 1889). Banting and Best adopted the same model of diabetes for their historical observation on the hypoglycemic properties of a crude pancreatic extract of insulin (Banting et al. 1922).

Alloxan and streptozotocin are the most prominent diabetogenic chemicals in diabetes research. Both chemical compounds are cytotoxic glucose analogs. Although their cytotoxicity is achieved via different pathways, their mechanisms of β-cell selective action are identical. In 1838, Wöhler and Liebig synthesized a pyrimidine derivative (Wöhler and Liebig 1838), which they later called alloxan (2,4,5,6-tetraoxohexahydropyrimidine). In 1943, alloxan became of interest in diabetes research when Dunn et al. (1943) reported that it could induce diabetes in animals as a result of the specific necrosis of the pancreatic β-cells. The resulting insulinopenia causes a state of experimental diabetes called "alloxan diabetes." The reduction product of alloxan, dialuric acid (Wöhler and Liebig 1838), has also been shown to be diabetogenic in animals, and to cause, like lipophilic alloxan derivatives such as butylalloxan, identical ultrastructural changes (Jörns et al. 1997).

Streptozotocin (2-deoxy-2-(3-methyl-3-nitrosoureido)-D-glucopyranose) is originally an antimicrobial agent and has also been used as a chemotherapeutic alkylating agent. In 1963, Rakieten et al. reported that streptozotocin was diabetogenic (Rakieten et al. 1963). Again, this insulinopenia syndrome, called "streptozotocin diabetes" (Schein et al. 1967), is caused by the specific necrosis of the pancreatic β-cells, and streptozotocin has been the agent of choice for the induction of diabetes in animals ever since.

After decades of research, a unifying explanation of how the β-cytotoxins cause functional deteriorations and degenerative changes in the β-cells, however, has been the subject of much debate, until Okamoto proposed a unifying concept as to the diabetogenicity of alloxan and streptozotocin (Okamoto 1981). In 1981, it was found that streptozotocin and alloxan caused DNA strand breaks in isolated rat islets (Yamamoto et al. 1981a). Considerable amounts of islet DNA were sedimented as a broad peak in the middle of the alkaline sucrose gradient with a concomitant decrease in undamaged DNA after only 5-min incubation with streptozotocin or alloxan. After 10- to 20-min incubation, islet DNA was almost completely fragmented. These results indicated that streptozotocin and alloxan produced strand breaks in islet DNA. Yamamoto et al. next prepared a nuclear fraction from rat islets, and assayed the activity of poly(ADP-ribose) polymerase (PARP), a nuclear enzyme that catalyzes the polymerization of the ADP-ribosyl moiety of NAD^+ to form poly(ADP-ribose); the enzyme was originally described by three independent groups in France and Japan (Chambon et al. 1966; Sugimura et al. 1967; Nishizuka et al. 1967) and has been implicated in DNA repair (Ueda and Hayaishi 1985). Both streptozotocin and alloxan were found to induce islet PARP activation with a peak at 10 min (Yamamoto et al. 1981a). The increase in nuclear PARP activity was associated with a concomitant decrease in cellular NAD^+, the substrate of PARP; the NAD^+ content of the islets showed a sharp drop within 20 min of incubation with either streptozotocin or alloxan. The diabetogenic doses of alloxan and streptozotocin induced in vivo DNA strand breaks and NAD^+ depletion in the islets (Yamamoto et al. 1981b). These results indicated that the β-cytotoxins cause DNA breaks to induce PARP activity, thereby depleting islet NAD^+. Furthermore, Uchigata et al. incubated isolated rat pancreatic islets in the presence of alloxan or streptozotocin with or without the addition of PARP inhibitors such as nicotinamide, 3-aminobenzamide, benzamide, 3-nitrobenzamide, 3-methoxybenzamide, theophylline, and 3-isobutyl-1-methylxanthine. The PARP inhibitors were found to reverse the alloxan- or streptozotocin-induced inhibition of proinsulin synthesis in a dose-dependent manner, and the stronger inhibitors exerted the protection at the lower concentrations (Uchigata et al. 1982). In vivo administrations of 3-aminobenzamide or nicotinamide to rats effectively protected against the alloxan- or streptozotocin-induced decrease in proinsulin synthesis (Uchigata et al. 1983).

It has been suggested that alloxan may work through the formation of the hydroxyl radical ($OH^{\cdot}$) (Heikkila et al. 1976), which is produced by the interaction between superoxide ($O_2^{-\cdot}$) and peroxide (H_2O_2) (Haber and Weiss 1934): $O_2^{-\cdot} + H_2O_2 \rightarrow OH^{\cdot} + OH^- + O_2$.

Superoxide dismutase and catalase catalyze the removal of $O_2^{-\cdot}$ and H_2O_2, respectively (McCord and Fridovich 1969), and hence, may inhibit the formation of $OH^{\bullet}$. Uchigata et al. showed that a combined administration of superoxide dismutase and catalase more effectively protects against alloxan-induced islet DNA strand breaks, as well as against proinsulin synthesis inhibition, than an administration of either of these radical scavenging enzymes alone (Uchigata et al. 1982). The results indicate that it is $OH^{\bullet}$ that is ultimately generated from alloxan to attack DNA. The concept was confirmed by Takasu et al. (1991). On the other hand, radical scavengers could not protect islet DNA against streptozotocin-induced islet DNA strand breaks (Uchigata et al. 1982). The breakage of DNA by streptozotocin is probably associated with its alkylating activity, as suggested with nitrosoureas (Erickson et al. 1977; LeDoux et al. 1986) and/or its nitric oxide generation activity (Kröncke et al. 1995).

From the experimental results described above, Okamoto proposed a basic model for the action of alloxan and streptozotocin in the induction of experimental diabetes (Okamoto 1981; Uchigata et al. 1982; Okamoto 1985). As shown in Fig. 6.1, the first step is the generation of free radicals by alloxan and

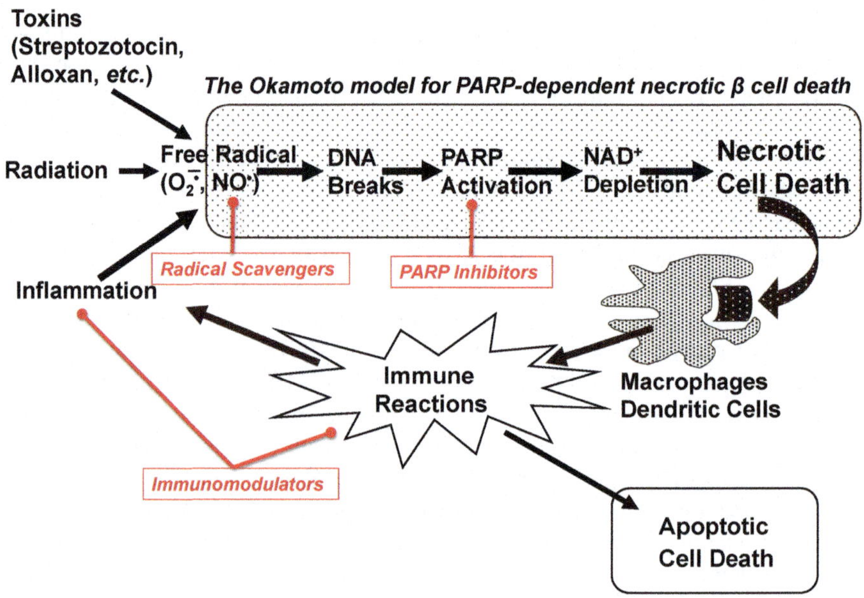

Fig. 6.1 The Okamoto model for PARP-dependent necrotic β-cell death (adapted from Okamoto and Takasawa 2003). The Okamoto model, originally proposed as a unifying model for β-cell damage and its prevention (Okamoto 1985; Okamoto et al. 1997; Takasawa and Okamoto 2002; Okamoto and Takasawa 2002, 2003), well explains both how autoimmunity for β-cell necrosis is initiated and how necrotic cell death is involved in various diseases in many tissues other than β-cells. Under physiological conditions, apoptotic cell death constitutively occurs for renewal and maintenance in animal bodies, whereas necrotic cells initiate and enhance (auto)immune reactions under pathological conditions. On the other hand, apoptotic cells recognized by macrophages/dendritic cells inhibit phlogistic (auto)immune responses. For the initiation of massive and pathological apoptotic cell death, necrotic cell death, which triggers autoimmune responses in macrophages/dentritic cells, is required (Sauter et al. 2000; Cocco and Ucker 2001). Based on the Okamoto model, prevention of β-cell necrosis is theoretically possible using radical scavengers (Nomikos et al. 1986; Roza et al. 1994; Abdel-Wahab and Abd-Allah 2000; Suarez-Pinzon et al. 2001; D'Aleo et al. 2009; Nagatani et al. 2011), PARP inhibitors (Yamamoto et al. 1981a; Yamada et al. 1982; Uchigata et al. 1983; Okamoto 1985; Masiello et al. 1990; Gale 1996; Knip et al. 2000; Reddy et al. 2001; Wada and Yagihashi 2004; Pandya et al. 2010), and immunomodulators (Mori et al. 1986; Toyota et al. 1986; Satoh et al. 1988; Seino et al. 1991; Takahashi et al. 1993). Recently, the term "parthanatos" has been proposed by the group of Dr. Dawson for PARP-dependent cell death (Harraz et al. 2008; Andrabi et al. 2008; Wang et al. 2011)

streptozotocin, which attack DNA to produce strand breaks. PARP then is activated to act to repair the DNA breaks, consuming cellular NAD^+. Since NAD^+ is the most abundant of cellular coenzymes and participates in a lot of biological reactions in mammalian cells, a severe reduction in intracellular NAD^+ to non-physiological levels may adversely affect β-cell functions, including proinsulin synthesis and insulin secretion. Protection of chemically induced β-cell necrosis was reported also in human islets (Hoorens and Pipeleers 1999). Therefore, in diabetes induction, the β-cells seem to be making a "suicide response" in their attempt to repair the damaged DNA. In fact, combined administration of streptozotocin or alloxan and PARP inhibitors such as nicotinamide, picolinamide, and 3-aminobenzamide prevented β-cytotoxin-induced diabetes and induced pancreatic β-cell tumors in rats (Yamagami et al. 1985; Takasawa et al. 1986).

Here, the problem arises as to why only pancreatic β-cells are specifically damaged by streptozotocin and alloxan. It has been conjectured that alloxan and streptozotocin have an affinity to the cell membranes of β-cells, because of their chemical structures (Cooperstein and Watkins 1981). ^{14}C-labeled alloxan or streptozotocin injected into mice was recovered in islets (Hammarström and Ullberg 1966; Tjälve et al. 1976). The NAD^+ content per DNA of pancreatic islets was approximately one half of that of the liver (Yamamoto et al. 1981b), and therefore, pancreatic β-cells may be more susceptible to reduction of NAD^+ levels. In fact, Malaisse et al. (1982) reported that the ability to provide protection against potent reactive radicals might be weak in islet cells, in view of the low glutathione peroxidase activity in islets. Recently, this concept was confirmed by the study of species differences in susceptibility of islets to streptozotocin between mouse and chick (Modak et al. 2007).

This model for the mechanism of action of alloxan and streptozotocin has received much attention because of its possible relevance to the effects of viruses and inflammation, especially those related to inflammatory damage of β-cells (Okamoto 1981, 1985; Okamoto et al. 1997; Takasawa and Okamoto 2002; Okamoto and Takasawa 2002, 2003), because biological events such as inflammation and virus infection, physical insult such as radiation, and chemical insult may independently or interactively produce β-cell DNA strand breaks (Fig. 6.1). Tsubouchi et al. have reported that a large dose of X-ray irradiation provoked necrosis of pancreatic islets in hamsters and that the hamsters exhibited a diabetic state (Tsubouchi et al. 1981). Cumulative β-cell damage induced by sub-diabetogenic doses of streptozotocin and by encephalomyocarditis or Coxsackie virus infection was shown to result in the development of diabetes (Toniolo et al. 1980). Recently, destruction of human islets by enterovirus infection was reported (Ylipaasto et al. 2012). Therefore, although type 1 diabetes can be caused by a variety of internal and external environmental factors, including immunologic abnormalities, inflammatory tissue damage, viruses, irradiation, and chemical insults, the final pathway may be one and the same, involving DNA damage, PARP activation, and NAD^+ depletion (Fig. 6.1). Accordingly, it may be possible to prevent type 1 diabetes by blocking immune reactions, scavenging free radicals, and inhibiting PARP. In effect, diabetes in NOD mice (Makino et al. 1980) is reported to be preventable by the treatment with immunosuppressive agents (Mori et al. 1986), an immunomodulator OK-432 (Okamoto 1976; Toyota et al. 1986), radical scavengers (Nomikos et al. 1986; Roza et al. 1994), and PARP inhibitors (Yamada et al. 1982; Nomikos et al. 1986). Concerning nitric oxide, Takamura et al. (1998) found that, in transgenic mice expressing type 2 nitric oxide synthase 2 (NOS2) constitutively in pancreatic β-cells, the β-cell mass is markedly reduced, resulting in the development of severe diabetes. NOS2 is usually induced by pro-inflammatory/inflammatory cytokines such as interleukin (IL)-1β, interferon (IFN)-γ, and tumor necrosis factor-α (Cetkovic-Cvrlje and Eizirik 1994; Akabane et al. 1995; Sekine et al. 2000; Thomas et al. 2002), and selective inhibitor of NOS2 (ONO-1714) attenuated insulitis in nonobese diabetic (NOD) mice (Kato et al. 2003).

PARP is one of the best known proteins with DNA-damage scanning activity, and poly(ADP-ribosyl)ation by PARP has been proposed to function in DNA repair by modifying architectural proteins proximal to DNA breaks, thus facilitating the opening of the condensed chromatin structure required for the recruitment of the repairing complex (Ueda and Hayaishi 1985). Paradoxically, in spite of this beneficial effect, PARP can induce necrotic cell death through NAD^+ depletion as described above (Fig. 6.1). In spite of the apparent Mr. Hyde-like side of this enzyme, it may represent an evolutionary strategy adopted by multicellular organisms to prevent the survival of cells that would otherwise transmit potentially dangerous genetic material/information. Many other tissues and cells such as those involved in cerebral ischemia (Eliasson et al. 1997; Plaschke et al. 2000; Ducrocq et al. 2000; Kauppinen et al. 2009; Alano et al. 2010; Matsuura et al. 2011), myocardial ischemia (Zingarelli et al. 1998; Bowes et al. 1999; Pieper et al. 2000; Szabó et al. 2004; Roesner et al. 2010), laryngeal injury (Stern et al. 1999), intestinal mucosal injury (Zingarelli et al. 1999; Jijon et al. 2000; Giannone et al. 2011), organ injury by hemorrhagic shock (Szabó et al. 1997; Oliver et al. 1999; Liaudet et al. 2000; Skarda et al. 2007), ischemic renal injury (Martin et al. 2000; del Moral et al. 2013), airway inflammation (Havranek et al. 2010), dyslipidemia-induced autonomic and vascular dysfunction (Hans et al. 2009), and diabetic myocardial and endothelial injury (Soriano et al. 2001; Pacher et al. 2002; Pacher and Szabó 2007) have been reported to die by the same mechanism as those in pancreatic β-cell death. In 1999, three independent groups in Germany, Japan, and the U.S. provided irrefutable support using PARP-deficient mice for the model (Burkart et al. 1999; Charron and Bonner-Weir 1999; Masutani et al. 1999; Pieper et al. 1999). The mice were extremely resistant to streptozotocin, and the β-cell death was prevented (Burkart et al. 1999; Charron and Bonner-Weir 1999; Masutani et al. 1999; Pieper et al. 1999; Li et al. 2013). Accordingly, the inhibition of PARP activity may be a possible therapeutic approach in a wide number of disorders other than diabetes.

Apoptosis of β*-Cells and its Regulators*

Apoptosis (programmed cell death) is an energy-dependent, genetically regulated cell suicide process that is evolutionarily conserved in multicellular organisms. Upon receiving an apoptotic signal, either at the mitochondria or on the cell surface, the cell undergoes a commitment to die and uses the machinery within it to undergo a highly orchestrated and characteristic process. In multicellular organisms, this process plays a critical role in all stages of life, from development, where apoptosis is important for tissue sculpting and remodeling, to maintenance of homeostasis during physiological turnover and the elimination of unwanted or harmful cells throughout the life of the organism (Thompson 1995; Jacobson et al. 1997). Histological evidence of apoptosis was discovered in the early 1970s, and the genetics of developmental apoptosis in the nematode *Caenorhabditis elegans* (Kerr et al. 1972; Marx 2002) provided the working genetic frame onto which mammalian genetic discoveries were made. Knockout mouse models for many of the components of the apoptotic signaling pathways have provided valuable insights into the mechanisms and the essential roles of each of the apoptotic genes, which are highly tissue- and context-specific.

The pancreatic β-cell plays an essential role in regulating glucose homeostasis. β-cell mass is dynamic and is tightly matched to meet the body's demand for insulin. During the neonatal period, the β-cell population increases rapidly, and this burst in expansion of β-cell mass is followed by a transient increase in β-cell death (Kassem et al. 2000). With increasing age, the rates of β-cell apoptosis and proliferation and/or neogenesis equilibrate at a frequency of 0.5 % under steady-state conditions (Bonner-Weir 2000).

Type 1 diabetes is a chronic autoimmune disease that affects 0.5 % of the population in the developed world and continues to increase in incidence. This disease is hallmarked by immune-mediated destruction of the pancreatic β-cells. Typically, at the time of diagnosis, patients with type 1 diabetes have an estimated 60–80 % reduction in β-cell mass (Notkins and Lernmark 2001). The current working model is that in genetically predisposed individuals T lymphocytes are aberrantly activated by the antigen-presenting cells (APCs) in the pancreatic draining lymph nodes. The activated T lymphocytes then circulate, target, and invade the islets, known as "insulitis." These activated T cells can then proceed to destroy the islets (Martin et al. 2001; Todd 2010; Todd et al. 2011). Many of the susceptibility genes for type 1 diabetes may act through the modulation of immune-related process, such as antigen presentation, expansion of self-reactive cells, and modification/regulation of the immune function (Concannon et al. 2009). The human leukocyte antigen (HLA) region in chromosome 6p21 explains more than 40 % of genetic risk of type 1 diabetes, suggesting that β-cell death in human type 1 diabetes is mainly immune-mediated apoptotic cell death. In this process, β-cell apoptosis initiated by pro-inflammatory cytokines such as IL-1β and IFN-γ is considered important (Zóka et al. 2013).

Type 2 diabetes, hallmarked by underlying insulin resistance, is also characterized by defects in glucose-responsive insulin secretion in addition to an eventual decline in β-cell mass (Rhodes 2005). As long as β-cells are able to compensate for the insulin resistance by enhancing insulin secretion and increasing β-cell mass, euglycemia can be maintained. However, in susceptible individuals, perhaps with genetic defects and/or exogenous insults, their β-cells may be unable to meet the body's demand for insulin, ultimately resulting in type 2 diabetes (Pimenta et al. 1995; Gunton et al. 2005; Marchetti et al. 2006; Florez 2008). As such, autopsy studies in which β-cell mass was quantified morphometrically in pancreata from patients with type 2 diabetes have shown β-cell apoptosis to account for the dramatic reduction of β-cell mass (Sakuraba et al. 2002; Butler et al. 2003; Leonardi et al. 2003; Yoon et al. 2003). Accumulating evidence indicates that β-cell loss (apoptosis) in type 2 diabetes results as a response to the combination of oxidative stress and endoplasmic reticulum (ER) stress (Montane et al. 2014); a great variety of stimuli, such as islet amyloid polypeptide (IAPP), cytokines, cellular cholesterol, or high glucose and lipid levels in the blood, can disturb ER homeostasis, leading to oxidative and ER stress, inflammation, and pancreatic β-cell apoptosis.

CD38–cADPR Signal System

Glucose increases the intracellular Ca^{2+} concentration in pancreatic β-cells to cause the secretion of insulin. Ashcroft et al. proposed that this increase in the Ca^{2+} concentration is provided extracellularly. That is, in the process of glucose metabolism, the millimolar concentrations of ATP produced inhibit the ATP-dependent potassium channel, causing membrane depolarization and the opening of the voltage-dependent Ca^{2+} channels (Ashcroft et al. 1984).

Cyclic ADP-ribose (cADPR) is a cyclic compound synthesized from NAD^{+}. This compound was first found in 1987 by Clapper et al. when studying Ca^{2+} release in sea urchin eggs (Clapper et al. 1987). However, the physiological significance of cADPR in mammalian systems was not then understood. According to the Okamoto model (Fig. 6.1), the decrease in the NAD^{+} level should cause the decrease in cADPR in β-cells. Thus, in 1993, Takasawa et al. proposed that insulin secretion by glucose occurs via cADPR-mediated Ca^{2+} mobilization from an intracellular Ca^{2+} pool, the endoplasmic reticulum (Takasawa et al. 1993a). To see whether the accumulation of cADPR is actually caused by glucose stimulation in pancreatic islets, normal Wistar rat and C57BL/6 J mouse islets were incubated with low and high glucose, and the cADPR content in the islets was assayed by radioimmunoassay with an anti-cADPR antibody. Incubation with high glucose caused the cADPR content of islets to increase within minutes, whereas the cADPR content of islets incubated with low glucose did not (Takasawa et al. 1998). The increase in the cADPR concentration in response to high glucose was confirmed using Balb/c mouse islets (An et al. 2001). Next, in pancreatic islet microsomes used as a cell-free system to study Ca^{2+} release, Takasawa et al. found that cADPR released Ca^{2+} from islet microsomes, as indicated by the prompt increase in fluo-3 fluorescence. The Ca^{2+} release by cADPR was completely inhibited by 8-amino-cADPR, an antagonist of cADPR (Takasawa et al. 1993a, 1998). However, inositol 1,4,5-trisphosphate (IP_3) did not cause the release of Ca^{2+} from islet microsomes although the islet microsomes remained responsive to cADPR. In microsomes from rat cerebellum, both IP_3 and cADPR caused the release of Ca^{2+}. Although heparin, an inhibitor of IP_3 binding to its receptor, blocked the IP_3-induced Ca^{2+} release from cerebellar microsomes, it did not block the cADPR-induced Ca^{2+} release from islet microsomes, indicating that islet endoplasmic reticulum responds to cADPR but not to IP_3. Although cerebellar microsomes respond to both cADPR and IP_3, cADPR induces Ca^{2+} release via a different mechanism than that utilized by IP_3. In digitonin-permeabilized pancreatic islets, cADPR as well as Ca^{2+} induced insulin secretion, but IP_3 did not. The combination of cADPR and Ca^{2+} did not induce significantly more insulin secretion than the addition of cADPR or Ca^{2+} alone, and the cADPR-induced insulin secretion was inhibited by the addition of EGTA. The cADPR-induced insulin secretion, therefore, appears to be mediated by Ca^{2+} mobilization from islet microsomes. Dose–response relationships between cADPR and insulin secretion from permeabilized islets were well fitted with those between cADPR and Ca^{2+} release from islet microsomes. Thus, it was concluded that glucose stimuli induce cADPR formation from NAD^{+} and that cADPR then mobilizes Ca^{2+} from the endoplasmic reticulum, serving as a second messenger for insulin secretion (Takasawa et al. 1993a).

The next issue concerns the mechanism by which the glucose stimulus induces the formation of cADPR. CD38, a 300-amino acid protein first recognized as a leukocyte antigen, was found to be expressed in a variety of tissues including pancreatic β-cells (Takasawa et al. 1993b; Koguma et al. 1994; Ebihara et al. 1997; Sasamori et al. 2004). CD38 has both ADP-ribosyl cyclase, synthesizing cADPR from NAD^{+}, and cADPR hydrolase to hydrolyze cADPR to ADP-ribose (Howard et al. 1993; Takasawa et al. 1993b; Koguma et al. 1994). Millimolar concentrations of ATP were found to inhibit the cADPR hydrolase activity, competing with the substrate (Takasawa et al. 1993b; Tohgo et al. 1994). Based on the competitive inhibition of the cADPR hydrolysis by ATP, cADPR and ATP appear to bind to the same site of CD38. By labeling the recombinant CD38 with an ATP analog, 5′-*p*-fluorosulfonylbenzoyladenosine, the binding site for ATP and/or cADPR was identified as the Lysine 129 of CD38 (Tohgo et al. 1997). Furthermore, the mutants did not bind cADPR, whereas they still used NAD^{+} as a substrate to form cADPR (Tohgo et al. 1997). These results indicate that ATP, produced in the process of glucose metabolism, competes with cADPR for the binding site, resulting in the inhibition of the cADPR hydrolase activity of CD38 and then in the accumulation of cADPR in β-cells.

cADPR is generally believed to activate a ryanodine receptor to release Ca^{2+} from the intracellular stores, the endoplasmic reticulum (Takasawa et al. 1993a; Hua et al. 1994; Mitchell et al. 2001). The type 2 ryanodine receptor is expressed in pancreatic islets (Takasawa et al. 1998, 2010). Experiments with islets revealed that cADPR does not bind directly to the ryanodine receptor, but acts on the receptor through a mediator such as FK506-binding protein (FKBP) 12.6 to release Ca^{2+}. The cellular target for FK506, one of the most widely used immunosuppressive agents, is thought to be FKBP12 and FKBP12.6. Rat FKBP12 is composed of 108 amino acids and is highly conserved among human, mouse, bovine, and rabbit FKBP12. Rat FKBP12.6 is also a 108-amino acid protein as are human and bovine FKBP12.6. Noguchi et al. (1997) isolated microsomes from rat islets, carried out immunoblot analyses, and found that rat islet microsomes contained FKBP12.6, but did not contain FKBP12. Interestingly, cADPR was found to bind to FKBP12.6 at a *Kd* value of 35 nM. The cADPR binding was inhibited by FK506, and neither structurally nor functionally related analogs of cADPR inhibited the cADPR binding to FKBP12.6. These results not only indicate that FKBP12.6 acts as a cADPR-binding protein, but also strongly suggest that cADPR is the physiological ligand of FKBP12.6 since FK506 does not normally exist in mammalian cells. As mentioned above, FKBP12.6 occurs in rat islet microsomes. When rat islet microsomes were treated with cADPR, FKBP12.6 dissociated from the microsomes and moved to the supernatant, releasing Ca^{2+} from the intracellular stores. The microsomes that were treated with FK506 or cADPR and were then devoid of FKBP12.6 did not show Ca^{2+} release by cADPR. These results and those from other experiments strongly suggest that, when cADPR binds to FKBP12.6 in the ryanodine receptor and causes the dissociation of FKBP12.6 from the ryanodine receptor to form the FKBP12.6-cADPR complex, the ryanodine receptor channel activity is thereby increased to release Ca^{2+} from the endoplasmic reticulum (Noguchi et al. 1997). When FK506 is present, cADPR cannot act on the ryanodine receptor to release Ca^{2+}, and the glucose-induced insulin secreting machinery ceases to function. Noguchi et al. (2008) confirmed that FKBP12.6 plays a role in glucose-induced insulin secretion downstream of ATP production, independently of ATP-sensitive K^+ channels, in pancreatic β-cells, by using FKBP12.6-deficient mice. In human, when FK506 was used as an immunosuppressant in kidney transplantation, hyperglycemia was observed in 20–35 % of the recipients (Pirsch et al. 1997).

It should also be noted that, in the presence of calmodulin (CaM), islet microsomes became sensitized to cADPR at much lower concentrations for Ca^{2+} release, and the Ca^{2+} release was greatly increased (Takasawa et al. 1995; Takasawa and Okamoto 2002). Inhibitors for CaM-dependent protein kinase II (CaM kinase II) completely abolished the glucose-induced insulin secretion as well as the cADPR-mediated and CaM-activated Ca^{2+} mobilization. Western blot analysis revealed that rat microsomes contain CaM kinase IIα but do not contain CaM. When the active 30-kDa chymotryptic fragment of CaM kinase II was added to the microsomes, fully activated cADPR-mediated Ca^{2+} release was observed in the absence of CaM (Takasawa et al. 1995; Takasawa and Okamoto 2002). These results strongly suggest that CaM kinase II is required to phosphorylate and activate the ryanodine receptor for the cADPR-mediated Ca^{2+} release. Takasawa et al. also found that cyclic AMP-dependent protein kinase (A-kinase) activates the cADPR-mediated

Ca^{2+} release from islet microsomes. In the absence of A-kinase, only a small amount of Ca^{2+} was released from the microsomes by low concentrations of cADPR (Takasawa and Okamoto 2002). On the other hand, when the catalytic subunit of A-kinase was added to the islet microsomes, the Ca^{2+} release was sensitized at lower concentrations of cADPR. As CaM kinase II was observed to be activated by glucose stimulation and several incretin peptide hormones such as VIP, PACAP, GLP-1, and GIP increase the intracellular concentrations of cyclic AMP to activate A-kinase, it is quite possible that the cADPR-mediated Ca^{2+} mobilization for insulin secretion is activated by CaM kinase II and A-kinase. Possibly, the activated kinases phosphorylate the ryanodine receptor to sensitize the Ca^{2+} channel for the cADPR signal. Recently, Takasawa et al. isolated a novel ryanodine receptor, which is generated from *type 2 ryanodine receptor* (*RyR2*) *gene* by alternative splicing of exons 4 and 75. The newly identified form of RyR2 is expressed in islets and neuronal cells and activated by cADPR, and has been proposed to act as an intracellular target for cADPR signaling, which has been demonstrated to be important for insulin secretion (Takasawa et al. 2010; Rossi and Sorrentino 2010).

Toward verifying the role of the CD38–cADPR signal system in the regulation of insulin secretion, Kato et al. created CD38 transgenic and knockout mice (Kato et al. 1995, 1999). The insulin secretion from the islets isolated from the CD38 transgenic mice was significantly higher than that from the control islets at high concentrations of glucose. When islets were exposed to ketoisocaproate, which, like glucose, is metabolized to form ATP, insulin secretion from transgenic islets was significantly higher than that of the control. However, with tolbutamide and KCl, the transgenic insulin secretion was not altered as compared to the control secretion. Tolbutamide blocks the ATP-sensitive K^+ channel and facilitates Ca^{2+} influx from extracellular sources through voltage-dependent Ca^{2+} channels, and KCl directly induces cell membrane depolarization resulting in Ca^{2+} influx. In glucose-tolerance tests, the serum insulin level of CD38-transgenic mice was significantly higher than that of control mice (Kato et al. 1995). On the other hand, the glucose-induced increase in the intracellular Ca^{2+} concentration was severely impaired in CD38 knockout mouse islets, and the glucose-induced insulin secretion was severely decreased (Kato et al. 1999). CD38 knockout mice showed impaired glucose tolerance, and the serum insulin level was lower than control. In order to see whether the observed phenotype could be rescued by the pancreatic β-cell, specific expression of CD38 cDNA was tested by crossbreeding the transgenic mice carrying a human CD38 cDNA under the rat insulin II promoter and the CD38 knockout mice. By intercrossing, CD38 knockout mice carrying the human CD38 transgene were generated. In effect, the human CD38 transgene ameliorated the glucose intolerance and the decreased insulin secretion. CD38 knockout mice did not show insulin resistance, suggesting that the observed phenotype is indeed caused by the absence of CD38 in pancreatic β-cells (Kato et al. 1999). The knockout islets, however, responded normally to the extracellular Ca^{2+} influx stimulants tolbutamide and KCl to secrete insulin (Kato et al. 1999). Thus, the CD38–cADPR signal system as well as the ATP-sensitive K^+ channel system appears to function in insulin secretion by glucose (Fig. 6.2).

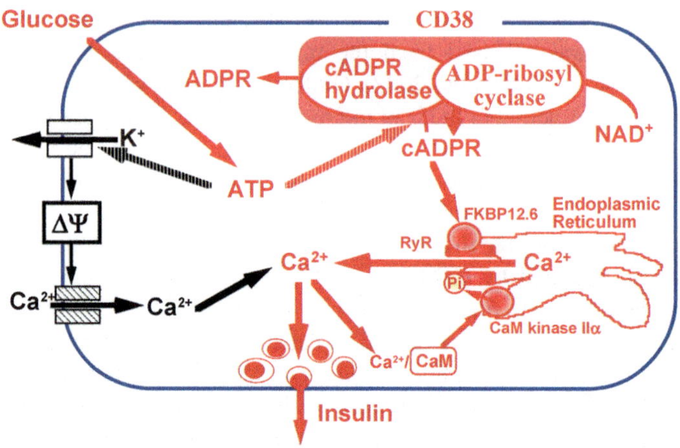

Fig. 6.2 The CD38–cADPR signal system in insulin secretion by glucose stimulation in pancreatic β-cells (adapted from Okamoto et al. 1997). The insulin secretion via the CD38–cADPR signal system is shown in *red*. cADPR binds to FKBP12.6, a component of ryanodine receptor (RyR) Ca^{2+} release channel complex, to release Ca^{2+}, dissociating FKBP12.6 from RyR (Noguchi et al. 1997). CaM kinase II phosphorylates RyR to sensitize and activate the Ca^{2+} channel (Pi, phosphorylation of RyR by CaM kinase IIα) (Takasawa et al. 1995; Suzuki et al. 2014). Ca^{2+}, released from intracellular store and/or supplied from extracellular sources, further activates CaM kinase II and amplifies the process. In this way, Ca^{2+}-induced Ca^{2+} release (CICR) can be explained. The conventional insulin secretion mechanism by Ca^{2+} influx from extracellular sources (Ashcroft et al. 1984) is shown in *black*

A wide body of evidence obtained in rats and mice supports this paradigm of insulin secretion based on the CD38–cADPR signal system. Recent results indicate that the CD38–cADPR signal system also functions in insulin secretion in man. As a result of a missense mutation (Arg140Trp) found in the CD38 gene in Japanese diabetic patients, the CD38 protein showed altered catalytic activities and a decreased production of cADPR (Nata et al. 1997; Yagui et al. 1998). The decreased function of the CD38 mutant may have contributed to the impairment of glucose-stimulated insulin secretion in type 2 diabetic patients. It is also significant that circulating anti-CD38 autoantibodies were detected in 10–14 % of Japanese (Ikehata et al. 1998) as well as Caucasian diabetic patients (Pupilli et al. 1999; Mallone et al. 2001; Antonelli et al. 2001, 2002; Marchetti et al. 2002). The autoantibody altered the in vitro enzymic activity of islet CD38 and insulin secretion. These findings provide further support for the concept that the CD38–cADPR signal system functions in insulin secretion in human.

Sleep apnea syndrome (SAS) is characterized by recurrent episodes of oxygen desaturation during sleep, the development of daytime sleepiness, and deterioration in the quality of life. Accumulating evidence suggests the association of intermittent hypoxia, a hallmark of SAS, and type 2 diabetes independently on body mass index and waist circumference. Recently, Ota et al. (2012) reported that intermittent hypoxia attenuated glucose-induced insulin secretion from pancreatic β-cells and that the mRNA level of CD38 in intermittent hypoxia-treated islets was significantly reduced, whereas the mRNA levels of glucose transporter 2, glucokinase, sulfonylurea receptor 1, L-type Ca^{2+} channel 1.2, insulin I, and insulin II were unchanged. Furthermore, overexpression of CD38 by introducing CD38 expression vector recovered the attenuation of glucose-induced insulin secretion by intermittent hypoxia.

The CD38–cADPR signal system for insulin secretion is different from the hypothesis of Ashcroft et al. (1984) in which Ca^{2+} influx from extracellular sources is involved in insulin secretion by glucose. In the signal system proposed by Berridge and Irvine, IP_3 induces Ca^{2+} release from the intracellular pool, the endoplasmic reticulum (Berridge and Irvine 1984). Accordingly, the CD38–cADPR signal system was the focus of intense debate (Islam et al. 1993; Takasawa et al. 1993c, 1998; Rutter et al. 1994; Webb et al. 1996; Islam and Berggren 1997; Okamoto et al. 1997). Discrepant results were reported in diabetic β-cells such as *ob/ob* mouse islets and RINm5F cells, which are often used for studying insulin secretion. However, the Ca^{2+} release responses of these diabetic β-cell microsomes differed greatly from those of normal islet microsomes (Takasawa et al. 1993a, b, c). Microsomes from normal C57BL mouse islets released Ca^{2+} in response to cADPR, but scarcely in response to IP_3. This response to cADPR was completely attenuated by the prior addition of 8-amino-cADPR. In *ob/ob* mouse islet microsomes, however, only a small amount of Ca^{2+} was released by cADPR, but much Ca^{2+} was released by IP_3 (Takasawa et al. 1998). RINm5F cell microsomes also responded well to IP_3 to release Ca^{2+}, but did not respond to cADPR. However, RINm5F cells are rat insulinoma-derived immortal cells that show almost no glucose-induced insulin secreting ability. Concerning intracellular Ca^{2+} release channels, the mRNA expression of RyR2, which is postulated to be a Ca^{2+} release channel for cADPR (Takasawa et al. 1998, 2010; Mitchell et al. 2001), was clearly detected in normal islets, but not in *ob/ob* islets. In contrast, IP_3 receptor (IP_3R1, IP_3R2, IP_3R4, and IP_3R5) mRNAs were not detected in normal islets, but were clearly detected in *ob/ob* islets, and although IP_3R3 mRNA was slightly detected in normal islets, the mRNA was significantly increased in *ob/ob* islets, well fitting with the observation that IP_3-induced Ca^{2+} mobilization preferentially worked in *ob/ob* islet microsomes. More importantly, the CD38 mRNA level was significantly decreased in *ob/ob* islets (Takasawa et al. 1998), and CD38 mRNA was not expressed in RINm5F cells (Takasawa et al. 1998). The decrease of CD38 mRNA in *ob/ob* islets may explain the low response in the cADPR content of β-cells by glucose stimulation (Takasawa et al. 1993a, b, c) because CD38 has both the ADP-ribosyl cyclase and the cADPR hydrolase activities, and the increase of cADPR by glucose stimulation is achieved by inhibition of the cADPR hydrolase activity of CD38. Decreased CD38 mRNA was also reported in islets of Goto–Kakizaki diabetic rats (Goto et al. 1976; Matsuoka et al. 1995), which show impaired glucose-induced insulin secretion. These results show that the CD38–cADPR signal system for insulin secretion acts under normal physiological conditions, but the IP_3 system becomes dominant in diabetic β-cells such as *ob/ob* mouse islets and RINm5F cells. In fact, it was more recently confirmed by Rutter et al. using aequorin chimera that MIN6 cells, which keep the ability of glucose-induced insulin section, show a dramatic Ca^{2+} mobilization in response to cADPR via the ryanodine receptor despite the fact that no response to IP_3 was observed (Mitchell et al. 2001; Varadi and Rutter 2002). Moreover, Balb/c mouse islets and MIN6 β-cells exhibited distinct increases in intracellular cADPR, Ca^{2+}, and insulin secretion by glucose stimulation (An et al. 2001; Ohta et al. 2011). They also showed the involvement of cADPR in the enhancement of insulin secretion by GLP-1 (Kim et al. 2008).

Regulators of β-cell regeneration

Background

Pancreatic β-cells, the site of insulin production in adult, have only a limited capacity for regeneration. In adult mouse and rat islets, the population of replicating β-cells was estimated to be only about 1 % of total β-cells. However, certain substances, such as glucose, essential amino acids, insulin, and growth hormone, were reported to stimulate some β-cell replication in fetal, newborn, or adult islets; nevertheless, the percentage of replicating β-cells was only about twice as high in β-cells treated with these substances than in control cells. According to the Okamoto model, streptozotocin and alloxan diabetes can be prevented by PARP inhibitors. In 1984, Yonemura et al. demonstrated that surgical diabetes could be ameliorated by using PARP inhibitors to induce the regeneration of pancreatic β-cells (Yonemura et al. 1984). Male Wistar rats underwent 90 % pancreatectomy and received daily intraperitoneal injection of nicotinamide or 3-aminobenzamide. The administration of PARP inhibitors ameliorated the surgical diabetes. The islets in the remaining pancreases of rats that had received PARP inhibitors for 3 months were extremely large, and almost the entire areas of the enlarged islets were stained for insulin (Fig. 6.3).

Reg Gene and Reg Protein as Regulator for β-Cell Regeneration

Pancreatic β-cells of the islets of *Langerhans* are the only cells that produce insulin in humans as well as in almost all animals, but they have a limited capacity for regeneration, which is a predisposing factor for the development of *diabetes mellitus*. Strategies for preventing β-cell destruction and influencing the replication and growth of the β-cell mass are therefore important for the prevention and/or treatment of diabetes (Takasawa and Okamoto 2002; Okamoto and Takasawa 2002). Terazono et al. isolated the regenerating islets and constructed a cDNA library. By differential screening of the regenerating islet-derived cDNA library,

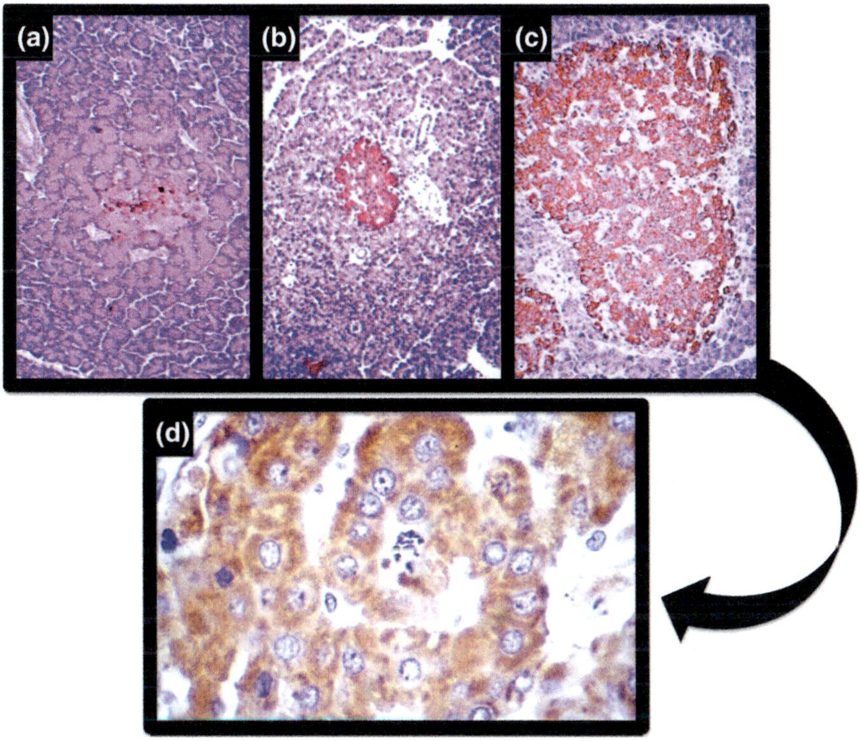

Fig. 6.3 Proliferation of insulin-positive islet cells in the remaining pancreas of 90 % depancreatized rats receiving daily injection of PARP inhibitor, nicotinamide. **a** Pancreatic tissue from 90 % depanreatized rat receiving daily injection of saline. **b** Pancreatic tissue from untreated control rat. **c** Pancreatic tissue from 90 % depancreatized rat receiving daily injection of nicotinamide. **d** The appearance of metaphase chromosomes indicates the proliferation of insulin-producing cells. PARP inhibitor upregulates a β-cell regeneration factor gene, *Reg*. Reg protein acts as an autocrine/paracrine growth factor for β-cell regeneration (Yonemura et al. 1984; Terazono et al. 1988; Watanabe et al. 1994; Akiyama et al. 2001; Shervani et al. 2004; Takasawa et al. 2006)

we found a novel gene expressed in regenerating islets. The cDNA had one large open reading frame encoding a 165-amino acid protein, and the deduced protein has a signal sequence. The novel gene was named *Reg*, that is, ***re****generating* ***g****ene*, with the implication that the gene may be involved in islet regeneration (Terazono et al. 1988). In order to see whether the gene product stimulates β-cell replication, Watanabe et al. administered recombinant rat Reg proteins to 90 % depancreatized rats and observed an increased [^{3}H]thymidine incorporation and frequent mitosis in islets of the remaining pancreas. On the 30th and 60th postoperative day, the fasting plasma glucose level of the rats that had received daily intraperitoneal injection of Reg protein was significantly lower than that of the 90 % depancreatized control rats. After 2 months, almost all the islets in the 90 % depancreatized control rats were destroyed, whereas the islets of the remaining pancreas in the Reg protein-treated rats were enlarged and almost entirely stained for insulin (Watanabe et al. 1994). Human REG protein administration also ameliorated diabetes in NOD mice with an increase in the β-cell mass (Gross et al. 1998). These results indicate that Reg protein stimulates the regeneration and/or growth of pancreatic β-cells, thereby ameliorating animal diabetes. Transgenic mice expressing *Reg* (*Reg I*) in β-cells showed increased [^{3}H]thymidine incorporation in the islets (Unno et al. 2002). In the *Reg I* transgene-carrying NOD mice, the development of diabetes was significantly retarded. On the other hand, *Reg I* knockout mice created by homologous recombination showed the decreased [^{3}H]thymidine incorporation in the islets. Further, when hyperplastic islets were induced by the injection of gold thioglucose, the islet sizes of *Reg I* knockout mice were significantly smaller than those from control *Reg*$^{+/+}$ mice (Unno et al. 2002). *Reg I* is induced in insulin-producing pancreatic β-cells by inflammatory stimulation such as by IL-6/glucocorticoids (Akiyama et al. 2001; Nakagawa et al. 2013) and acts as an autocrine/paracrine growth factor for β-cell regeneration via a cell surface Reg receptor (Kobayashi et al. 2000; Takasawa et al. 2006) to ameliorate experimental diabetes (Watanabe et al. 1994; Gross et al. 1998; Unno et al. 2002; Shervani et al. 2004).

Reg Family Genes

The *Reg* and *Reg*-related genes were isolated and revealed to constitute a multigene family, the *Reg* gene family (Okamoto and Takasawa 2002; Nata et al. 2004). Based on the primary structures of the Reg proteins, the members of the family are grouped into four subclasses: types I, II, III, and IV (Fig. 6.4). In human, five *REG* family genes, i.e., *REG Iα* (Terazono et al. 1988; Watanabe et al. 1990), *REG Iβ* (Moriizumi et al. 1994), *REG*-related sequence (*RS*) (Watanabe et al. 1990), *HIP* (Lasserre et al. 1994)/*PAP* (Dusetti et al. 1994), and *REG III* (Nata et al. 2004) are randomly ordered in the 95-kbp region of chromosome 2p12 (Miyashita et al. 1995; Nata et al. 2004) whereas *REG IV* locates on chromosome 1p12-13.1 (Hartupee et al. 2001; Kämäräinen et al. 2003). In mouse genome, all the *Reg* family genes except for *Reg IV*, i.e., *Reg I, Reg II, Reg IIIα, Reg IIIβ, Reg IIIγ*, and *Reg IIIδ,* were mapped to a contiguous 75-kbp region of chromosome 6C (Narushima et al. 1997; Abe et al. 2000), whereas *Reg IV* was mapped on chromosome 3. Type I (and Type II) Reg proteins are expressed in regenerating islets and involved in β-cell regeneration (Omori et al. 2003; Shervani et al. 2004; Takasawa et al. 2006; Lu et al. 2006; Planas et al. 2006; Ota et al. 2013; Calderari et al. 2014; Aida et al. 2014).

Reg Protein Signal

To determine what intracellular signal transduction events are induced in pancreatic β-cells by Reg protein, we tested several signal transduction pathways

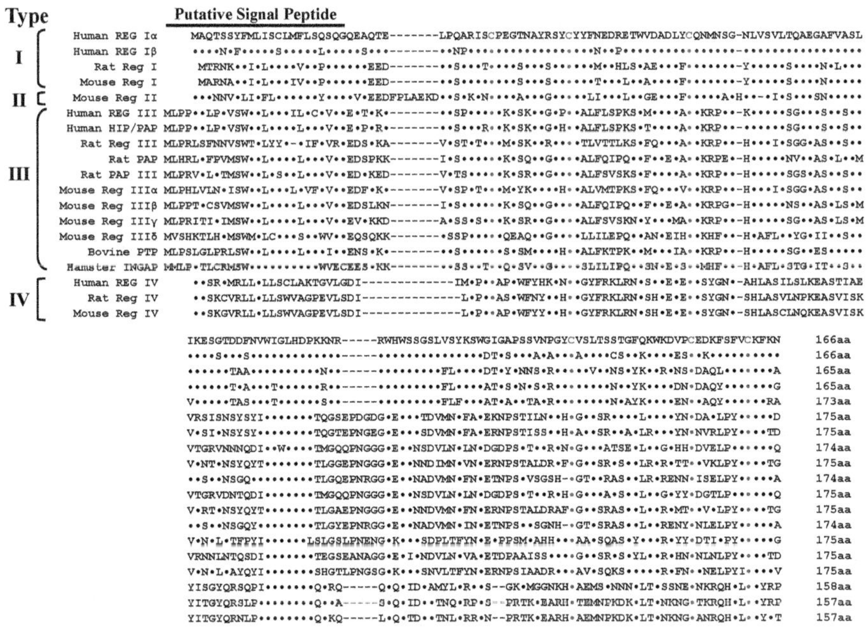

Fig. 6.4 Alignment of amino acid sequences of the Reg family. Based on the primary structure of the encoded proteins [adapted from (Okamoto and Takasawa 2002; Kämäräinen et al. 2003; Nata et al. 2004; Namikawa et al. 2005)], the members of the *Reg* gene family are grouped into four subgroups: types I, II, III, and IV. *Dots* indicate amino acids identical to human REG Iα. *Dashes* indicate gaps for maximal alignment. Six conserved cysteines in the mature proteins are indicated by *red*. Types I, II, and III *Reg* genes are clustered in a restricted region of the same chromosome (Miyashita et al. 1995; Narushima et al. 1997; Abe et al. 2000; Nata et al. 2004). Human *REG*-related sequence (*RS*) is omitted from the alignment because human *RS* is a pseudogene (Watanabe et al. 1990)

in pancreatic β-cells by Reg protein stimulation using PathDetect in vivo *trans*-reporting systems and found that ATF-2, but not CREB, cJun, Elk1, CHOP, or cFos, was activated by Reg protein stimulation (Takasawa et al. 2006). The ATF-2 activation was also observed by the cotransfection of the Reg receptor expression plasmid instead of Reg protein addition to the medium. Phosphorylation of ATF-2 at Thr-71 was increased by the stimulation of Reg protein. These results indicate that the Reg–Reg receptor system activates ATF-2, suggesting that genes under the control of ATF-2 play an important role in the cell cycle progression in pancreatic β-cells.

We investigated whether the target of ATF-2 is the cyclin D1 promoter in Reg protein-activated pancreatic β-cells by using a reporter plasmid containing the rat cyclin D1 promoter in transiently transfected RINm5F β-cells. The cyclin D1 promoter activation by Reg protein was induced in a dose-dependent manner. Furthermore, the cotransfection of an ATF-2 expression vector with the cyclin D1 reporter construct also increased the activity of the cyclin D1 promoter.

The increases in the cyclin D1 promoter were almost the same as the increases in phospho-ATF-2. In electrophoretic mobility assay using cyclin D1 promoter sequence, Reg protein-stimulated β-cell nuclear proteins formed a specific complex with the ATF-2 binding site. Addition of the antibody against ATF-2 resulted in the formation of a supershift, suggesting the presence of ATF-2 in the complex. The essential involvement of ATF-2 activation in the Reg protein-induced activation of the cyclin D1 promoter was further confirmed by the inhibition of the Reg protein-induced activation of the cyclin D1 promoter with the cotransfection of the dominant-negative ATF-2 expression plasmid and by physical in vivo interaction of ATF-2 and cyclin D1 promoter in chromatin immunoprecipitation (ChIP) assay (Takasawa et al. 2006).

To determine the intracellular signaling pathway(s) by which Reg protein induced the cyclin D1 promoter activity via the ATF-2 activation, several chemical inhibitors were employed. The addition of PI(3)K inhibitors LY294002 and wortmannin attenuated the Reg protein-induced ATF-2 phosphorylation/activation. SB203580, a p38 mitogen-activated protein (MAP) kinase inhibitor, PD98059, an inhibitor of the MAP kinase/ERK kinase-ERK pathway, KN-62 and KN-93, inhibitors of Ca^{2+}/CaM kinase II, K-252a, an inhibitor of CaM kinase IV, and H-89, the A-kinase inhibitor, showed no inhibitory effect on the cyclin D1 promoter activity in β-cells. In addition, overexpression of MEKK, MEK1, MEK3, and the catalytic subunit of A-kinase failed to increase the cyclin D1 promoter activity. The Reg protein-induced cyclin D1 promoter activity was actually inhibited by the addition of LY294002. These results likely indicate that an LY294002-sensitive pathway involving PI(3)K mediates the signal transduction by Reg protein. Moreover, the immunoprecipitated PI(3)K phosphorylated ATF-2 in a time-dependent and dose-dependent fashion, indicating that PI(3)K directly phosphorylates ATF-2 to activate the cyclin D1 promoter (Takasawa et al. 2006).

We produced *Reg I* knockout mice by homologous recombination (Unno et al. 2002). In *Reg I* knockout mice, no Reg I protein expression was detected in the pancreas and therefore no Reg I was secreted into the islet culture medium (Unno et al. 2002; Takasawa et al. 2006). We isolated pancreatic islets from *Reg I* knockout and wild-type control mice and compared their 5'-bromo-2'-deoxyuridine (BrdU) incorporation. The BrdU incorporation in *Reg I* knockout islets without the addition of Reg I protein in the culture medium was reduced. We then compared the levels of phospho-ATF-2, cyclin D1 protein, and phospho-Rb in Reg I-deficient mouse islets with those in normal littermates to determine whether Reg I induces the cyclin D1 gene for the cell cycle progression via the activation of PI(3)K/ATF-2. Wild-type islets secreted Reg I in the culture medium, and the phospho-Rb level in the islets was much higher than that in *Reg I* knockout islets. This well fitted the decreased BrdU incorporation in *Reg I* knockout islets. As expected, both the levels of phospho-ATF-2 and cyclin D1 were decreased in *Reg I* knockout islets, whereas the levels of other proteins such as PARP (Akiyama et al. 2001; Okamoto and Takasawa 2002) and CD38 (Takasawa et al. 1993b; Koguma et al. 1994; Kato et al. 1999; Okamoto and Takasawa 2002) as controls were unchanged.

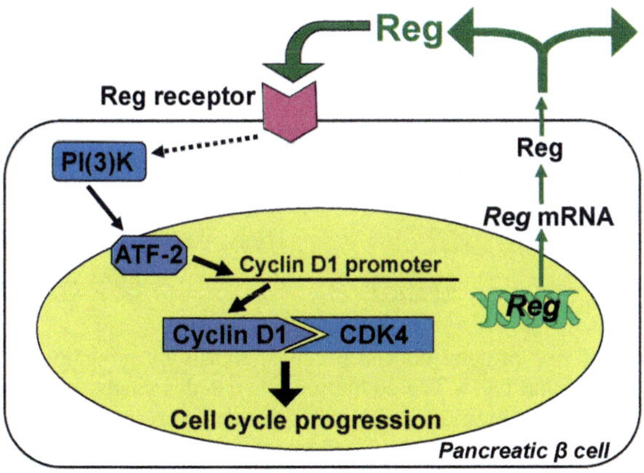

Fig. 6.5 A putative role of Reg protein in β-cell regeneration through the PI(3)K/ATF-2-cyclin D1 signaling pathway. Upon Reg protein stimulation, PI(3)K becomes activated. ATF-2 is phosphorylated in a PI(3)K-dependent manner, and phosphorylated ATF-2 binds cyclin D1 gene promoter through the ATF-2 binding sequence (-57 to -52 of rat cyclin D1 promoter) to activate the cyclin D1 promoter. The increase of cyclin D1 activates CDK4 and the subsequent G1-S progression of the cell cycle of pancreatic β-cells (Takasawa et al. 2006). Reg protein synthesized from *Reg* gene is secreted from β-cells and acts as an autocrine/paracrine growth factor via Reg receptor (Kobayashi et al. 2000)

These results show conclusively that the activation of the cyclin D1 gene promoter in response to Reg I protein stimulation is mediated by the PI(3)K/ATF-2 signal transduction pathway in pancreatic β-cells (Fig. 6.5). The growth of pancreatic β-cells is determined by the number of β-cells entering the cell cycle rather than by changes in the rate of the cycle (Hellerström and Swenne 1991; Yonemura et al. 1984; Okamoto 1985; Rane et al. 1999; Okamoto and Takasawa 2002; Unno et al. 2002). Reg, an autocrine/paracrine pancreatic β-cell regeneration factor, appears to stimulate the cell cycle progression (Watanabe et al. 1994; Gross et al. 1998; Kobayashi et al. 2000; Levine et al. 2000; Akiyama et al. 2001; Unno et al. 2002; Shervani et al. 2004), although some controversial data have been reported (Smith et al. 1994; Moriscot et al. 1996). Progression from the G1 to S phase of the cell cycle requires the activation of cyclin-dependent kinase (CDK) 4, and the CDK4 activation is controlled by the complex formation with its catalytic partner, cyclin D1. The expression of the cyclin D1 is controlled largely and perhaps entirely by extracellular signals. Reg protein activated the cyclin D1 promoter for the cell cycle progression, and ATF-2 is an essential transcription factor in the process of Reg protein-induced cyclin D1 promoter activation. The PI(3)K activation is involved in the ATF-2 activation for cyclin D1 expression. It has been reported that disruption of the CDK4 gene resulted in the development of insulin-deficient diabetes due to a reduction in the number of pancreatic β-cells and that the expression of a mutant CDK4, which escaped from the inhibitory regulation of CDK4, caused islets to become hyperplastic (Rane et al. 1999). The results of the CDK4 disruption well explain the results that the islets from the *Reg I* knockout

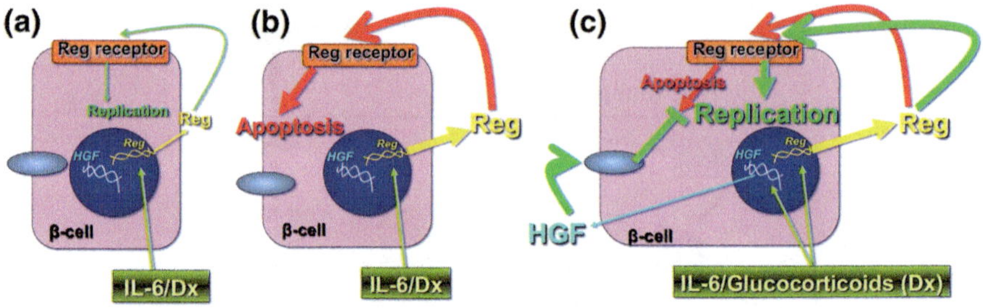

Fig. 6.6 Prevention of Reg protein-induced β-cell apoptosis by induction of *HGF* gene transcription by IL-6/dexamethasone. **a** The addition of IL-6 + dexamethasone induces *Reg* (*Reg I*) gene expression in β-cells to promote β-cell proliferation (Akiyama et al. 2001). **b** High concentrations of Reg protein (Reg I) induce β-cell apoptosis (Kobayashi et al. 2000). **c** The addition of IL-6 + dexamethasone also induces *HGF* gene expression via STAT3 activation in β-cells. HGF attenuates the high-concentration Reg protein-induced apoptosis in β-cells (Nakagawa et al. 2013). Thus, the combined addition of IL-6 + dexamethasone induces both *Reg* and *HGF* expressions to promote proliferation of pancreatic β-cells effectively. In fact, simultaneous induction of *IL-6*, *Reg I* (*REG Iα*), and *HGF* mRNA in β-cells along with β-cell proliferation was recently reported in intermittent hypoxia-induced β-cell proliferation (Ota et al. 2013)

mice showed reduced BrdU incorporation (Takasawa et al. 2006) and that the islet hyperplasia induced by the gold thioglucose treatment was attenuated by *Reg I* gene disruption (Unno et al. 2002) because Reg protein induces a regulatory subunit of CDK4, cyclin D1, for the cell cycle progression. Reg protein was also reported to work as a growth factor for Schwann cells (Livesey et al. 1997), motoneurons (Nishimune et al. 2000), gastrointestinal epithelium (Fukui et al. 1998; Kazumori et al. 2000), heart muscle (Kiji et al. 2005), and skeletal muscle (Klasan et al. 2014), and cyclin D1 was suggested to be a key molecule for the regenerative growth of Schwann cells (Kim et al. 2000, 2001). Therefore, the signal transduction mechanism in the Reg protein-induced regenerative growth of other tissues including Schwann cells may be the same as that of pancreatic β-cells (Fig. 6.6).

Role of HGF in β-Cell Regeneration

HGF Attenuates the Apoptosis Induced by the High Concentrations of Reg I Protein

We previously reported that high concentrations of Reg I protein (300–1000 nM) induced the apoptosis of RINm5F β-cells (Kobayashi et al. 2000). Higher serum concentrations of Reg family protein (over 10 μM) were reported in pathological conditions such as pancreatitis (Satomura et al. 1995; Viterbo et al. 2009). A single addition of IL-6, dexamethasone, nicotinamide, nor 3-aminobenzamide alone, did not inhibit the apoptosis induced by 1000 nM Reg I protein (Nakagawa et al. 2013). On the other hand, the combined addition of IL-6 and dexamethasone significantly inhibited the apoptosis. The addition of PARP inhibitors, nicotinamide or 3-aminobenzamide, to IL-6 + dexamethasone was ineffective. Moreover, the addition of 2.5 and 25 ng/ml hepatocyte growth factor (HGF) showed an anti-apoptotic activity against the high-concentration Reg I-induced apoptosis as well as the combined addition of IL-6 + dexamethasone. In the presence of 1000 nM Reg I protein, the viable cell numbers were increased by the addition of IL-6 + dexamethasone as well as by the addition of HGF. The combined addition of IL-6 + dexamethasone and HGF did not increase significantly more WST-1 cleavage than the addition of IL-6 + dexamethasone or HGF alone (Nakagawa et al. 2013). These results suggest that HGF is induced and secreted from pancreatic β-cells by the combined addition of IL-6 + dexamethasone to protect the Reg I-induced apoptosis.

Induction of *HGF* by the Combined Addition of IL-6 + dexamethasone

The combined addition of IL-6 + dexamethasone induced the expression of *HGF* mRNA, whereas the addition of IL-6, dexamethasone, nor nicotinamide alone, did not induce the *HGF* mRNA expression (Nakagawa et al. 2013). A little increase of the mRNA was observed by the combined addition of IL-6 + dexamethasone + nicotinamide. In the RINm5F culture medium, significantly higher concentrations of HGF were detected by ELISA with the combined addition of IL-6 + dexamethasone. The combined addition of IL-6 + dexamethasone induced the *HGF* mRNA expression in rat primary cultured islets. These results indicate that pancreatic β-cells express *HGF* mRNA and secrete HGF protein in response to the inflammatory stimuli such as the combined stimulation of IL-6 and glucocorticoids.

Activation of the *HGF* Promoter by IL-6 + dexamethasone

In order to determine whether the increase of *HGF* mRNA was caused by the activation of transcription, a 1395-bp fragment containing 1336 bp of the promoter region of the rat *HGF* gene was fused to the luciferase gene and transfected it into RIN5mF cells. The combined addition of IL-6 + dexamethasone markedly increased the relative luciferase activity, whereas IL-6, dexamethasone, nicotinamide, 3-aminobenzamide alone, IL-6 + nicotinamide, IL-6 + 3-aminobenzamide, dexamethasone + nicotinamide nor dexamethasone +3-aminobenzamide was ineffective (Nakagawa et al. 2013). The addition of PARP inhibitors to the combination of IL-6 + dexamethasone did not additionally increase the promoter activity. These promoter assays revealed that the induction of the *HGF* mRNA occurred at the transcriptional level. We previously reported that the combined addition of IL-6 + dexamethasone induced the rat *Reg I* gene activation in RINm5F cells (Akiyama et al. 2001). To investigate whether the *HGF* promoter was activated by Reg I protein secreted from RINm5F cells to surrounding culture medium, we added increasing concentrations of Reg I protein (3.3, 33, and 333 nM) in the RINm5F culture medium and analyzed the *HGF* promoter activity. The addition of Reg I protein did not activate the *HGF* transcription. These results indicate that the *HGF* promoter activation by the combined addition of IL-6 + dexamethasone is not mediated by Reg I protein in the medium and that the promoters for *HGF* and *Reg I* are independently activated by the combined addition of IL-6 + dexamethasone in pancreatic β-cells.

Localization of IL-6/dexamethasone-Responsive Region in the *HGF* Promoter

In order to identify the region necessary for the induction of the *HGF* gene by IL-6 + dexamethasone, progressive deletions of the *HGF* promoter were performed. The deletion down to position −96 did not alter significantly the expression of the reporter gene induced by the combined addition of IL-6 + dexamethasone, but an additional deletion to nucleotide −92 caused a remarkable decrease of promoter activity (Nakagawa et al. 2013), indicating that the region from −96 to −92 is essential for the IL-6 + dexamethasone-sensitive *HGF* promoter activities. The 5'-upstream region of the rat *HGF* gene (−148 to −1) was highly conserved in human (94.6 %), mouse (98.0 %), and chicken (79.2 %) *HGF* genes. A computer-aided search for sequences similar to known *cis*-acting elements revealed that there were four possible binding sites for transcription factors: signal transducer activator of the transcription family (STATx), nuclear respiratory factor 2 (Nrf-2), GATA transcription factor (GATA), and heat shock factor 2 (HSF-2).

STAT3 is a Key Factor for *HGF* Transcription

To map out the *cis*-elements of *HGF* promoter that are responsible for the IL-6 + dexamethasone-induced *HGF* transcription, site-directed mutagenesis of the possible transcription factor binding sites was conducted within the luciferase construct containing *HGF* promoter region up to −1336 from transcription initiation site. STAT-M1, which altered STAT binding motif, STAT-M2, which altered binding motifs of STAT and Nrf-2, and GATA-M, which destroyed binding motifs of GATA and HSF-2, were constructed. STAT-M1 and STAT-M2 showed remarkable reductions in promoter activities, and GATA-M showed almost the same promoter activity as that of −1136 construct. Comparison between STAT-M1 and STAT-M2 showed that STAT-M2 did not cause a further decrease of the *HGF* promoter activity, despite the additional mutation in Nrf-2 binding site, indicating that the STAT binding element is the most important for the *HGF* transcriptional activation by the combined stimulation of IL-6 + dexamethasone (Nakagawa et al. 2013).

To investigate specific binding of STAT transcription factor (STAT3) to potential STAT binding element in vivo, we used ChIP assay to identify the binding state of STAT3 to the predicted STAT3 binding element in the *HGF* promoter in rat RINm5F and human 1.1B4 β-cells after IL-6 + dexamethasone treatment. The in vivo physical interaction of STAT3 with the rat *HGF* promoter in response to the IL-6 + dexamethasone stimulation was demonstrated. Further increment of STAT3 binding by IL-6 + dexamethasone + nicotinamide was not observed, which is completely consistent with the results of promoter assays. The induction of *HGF* mRNA by IL-6 + dexamethasone was also demonstrated in human 1.1B4 β-cells by real-time RT-PCR. Furthermore, in vivo physical interaction of STAT3 with the STAT binding element of the human *HGF* promoter was confirmed in human 1.1B4 pancreatic β-cell line (Nakagawa et al. 2013).

These results strongly suggest that the combined stimulation of IL-6 and glucocorticoids induces the activation of both *Reg* and *HGF* genes and that the anti-apoptotic effects of HGF against the high-concentration Reg I-induced apoptosis may help β-cell regeneration by Reg I protein. Requirement of both IL-6 and dexamethasone for STAT3 activation was also reported in rat serine protease inhibitor-3 gene transcription (Kordula and Travis 1996). However, in most cases, IL-6 induces STAT(3) activation and glucocorticoids such as dexamethasone suppress STAT(3) activation (Mihara et al. 2012). Although why glucocorticoid is required for STAT3 activation in addition to IL-6 remains unknown, simultaneous transcriptional activation of *HGF* and *Reg I* by the combined addition of IL-6 and dexamethasone in pancreatic β-cells may be a novel β-cell regeneration strategy/therapy.

References

Abdel-Wahab MH, Abd-Allah AR (2000) Possible protective effect of melatonin and/or desferrioxamine against streptozotocin-induced hyperglycaemia in mice. Pharmacol Res 41(5):533–537

Abe M, Nata K, Akiyama T, Shervani NJ, Kobayashi S, Tomioka-Kumagai T, Ito S, Takasawa S, Okamoto H (2000) Identification of a novel *Reg* family gene, *Reg IIIδ*, and mapping of all three types of *Reg* family gene in a 75 kilobase mouse genomic region. Gene 246(102):111–122

Aida K, Saitoh S, Nishida Y, Yokota S, Ohno S, Mao X, Akiyama D, Tanaka S, Awata T, Shimada A, Oikawa Y, Shimura H, Furuya F, Takizawa S, Ichijo M, Ichijo S, Itakura J, Fujii H, Hashiguchi A, Takasawa S, Endo T, Kobayashi T (2014) Distinct cell clusters touching islet cells induce islet cell replication in association with over-expression of regenerating gene (REG) protein in fulminant Type 1 diabetes. PLoS ONE 9(4):e95110

Akabane A, Kato I, Takasawa S, Unno M, Yonekura H, Yoshimoto T, Okamoto H (1995) Nicotinamide inhibits IRF-1 mRNA induction and prevents IL-1β-induced nitric oxide synthase expression in pancreatic β-cells. Biochem Biophys Res Commun 215(2):524–530

Akiyama T, Takasawa S, Nata K, Kobayashi S, Abe M, Shervani NJ, Ikeda T, Nakagawa K, Unno M, Matsuno S, Okamoto H (2001) Activation of *Reg* gene, a gene for insulin-producing β-cell regeneration: poly(ADP-ribose) polymerase binds *Reg* promoter and regulates the transcription by autopoly(ADP-ribosyl)ation. Proc Natl Acad Sci USA 98(1):48–53

Alano CC, Garnier P, Ying W, Higashi Y, Kauppinen TM, Swanson RA (2010) NAD^+ depletion is necessary and sufficient for poly(ADP-ribose) polymerase-1-mediated neuronal death. J Neurosci 30(8):2967–2978

An NH, Han MK, Um C, Park BH, Park BJ, Kim HK, Kim UH (2001) Significance of ecto-cyclase activity of CD38 in insulin secretion of mouse pancreatic islet cells. Biochem Biophys Res Commun 282(3):781–786

Andrabi SA, Dawson TM, Dawson VL (2008) Mitochondrial and nuclear cross talk in cell death: parthanatos. Ann NY Acad Sci 1147:233–241

Antonelli A, Baj G, Marchetti P, Fallahi P, Surico N, Pupilli C, Malavasi F, Ferrannini E (2001) Human anti-CD38 autoantibodies raise intracellular calcium and stimulate insulin release in human pancreatic islets. Diabetes 50(5):985–991

Antonelli A, Tuomi T, Nannipieri M, Fallahi P, Nesti C, Okamoto H, Groop L, Ferrannini E (2002) Autoimmunity to CD38 and GAD in Type I and Type II diabetes: CD38 and HLA genotypes and clinical phenotypes. Diabetologia 45(9):1298–12306

Ashcroft FM, Harrison DE, Ashcroft SJH (1984) Glucose induces closure of single potassium channels in isolated rat pancreatic β-cells. Nature 312(5993):446–448

Banting FG, Best CH, Collip JB, Campbell WR, Fletcher AA (1922) Pancreatic extracts in the treatment of diabetes mellitus: preliminary report. Can Med Assoc J 12:141–146

Berridge MJ, Irvine RF (1984) Inositol trisphosphate, a novel second messenger in cellular signal transduction. Nature 312(5992):315–321

Bonner-Weir S (2000) Perspective: postnatal pancreatic β-cell growth. Endocrinology 141(6):1926–1929

Bowes J, McDonald MC, Piper J, Thiemermann C (1999) Inhibitors of poly (ADP-ribose) synthetase protect rat cardiomyocytes against oxidant stress. Cardiovasc Res 41(1):126–134

Burkart V, Wang Z-Q, Radons J, Heller B, Herceg Z, Stingl L, Wagner EF, Kolb H (1999) Mice lacking the poly(ADP-ribose) polymerase gene are resistant to pancreatic beta-cell destruction and diabetes development induced by streptozocin. Nature Med 5(3):314–319

Butler AE, Janson J, Bonner-Weir S, Ritzel R, Rizza RA, Butler PC (2003) β-Cell deficit and increased β-cell apoptosis in humans with Type 2 diabetes. Diabetes 52(1):102–110

Calderari S, Irminger J-C, Giroix M-H, Ehses JA, Gangnerau M-N, Coulaud J, Rickenbach K, Gauguier D, Halban P, Serradas P, Homo-Delarche F (2014) *Regenerating 1* and *3b* gene expression in the pancreas of type 2 diabetic Goto-Kakizaki (GK) rats. PLoS ONE 9(2):e90045

Cetkovic-Cvrlje M, Eizirik DL (1994) TNF-α and IFN-γ potentiate the deleterious effects of IL-1β on mouse pancreatic islets mainly via generation of nitric oxide. Cytokine 6(4):399–406

Chambon P, Weill JD, Doly J, Strosser MT, Mandel P (1966) On the formation of a novel adenylic compound by enzymatic extracts of liver nuclei. Biochem Biophys Res Commun 25(6):638–643

Charron MJ, Bonner-Weir S (1999) Implicating PARP and NAD^+ depletion in type I diabetes. Nature Med 5(3):269–270

Clapper DL, Walseth TF, Dargie PJ, Lee HC (1987) Pyridine nucleotide metabolites stimulate calcium release from sea urchin egg microsomes desensitized to inositol trisphosphate. J Biol Chem 262(20):9561–9568

Cocco RE, Ucker DS (2001) Distinct modes of macrophage recognition for apoptotic and necrotic cells are not specified exclusively by phosphatidylserine exposure. Mol Biol Cell 12(4):919–930

Concannon P, Rich SS, Nepom GT (2009) Genetics of type 1A diabetes. N Engl J Med 360(16):1646–1654

Cooperstein SJ, Watkins D (1981) Actions of toxic drugs on islet cells. In: Cooperstein SJ, Watkins D (eds) The islets of Langerhans. Academic Press, New York, pp 387–425

D'Aleo V, Del Guerra S, Martano M, Bonamassa B, Canistro D, Soleti A, Valgimigli L, Paolini M, Filipponi F, Boggi U, Del Prato S, Lupi R (2009) The non-peptidyl low molecular weight radical scavenger IAC protects human pancreatic islets from lipotoxicity. Mol Cell Endocrinol 309(1–2):63–66

del Moral RM, Gómez-Morales M, Hernández-Cortés P, Aguilar D, Caballero T, Aneiros-Fernández J, Caba-Molina M, Rodríguez-Martínez MD, Peralta A, Galindo-Moreno P, Osuna A, Oliver FJ, del Moral RG, O'Valle F (2013) PARP inhibition attenuates histopathological lesion in ischemia/reperfusion renal mouse model after cold prolonged ischemia. Sci World J 2013:486574

Ducrocq S, Benjelloun N, Plotkine M, Ben-Ari Y, Charriaut-Marlangue C (2000) Poly(ADP-ribose) synthase inhibition reduces ischemic injury and inflammation in neonatal rat brain. J Neurochem 74(6):2504–2511

Dunn JS, Sheehan HL, McLetchie NGB (1943) Necrosis of islets of Langerhans produced experimentally. Lancet 241(6242):484–487

Dusetti NJ, Frigerio JM, Fox MF, Swallow DM, Dagorn JC, Iovanna JL (1994) Molecular cloning, genomic organization, and chromosomal localization of the human pancreatitis-associated protein (PAP) gene. Genomics 19(1):108–114

Ebihara S, Sasaki T, Hida W, Kikuchi Y, Oshiro T, Shimura S, Takasawa S, Okamoto H, Nishiyama A, Akaike N, Shirato K (1997) Role of cyclic ADP-ribose in ATP-activated potassium currents in alveolar macrophages. J Biol Chem 272(25):16026–16029

Eliasson MJ, Sampei K, Mandir AS, Hurn PD, Traystman RJ, Bao J, Pieper A, Wang ZQ, Dawson TM, Snyder SH, Dawson VL (1997) Poly(ADP-ribose) polymerase gene disruption renders mice resistant to cerebral ischemia. Nat Med 3(10):1089–1095

Erickson LC, Bradley MO, Kohn KW (1977) Strand breaks in DNA from normal and transformed human cells treated with 1,3-bis(2-chloroethyl)-1-nitrosourea. Cancer Res 37(10):3744–3750

Florez JC (2008) Newly identified loci highlight beta cell dysfunction as a key cause of type 2 diabetes: where are the insulin resistance genes? Diabetologia 51(7):1100–1110

Fukui H, Kinoshita Y, Maekawa T, Okada A, Waki S, Hassan S, Okamoto H, Chiba T (1998) Regenerating gene protein may mediate gastric mucosal proliferation induced by hypergastrinemia in rats. Gastroenterology 115(6):1483–1493

Gale EA (1996) Molecular mechanisms of beta-cell destruction in IDDM: the role of nicotinamide. Horm Res 45(suppl 1):39–43

Galluzzi L, Vitale I, Abrams JM, Alnemri ES, Baehrecke EH, Blagosklonny MV, Dawson TM, Dawson VL, El-Deiry WS, Fulda S, Gottlieb E, Green DR, Hengartner MO, Kepp O, Knight RA, Kumar S, Lipton SA, Lu X, Madeo F, Malorni W, Mehlen P, Nuñez G, Peter ME, Piacentini M, Rubinsztein DC, Shi Y, Simon HU, Vandenabeele P, White E, Yuan J, Zhivotovsky B, Melino G, Kroemer G (2012) Molecular definitions of cell death subroutines: recommendations of the Nomenclature Committee on Cell Death 2012. Cell Death Differ 19(1):107–120

Giannone PJ, Alcamo AA, Schanbacher BL, Nankervis CA, Besner GE, Bauer JA (2011) Poly(ADP-ribose) polymerase-1: a novel therapeutic target in necrotizing enterocolitis. Pediatr Res 70(1):67–71

Goto Y, Kakizaki M, Masaki N (1976) Production of spontaneous diabetic rats by repetition of selective breeding. Tohoku J Exp Med 119(1):85–90

Gross DJ, Weiss L, Reibstein I, van den Brand J, Okamoto H, Clark A, Slavin S (1998) Amelioration of diabetes in nonobese diabetic mice with advanced disease by linomide-induced immunoregulation combined with Reg protein treatment. Endocrinology 139(5):2369–2374

Gunton JE, Kulkarni RN, Yim S, Okada T, Hawthorne WJ, Tseng YH, Roberson RS, Ricordi C, O'Connell PJ, Gonzalez FJ, Kahn CR (2005) Loss of ARNT/HIF1β mediates altered gene expression and pancreatic-islet dysfunction in human type 2 diabetes. Cell 122:337–349

Haber F, Weiss J (1934) The catalytic decomposition of hydrogen peroxide by iron salts. Proc R Soc Lond A147:332–351

Hammarström L, Ullberg S (1966) Specific uptake of labeled alloxan in the pancreatic islets. Nature 212:708–709

Hans CP, Feng Y, Naura AS, Zerfaoui M, Rezk BM, Xia H, Kaye AD, Matrougui K, Lazartigues E, Boulares AH (2009) Protective effects of PARP-1 knockout on dyslipidemia-induced autonomic and vascular dysfunction in ApoE mice: effects on eNOS and oxidative stress. PLoS ONE 4(10):e7430

Harraz MM, Dawson TM, Dawson VL (2008) Advances in neuronal cell death 2007. Stroke 39(2):286–288

Hartupee JC, Zhang H, Bonaldo MF, Soares MB, Dieckgaefe BK (2001) Isolation and characterization of a cDNA encoding a novel member of the human regenerating protein family: Reg IV. Biochim Biophys Acta 1518(3):287–293

Havranek T, Aujla PK, Nickola TJ, Rose MC, Scavo LM (2010) Increased poly(ADP-ribose) polymerase (PARP)-1 expression and activity are associated with inflammation but not goblet cell metaplasia in murine models of allergen-induced airway inflammation. Exp Lung Res 36(7):381–389

Heikkila RE, Winston B, Cohen G, Barden H (1976) Alloxan-induced diabetes—evidence for hydroxyl radical as a cytotoxic intermediate. Biochem Pharmacol 25(9):1085–1092

Hellerström C, Swenne I (1991) Functional maturation and proliferation of fetal pancreatic β-cells. Diabetes 40(Suppl. 2):89–93

Hoorens A, Pipeleers D (1999) Nicotinamide protects human beta cells against chemically-induced necrosis, but not against cytokine-induced apoptosis. Diabetologia 42(1):55–59

Howard M, Grimaldi JC, Bazan JF, Lund FE, Santos-Argumedo L, Parkhouse RM, Walseth TF, Lee HC (1993) Formation and hydrolysis of cyclic ADP-ribose catalyzed by lymphocyte antigen CD38. Science 262(5136):1056–1059

Hua SY, Tokimasa T, Takasawa S, Furuya Y, Nohmi M, Okamoto H, Kuba K (1994) Cyclic ADP-ribose modulates Ca^{2+} release channels for activation by physiological Ca^{2+} entry in bullfrog sympathetic neurons. Neuron 12(5):1073–1079

Ikehata F, Satoh J, Nata K, Tohgo A, Nakazawa T, Kato I, Kobayashi S, Akiyama T, Takasawa S, Toyota T, Okamoto H (1998) Autoantibodies against CD38 (ADP-ribosyl cyclase/cyclic ADP-ribose hydrolase) that impair glucose-induced insulin secretion in noninsulin-dependent diabetes patients. J Clin Invest 102(2):395–401

Islam MS, Berggren PO (1997) Cyclic ADP-ribose and the pancreatic beta cell: where do we stand? Diabetologia 40(12):1480–1484

Islam MS, Larsson O, Berggren PO (1993) Cyclic ADP-ribose in β-cells. Science 262(5133):584–586

Jacobson MD, Weil M, Raff MC (1997) Programmed cell death in animal development. Cell 88(3):347–354

Jijon HB, Churchill T, Malfair D, Wessler A, Jewell LD, Parsons HG, Madsen KL (2000) Inhibition of poly(ADP-ribose) polymerase attenuates inflammation in a model of chronic colitis. Am J Physiol Gastrointest Liver Physiol 279(3):G641–G651

Jörns A, Munday R, Tiedge M, Lenzen S (1997) Comparative toxicity of alloxan, N-alkylalloxans and ninhydrin to isolated pancreatic islets *in vitro*. J Endocrinol 155(2):283–293

Kämäräinen M, Heiskala K, Knuutila S, Heiskala M, Winqvist O, Andersson LC (2003) RELP, a novel human REG-like protein with up-regulated expression in inflammatory and metaplastic gastrointestinal mucosa. Am J Pathol 163(1):11–20

Kassem SA, Ariel I, Thornton PS, Scheimberg I, Glaser B (2000) β-cell proliferation and apoptosis in the developing normal human pancreas and in hyperinsulinism of infancy. Diabetes 49(8):1325–1333

Kato I, Takasawa S, Akabane A, Tanaka O, Abe H, Takamura T, Suzuki Y, Nata K, Yonekura H, Yoshimoto T, Okamoto H (1995) Regulatory role of CD38 (ADP-ribosyl cyclase/cyclic ADP-ribose hydrolase) in insulin secretion by glucose in pancreatic β-cells. Enhanced insulin secretion in CD38-expressing transgenic mice. J Biol Chem 270(50):30045–30050

Kato I, Yamamoto Y, Fujimura M, Noguchi N, Takasawa S, Okamoto H (1999) CD38 disruption impairs glucose-induced increases in cyclic ADP-ribose, $[Ca^{2+}]i$, and insulin secretion. J Biol Chem 274(4):1869–1872

Kato Y, Miura Y, Yamamoto N, Ozaki N, Oiso Y (2003) Suppressive effects of a selective inducible nitric oxide synthase (iNOS) inhibitor on pancreatic beta-cell dysfunction. Diabetologia 46(9):1228–1233

Kauppinen TM, Suh SW, Berman AE, Hamby AM, Swanson RA (2009) Inhibition of poly(ADP-ribose) polymerase suppresses inflammation and promotes recovery after ischemic injury. J Cereb Blood Flow Matab 29(4):820–829

Kazumori H, Ishirara S, Hoshino E, Kawashima K, Moriyama N, Suetsugu H, Sato H, Adachi K, Fukuda R, Watanabe M, Takasawa S, Okamoto H, Fukui H, Chiba T, Kinoshita Y (2000) Neutrophil chemoattractant 2β regulates expression of the *Reg* gene in injured gastric mucosa in rats. Gastroenterology 119(6):1610–1622

Kerr JF, Wyllie AH, Currie AR (1972) Apoptosis: a basic biological phenomenon with wide-ranging implications in tissue kinetics. Br J Cancer 26(4):239–257

Kiji T, Dohi Y, Takasawa S, Okamoto H, Nonomura A, Taniguchi S (2005) Activation of regenerating gene *Reg* in rat and human hearts in response to acute stress. Am J Physiol Heart Circ Physiol 289(1):H277–H284

Kim HA, Pomeroy SL, Whoriskey W, Pawlitzky I, Benowitz LI, Sicinski P, Stiles CD, Roberts TM (2000) A developmentally regulated switch directs regenerative growth of Schwann cells through cyclin D1. Neuron 26(2):405–416

Kim HA, Ratner N, Roberts TM, Stiles CD (2001) Schwann cell proliferative responses to cAMP and Nf1 are mediated by cyclin D1. J Neurosci 21(4):1110–1116

Kim BJ, Park KH, Yim CY, Takasawa S, Okamoto H, Im MJ, Kim UH (2008) Generation of nicotinic acid adenine dinucleotide phosphate and cyclic ADP-ribose by glucagon-like peptide-1 evokes Ca^{2+} signal that is essential for insulin secretion in mouse pancreatic islets. Diabetes 57(4):868–878

Klasan GS, Ivanac D, Erzen DJ, Picard A, Takasawa S, Peharec S, Arbanas J, Girotto D, Jerkovic R (2014) Reg3G gene expression in regenerating skeletal muscle and corresponding nerve. Muscle Nerve 49(1):61–68

Knip M, Douek IF, Moore WP, Gillmor HA, McLean AE, Bingley PJ, Gale EA, European Nicotinamide Diabetes Intervention Trial Group (2000) Safety of high-dose nicotinamide: a review. Diabetologia 43(11):1337–1345

Kobayashi S, Akiyama T, Nata K, Abe M, Tajima M, Shervani NJ, Unno M, Matsuno S, Sasaki H, Takasawa S, Okamoto H (2000) Identification of a receptor for Reg (Regenerating gene) protein, a pancreatic β-cell regeneration factor. J Biol Chem 275(15):10723–10726

Koguma T, Takasawa S, Tohgo A, Karasawa T, Furuya Y, Yonekura H, Okamoto H (1994) Cloning and characterization of cDNA encoding rat ADP-ribosyl cyclase/cyclic ADP-ribose hydrolase (homologue to human CD38) from islets of Langerhans. Biochim Biophys Acta 1223(1):160–162

Kordula T, Travis J (1996) The role of Stat and C/EBP transcription factors in the synergistic activation of rat serine protease inhibitor-3 gene by interleukin-6 and dexamethasone. Biochem J 313(Pt 3):1019–1027

Kroemer G, El-Deiry WS, Golstein P, Peter ME, Vaux D, Vandenabeele P, Zhivotovsky B, Blagosklonny MV, Malorni W, Knight RA, Piacentini M, Nagata S, Melino G (2005) Classification of cell death: recommendations of the Nomenclature Committee on Cell Death. Cell Death Differ 12(suppl 2):1463–1467

Kröncke KD, Fehsel K, Sommer A, Rodriguez ML, Kolb-Bachofen V (1995) Nitric oxide generation during cellular metabolization of the diabetogenic N-methyl-N-nitroso-urea streptozotozin contributes to islet cell DNA damage. Biol Chem Hoppe Seyler 376(3):179–185

Lasserre C, Simon MT, Ishikawa H, Diriong S, Nguyen VC, Christa L, Vernier P, Brechot C (1994) Structural organization and chromosomal localization of a human gene (*HIP/PAP*) encoding a C-type lectin overexpressed in primary liver cancer. Eur J Biochem 224(1):29–38

LeDoux SP, Woodley SE, Patton NJ, Wilson GL (1986) Mechanisms of nitrosourea-induced beta-cell damage. Alterations in DNA. Diabetes 35(8):866–872

Leonardi O, Mints G, Hussain MA (2003) Beta-cell apoptosis in the pathogenesis of human type 2 diabetes mellitus. Eur J Endocrinol 148(2):99–102

Levine JL, Patel KJ, Zheng Q, Shuldiner AR, Zenilman ME (2000) A recombinant rat regenerating protein is mitogenic to pancreatic derived cells. J Surg Res 89(1):60–65

Li B, Luo C, Chowdhury S, Gao Z-H, Liu J-L (2013) Parp1 deficient mice are protected from streptozotocin-induced diabetes but not caerulein-induced pancreatitis, independent of the induction of Reg family genes. Regul Pept 186:83–91

Liaudet L, Soriano FG, Szabó E, Virág L, Mabley JG, Salzman AL, Szabo C (2000) Protection against hemorrhagic shock in mice genetically deficient in poly(ADP-ribose)polymerase. Proc Natl Acad Sci USA 97(18):10203–10208

Livesey FJ, O'Brien JA, Li M, Smith AG, Murphy LJ, Hunt SP (1997) A Schwann cell mitogen accompanying regeneration of motor neurons. Nature 390(6660):614–618

Lu Y, Ponton A, Okamoto H, Takasawa S, Herrera PL, Liu JL (2006) Activation of the Reg family genes by pancreatic-specific IGF-I gene deficiency and after streptozotocin-induced diabetes in mouse pancreas. Am J Physiol Endocrinol Metab 291(1):E50–E58

Makino S, Kunimoto K, Muraoka Y, Mizushima Y, Katagiri K, Tochino Y (1980) Breeding of a non-obese diabetic strain of mice. Jikken Dobutsu 29(1):1–13

Malaisse WJ, Malaisse-Lagae F, Sener A, Pipeleers DG (1982) Determinants of the selective toxicity of alloxan to the pancreatic B cell. Proc Natl Acad Sci USA 79(3):927–930

Mallone R, Ortolan E, Baj G, Funaro A, Giunti S, Lillaz E, Saccucci F, Cassader M, Cavallo-Perin P, Malavasi F (2001) Autoantibody response to CD38 in Caucasian patients with type 1 and type 2 diabetes: immunological and genetic characterization. Diabetes 50(4):752–762

Marchetti P, Antonelli A, Lupi R, Marselli L, Fallahi P, Nesti C, Baj G, Ferrannini E (2002) Prolonged in vitro exposure to autoantibodies against CD38 impairs the function and survival of human pancreatic islets. Diabetes 51(suppl 3):S474–S477

Marchetti P, Del Prato S, Lupi R, Del Guerra S (2006) The pancreatic beta-cell in human Type 2 diabetes. Nutr Metab Cardiovasc Dis 16(Suppl 1):S3–S6

Martin DR, Lewington AJ, Hammerman MR, Padanilam BJ (2000) Inhibition of poly(ADP-ribose) polymerase attenuates ischemic renal injury in rats. Am J Physiol Integr Comp Physiol 279(5):R1834–R1840

Martin S, van den Engel NK, Vinke A, Heidenthal E, Schulte K, Kolb H (2001) Dominant role of intercellular adhesion molecule-1 in the pathogenesis of autoimmune diabetes in non-obese diabetic mice. J Autoimmun 17(2):109–117

Marx J (2002) Nobel Prize in Physiology or Medicine. Tiny worm takes a star turn. Science 298(5593):526

Masiello P, Novelli M, Fierabracci V, Bergamini E (1990) Protection by 3-aminobenzamide and nicotinamide against streptozotocin-induced beta-cell toxicity in vivo and in vitro. Res Commun Chem Pathol Pharmacol 69(1):17–32

Masutani M, Suzuki H, Kamada N, Watanabe M, Ueda O, Nozaki T, Jishage K, Watanabe T, Sugimoto T, Nakagama H, Ochiya T, Sugimura T (1999) Poly(ADP-ribose) polymerase gene disruption conferred mice resistant to streptozotocin-induced diabetes. Proc Natl Acad Sci USA 96(5):2301–2304

Matsuoka T, Kajimoto Y, Watada H, Umayahara Y, Kubota M, Kawamori R, Yamasaki Y, Kamada T (1995) Expression of CD38 gene, but not of mitochondrial glycerol-3-phosphate dehydrogenase gene, is impaired in pancreatic islets of GK rats. Biochem Biophys Res Commun 214(1):239–246

Matsuura S, Egi Y, Yuki S, Horikawa T, Satoh H, Akira T (2011) MP-124, a novel poly(ADP-ribose) polymerase-1 (PARP-1) inhibitor, ameliorates ischemic brain damage in a non-human primate model. Brain Res 1410:122–131

McCord JM, Fridovich I (1969) Superoxide dismutase. An enzymic function for erythrocuprein (Hemocuprein). J Biol Chem 244(22):6049–6055

Mihara M, Hashizume M, Yoshida H, Suzuki M, Shiina M (2012) IL-6/IL-6 receptor system and its role in physiological and pathological conditions. Clin Sci (Lond) 122(4):143–159

Mitchell KJ, Pinton P, Varadi A, Tacchetti C, Ainscow EK, Pozzan T, Rizzuto R, Rutter GA (2001) Dense core secretory vesicles revealed as a dynamic Ca^{2+} store in neuroendocrine cells with a vesicle-associated membrane protein aequorin chimaera. J Cell Biol 155(1):41–51

Miyashita H, Nakagawara K, Mori M, Narushima Y, Noguchi N, Moriizumi S, Takasawa S, Yonekura H, Takeuchi T, Okamoto H (1995) Human *REG* family genes are tandemly ordered in a 95-kilobase region of chromosome 2p12. FEBS Lett 377(3):429–433

Modak MA, Datar SP, Bhonde RR, Ghaskadbi SS (2007) Differential susceptibility of chick and mouse islets to streptozotocin and its co-relation with islet antioxidant status. J Comp Physiol B 177(2):247–257

Montane J, Cadavez L, Novials A (2014) Stress and the inflammatory process: a major cause of pancreatic cell death in type 2 diabetes. Diabtes Metab Syndr Obes 7:25–34

Mori Y, Suko M, Okudaira H, Matsuba I, Tsuruoka A, Sasaki A, Yokoyama H, Tanase T, Shida T, Nishimura M, Terada E, Ikeda Y (1986) Preventive effects of cyclosporine on diabetes in NOD mice. Diabetologia 29(4):244–247

Moriizumi S, Watanabe T, Unno M, Nakagawara K, Suzuki Y, Miyashita H, Yonekura H, Okamoto H (1994) Isolation, structural determination and expression of a novel *reg* gene, human *reg Iβ*. Biochim Biophys Acta 1217(2):199–202

Moriscot C, Renaud W, Bouvier R, Figarella-Branger D, Figarella C, Guy-Crotte O (1996) Absence of correlation between reg and insulin gene expression in pancreas during fetal development. Pediatr Res 39(2):349–353

Nagatani S, Sudo T, Murakami Y, Uemura K, Hiyama E, Sueda T (2011) Edaravone, a free radical scavenger, promotes engraftment of intraportally transplanted islet cells. Pancreas 40(1):126–130

Nakagawa K, Takasawa S, Nata K, Yamauchi A, Itaya-Hironaka A, Ota H, Yoshimoto K, Sakuramoto-Tsuchida S, Miyaoka T, Takeda M, Unno M, Okamoto H (2013) Prevention of Reg I-induced β-cell apoptosis by IL-6/dexamethasone through activation of *HGF* gene regulation. Biochim Biophys Acta 1833(12):2988–2995

Namikawa K, Fukushima M, Murakami K, Suzuki A, Takasawa S, Okamoto H, Kiyama H (2005) Expression of Reg/PAP family members during motor nerve regeneration in rat. Biochem Biophys Res Commun 332(1):126–134

Narushima Y, Unno M, Nakagawara K, Mori M, Miyashita H, Suzuki Y, Noguchi N, Takasawa S, Kumagai T, Yonekura H, Okamoto H (1997) Structure, chromosomal localization and expression of mouse genes encoding type III *Reg*, *Reg IIIα*, *Reg IIIβ*. Reg IIIγ. Gene 185(2):159–168

Nata K, Takamura T, Karasawa T, Kumagai T, Hashioka W, Tohgo A, Yonekura H, Takasawa S, Nakamura S, Okamoto H (1997) Human gene encoding CD38 (ADP-ribosyl cyclase/cyclic ADP-ribose hydrolase): organization, nucleotide sequence and alternative splicing. Gene 186(2):285–292

Nata K, Liu Y, Xu L, Ikeda T, Akiyama T, Noguchi N, Kawaguchi S, Yamauchi A, Takahashi I, Shervani NJ, Onogawa T, Takasawa S, Okamoto H (2004) Molecular cloning, expression and chromosomal localization of a novel human *REG* family gene. REG III. Gene 340(1):161–170

Nishimune H, Vasseur S, Wiese S, Birling MC, Holtmann B, Sendtner M, Iovanna JL, Henderson CE (2000) Reg-2 is a motoneuron neurotrophic factor and a signalling intermediate in the CNTF survival pathway. Nature Cell Biol 2(12):906–914

Nishizuka Y, Ueda K, Nakazawa K, Hayaishi O (1967) Studies on the polymer of adenosine diphosphate ribose. I. Enzymic formation from nicotinamide adenine dinucleotide in mammalian nuclei. J Biol Chem 242(13):3164–3171

Noguchi N, Takasawa S, Nata K, Tohgo A, Kato I, Ikehata F, Yonekura H, Okamoto H (1997) Cyclic ADP-ribose binds to FK506-binding protein 12.6 to release Ca^{2+} from islet microsomes. J Biol Chem 272(6):3133–3136

Noguchi N, Yoshikawa T, Ikeda T, Takahashi I, Shervani NJ, Uruno A, Yamauchi A, Nata K, Takasawa S, Okamoto H, Sugawara A (2008) FKBP12.6 disruption impairs glucose-induced insulin secretion. Biochem Biophys Res Commun 371(4):735–740

Nomikos IN, Prowse SJ, Carotenuto P, Lafferty KJ (1986) Combined treatment with nicotinamide and desferrioxamine prevents islet allograft destruction in NOD mice. Diabetes 35(11):1302–1304

Notkins AL, Lernmark A (2001) Autoimmune type 1 diabetes: resolved and unresolved issues. J Clin Invest 108(9):1247–1252

Ohta Y, Kitanaka A, Mihara K, Imataki O, Ohnishi H, Tanaka T, Taminato T, Kubota Y (2011) Expression of CD38 with intracellular enzymatic activity: a possible explanation for the insulin release induced by intracellular cADPR. Mol Cell Biochem 352(1–2):293–299

Okamoto H (1981) Regulation of proinsulin synthesis in pancreatic islets and a new aspect to insulin-dependent diabetes. Mol Cell Biochem 37(1):43–61

Okamoto H (1985) Molecular basis of experimental diabetes: degeneration, oncogenesis and regeneration of pancreatic B-cells of islets of Langerhans. BioEssays 2:15–21

Okamoto H Sr (1976) Antitumor activity of streptolysin S-forming streptococci. In: Bernheimer AW (ed) Mechanisms in bacterial toxicology. Wiley, New York, pp 237–257

Okamoto H, Takasawa S (2002) Recent advances in the Okamoto model: the CD38-cyclic ADP-ribose signal system and the regenerating gene protein (Reg)-Reg receptor system in β-cells. Diabetes 51(suppl 3):S462–S473

Okamoto H, Takasawa S (2003) Recent advances in physiological and pathological significance of NAD^+ metabolites: roles of poly(ADP-ribose) and cyclic ADP-ribose in insulin secretion and diabetogenesis. Nutr Res Rev 16(2):253–266

Okamoto H, Takasawa S, Nata K (1997) The CD38-cyclic ADP-ribose signaling system in insulin secretion: molecular basis and clinical implications. Diabetologia 40(12):1485–1491

Oliver FJ, Ménissier-de Murcia J, Nacci C, Decker P, Andriantsitohaina R, Muller S, de la Rubia G, Stoclet JC, de Murcia G (1999) Resistance to endotoxic shock as a consequence of defective NF-κB activation in poly (ADP-ribose) polymerase-1 deficient mice. EMBO J 18(16):4446–4454

Omori K, Matsuda T, Ferreri K, Todorov I, Okamoto H, Takasawa S, Kandeel F, Mullen Y (2003) Induction of *REG* gene expression, *in vitro*, in human islets. Transplantation 76(4):S59

Ota H, Tamaki S, Itaya-Hironaka A, Yamauchi A, Sakuramoto-Tsuchida S, Morioka T, Takasawa S, Kimura H (2012) Attenuation of glucose-induced insulin secretion by intermittent hypoxia via down-regulation of CD38. Life Sci 90(5–6):206–211

Ota H, Itaya-Hironaka A, Yamauchi A, Sakuramoto-Tsuchida S, Miyaoka T, Fujimura T, Tsujinaka H, Yoshimoto K, Nakagawara K, Tamaki S, Takasawa S, Kimura H (2013) Pancreatic β-cell proliferation by intermittent hypoxia via up-regulation of *Reg* family genes and *HGF* gene. Life Sci 93(18–19):664–672

Pacher P, Szabó C (2007) Role of poly(ADP-ribose) polymerase 1 (PARP-1) in cardiovascular diseases: the therapeutic potential of PARP inhibitors. Cardiovasc Drug Rev 25(3):235–260

Pacher P, Liaudet L, Soriano FG, Mabley JG, Szabó É, Szabó C (2002) The role of poly(ADP-ribose) polymerase activation in the development of myocardial and endothelial dysfunction in diabetes. Diabetes 51(2):514–521

Pandya KG, Patel MR, Lau-Cam CA (2010) Comparative study of the binding characteristics to and inhibitory potencies towards PARP and in vivo antidiabetogenic potencies of taurine, 3-aminobenzamide and nicotinamide. J Biomed Sci 17(suppl 1):S16

Pieper AA, Brat DJ, Krug DK, Watkins CC, Gupta A, Blackshaw S, Verma A, Wang Z-Q, Snyder SH (1999) Poly(ADP-ribose) polymerase-deficient mice are protected from streptozotocin-induced diabetes. Proc Natl Acad Sci USA 96(6):3059–3064

Pieper AA, Walles T, Wei G, Clements EE, Verma A, Snyder SH, Zweier JL (2000) Myocardial postischemic injury is reduced by polyADPripose polymerase-1 gene disruption. Mol Med 6(4):271–282

Pimenta W, Korytkowski M, Mitrakou A, Jenssen T, Yki-Jarvinen H, Evron W, Dailey G, Gerich J (1995) Pancreatic beta-cell dysfunction as the primary genetic lesion in NIDDM. JAMA 273(23):1855–1861

Pirsch JD, Miller J, Deierhoi MH, Vincenti F, Filo RS (1997) A comparison of tacrolimus (FK506) and cyclosporine for immunosuppression after cadaveric renal transplantation. FK506 Kidney Transplant Study Group. Transplantation 63(7):977–983

Planas R, Alba A, Carrillo J, Puertas MC, Ampudia R, Pastor X, Okamoto H, Takasawa S, Gurr W, Pujol-Borrell R, Verdaguer J, Vives-Pi M (2006) *Reg* (regenerating) gene overexpression in islets from non-obese diabetic mice with accelerated diabetes: role of IFNβ. Diabetologia 49(10):2379–2387

Plaschke K, Kopitz J, Weigand MA, Martin E, Bardenheuer HJ (2000) The neuroprotective effect of cerebral poly(ADP-ribose)polymerase inhibition in a rat model of global ischemia. Neurosci Lett 284(1–2):109–112

Pupilli C, Giannini S, Marchetti P, Lupi R, Antonelli A, Malavasi F, Takasawa S, Okamoto H, Ferrannini E (1999) Autoantibodies to CD38 (ADP-ribosyl cyclase/cyclic ADP-ribose hydrolase) in Caucasian patients with diabetes: effects on insulin release from human islets. Diabetes 48(12):2309–2315

Rakieten N, Rakieten ML, Nadkarni MR (1963) Studies on the diabetogenic action of streptozotocin (NSC-37917). Cancer Chemother Rep 29:91–98

Rane SG, Dubus P, Mettus RV, Galbreath EJ, Boden G, Reddy EP, Barbacid M (1999) Loss of Cdk4 expression causes insulin-deficient diabetes and Cdk4 activation results in beta-islet cell hyperplasia. Nature Genet 22(1):44–52

Reddy S, Karanam M, Robinson E (2001) Prevention of cyclophosphamide-induced accelerated diabetes in the NOD mouse by nicotinamide or a soy protein-based infant formula. Int J Exp Diabetes Res 1(4):299–313

Rhodes CJ (2005) Type 2 diabetes—a matter of β-cell life and death? Science 307(5708):380–384

Roesner JP, Mersmann J, Bergt S, Bohnenberg K, Barthuber C, Szabo C, Nöldge-Schomburg GE, Zacharowski K (2010) Therapeutic injection of PARP inhibitor INO-1001 preserves cardiac function in porcine myocardial ischemia and reperfusion without reducing infarct size. Shock 33(5):507–512

Rossi D, Sorrentino V (2010) The multiple alternatives of intracellular calcium signaling: a functionally distinct RyR splicing variant in pancreatic islets. Islets 2(6):383–385

Roza AM, Slakey DP, Pieper GM, Van Ye TM, Moore-Hilton G, Komorowski RA, Johnson CP, Hedlund BE, Adams MB (1994) Hydroxyethyl starch deferoxamine, a novel iron chelator, delays diabetes in BB rats. J Lab Clin Med 123(4):556–560

Rutter GA, Theler JM, Li G, Wolheim CB (1994) Ca^{2+} stores in insulin-secreting cells: lack of effect of cADP ribose. Cell Calcium 16(2):71–80

Sakuraba H, Mizukami H, Yagihashi N, Wada R, Hanyu C, Yagihashi S (2002) Reduced beta-cell mass and expression of oxidative stress-related DNA damage in the islet of Japanese Type II diabetic patients. Diabetologia 45(1):85–96

Sasamori K, Sasaki T, Takasawa S, Tamada T, Nara M, Irokawa T, Shimura S, Shirato K, Hattori T (2004) Cyclic ADP-ribose, a putative Ca^{2+}-mobilizing second messenger, operates in submucosal gland acinar cells. Am J Physiol Lung Cell Mol Physiol 287(1):L69–L78

Satoh J, Shintani S, Oya K, Tanaka S, Nobunaga T, Toyota T, Goto Y (1988) Treatment with streptococcal preparation (OK-432) suppresses anti-islet autoimmunity and prevents diabetes in BB rats. Diabetes 37(9):1188–1194

Satomura Y, Sawabu N, Mouri I, Yamakawa O, Watanabe H, Motoo Y, Okai T, Ito T, Kaneda K, Okamoto H (1995) Measurement of serum PSP/reg-protein concentration in various diseases with a newly developed enzyme-linked immunosorbent assay. J Gastroenterol 30(5):643–650

Sauter B, Albert ML, Francisco L, Larsson M, Somersan S, Bhardwaj N (2000) Consequences of cell death: exposure to necrotic tumor cells, but not primary tissue cells or apoptotic cells, induces the maturation of immunostimulatory dendritic cells. J Exp Med 191(3):423–434

Schein PS, Cooney DA, Vernon ML (1967) The use of nicotinamide to modify the toxicity of streptozotocin diabetes without loss of antitumor activity. Cancer Res 27(12):2324–2332

Seino H, Satoh J, Shintani S, Takahashi K, Zhu XP, Masuda T, Nobunaga T, Saito M, Terano Y, Toyota T (1991) Inhibition of autoimmune diabetes in NOD mice with serum from streptococcal preparation (OK-432)-injected mice. Clin Exp Immunol 86(3):413–418

Sekine N, Ishikawa T, Okazaki T, Hayashi M, Wollheim CB, Fujita T (2000) Synergistic activation of NF-κB and inducible isoform of nitric oxide synthase induction by interferon-γ and tumor necrosis factor-α in INS-1 cells. J Cell Physiol 184(1):46–57

Shervani NJ, Takasawa S, Uchigata Y, Akiyama T, Nakagawa K, Noguchi N, Takada H, Takahashi I, Yamauchi A, Ikeda T, Iwamoto Y, Nata K, Okamoto H (2004) Autoantibodies to REG, a beta-cell regeneration factor, in diabetic patients. Eur J Clin Invest 34(11):752–758

Skarda DE, Putt KS, Hergenrother PJ, Mulier KE, Beilman GJ (2007) Increased poly(ADP-ribose) polymerase activity during porcine hemorrhagic shock is transient and predictive of mortality. Resuscitation 75(1):135–144

Smith FE, Bonner-Weir S, Leahy JL, Laufgraben MJ, Ogawa Y, Rosen KM, Villa-Komaroff L (1994) Pancreatic Reg/pancreatic stone protein (PSP) gene expression does not correlate with beta-cell growth and regeneration in rats. Diabetologia 37(10):994–999

Soriano GF, Virág L, Jagtap P, Szabó É, Mabley JG, Liaudet J, Marton A, Hoyt DG, Murthy KGK, Salzman AL, Southan GJ, Szabó C (2001) Diabetic endothelial dysfunction: the role of poly(ADP-ribose) polymerase activation. Nat Med 7(1):108–113

Stern Y, Salzman A, Cotton RT, Zingarelli B (1999) Protective effect of 3-aminobenzamide, an inhibitor of poly (ADP-ribose) synthetase, against laryngeal injury in rats. Am J Respir Crit Care Med 160(5 Pt 1):1743–1749

Suarez-Pinzon WL, Mabley JG, Strynadka K, Power RF, Szabó C, Rabinovitch A (2001) An inhibitor of inducible nitric oxide synthase and scavenger of peroxynitrite prevents diabetes development in NOD mice. J Autoimmun 16(4):449–455

Sugimura T, Fujimura S, Hasegawa S, Kawamura Y (1967) Polymerization of the adenosine 5'-diphosphate ribose moiety of NAD by rat liver nuclear enzyme. Biochim Biophys Acta 138(2):438–441

Suzuki H, Takasawa S, Usui I, Ishii Y, Kato I, Okamoto H, Kobayashi M, Sasahara M, Tobe K (2014) The Ca^{2+}/calmodulin-dependent protein kinase IIα (Thr286Asp) transgenic mice: a novel mouse model of severe insulin-dependent diabetes. Austin J Endocrinol Diabet 1(1):9

Szabó C, Cuzzocrea S, Zingarelli B, O'Connor M, Salzman AL (1997) Endothelial dysfunction in a rat model of endotoxic shock. Importance of the activation of poly (ADP-ribose) synthetase by peroxynitrite. J Clin Invest 100(3):723–735

Szabó G, Liaudet L, Hagl S, Szabó C (2004) Poly(ADP-ribose) polymerase activation in the reperfused myocardium. Cardiovasc Res 61(3):471–480

Takahashi K, Satoh J, Seino H, Zhu XP, Sagara M, Masuda T, Toyota T (1993) Prevention of type I diabetes with lymphotoxin in BB rats. Clin Immunol Immunopathol 69(3):318–323

Takamura T, Kato I, Kimura N, Nakazawa T, Yonekura H, Takasawa S, Okamoto H (1998) Transgenic mice overexpressing type 2 nitric-oxide synthase in pancreatic β-cells develop insulin-dependent diabetes without insulitis. J Biol Chem 273(5):2493–2496

Takasawa S, Okamoto H (2002) Pancreatic β-cell death, regeneration and insulin Secretion: roles of poly(ADP-ribose) polymerase and cyclic ADP-ribose. Int J Exp Diabetes Res 3(2):79–96

Takasawa S, Yamamoto H, Terazono K, Okamoto H (1986) Novel gene activated in rat insulinomas. Diabetes 35(10):1178–1180

Takasawa S, Nata K, Yonekura H, Okamoto H (1993a) Cyclic ADP-ribose in insulin secretion from pancreatic β-cells. Science 259(5093):370–373

Takasawa S, Nata K, Yonekura H, Okamoto H (1993b) Cyclic ADP-ribose in β-cells: Response. Science 262(5133):585

Takasawa S, Tohgo A, Noguchi N, Koguma T, Nata K, Sugimoto T, Yonekura H, Okamoto H (1993c) Synthesis and hydrolysis of cyclic ADP-ribose by human leukocyte antigen CD38 and inhibition of the hydrolysis by ATP. J Biol Chem 268(35):26052–26054

Takasawa S, Ishida A, Nata K, Nakagawa K, Noguchi N, Tohgo A, Kato I, Yonekura H, Fujisawa H, Okamoto H (1995) Requirement of calmodulin-dependent protein kinase II in cyclic ADP-ribose-mediated intracellular Ca^{2+} mobilization. J Biol Chem 270(51):30257–30259

Takasawa S, Akiyama T, Nata K, Kuroki M, Tohgo A, Noguchi N, Kobayashi S, Kato I, Katada T, Okamoto H (1998) Cyclic ADP-ribose and inositol 1,4,5-trisphosphate as alternate second messengers for intracellular Ca^{2+} mobilization in normal and diabetic β-cells. J Biol Chem 273(5):2497–2500

Takasawa S, Ikeda T, Akiyama T, Nata K, Nakagawa K, Shervani NJ, Noguchi N, Murakami-Kawaguchi S, Yamauchi A, Takahashi I, Tomioka-Kumagai T, Okamoto H (2006) Cyclin

D1 activation through ATF-2 in Reg-induced pancreatic β-cell regeneration. FEBS Lett 580(2):585–591

Takasawa S, Kuroki M, Nata K, Noguchi N, Ikeda T, Yamauchi A, Ota H, Itaya-Hironaka A, Sakuramoto-Tsuchida S, Takahashi I, Yoshikawa T, Shimosegawa T, Okamoto H (2010) A novel ryanodine receptor expressed in pancreatic islets by alternative splicing from type 2 ryanodine receptor gene. Biochem Biophys Res Commun 397(2):140–145

Takasu N, Asawa T, Komiya I, Nagasawa Y, Yamada T (1991) Alloxan-induced DNA strand breaks in pancreatic islets. Evidence for H_2O_2 as an intermediate. J Biol Chem 266(4):2112–2114

Terazono K, Yamamoto H, Takasawa S, Shiga K, Yonemura Y, Tochino Y, Okamoto H (1988) A novel gene activated in regenerating islets. J Biol Chem 263(5):2111–2114

Thomas HE, Darwiche R, Corbett JA, Kay TW (2002) Interleukin-1 plus γ-interferon-induced pancreatic β-cell dysfunction is mediated by β-cell nitric oxide production. Diabetes 51(2):311–316

Thompson CB (1995) Apoptosis in the pathogenesis and treatment of disease. Science 267(5203):1456–1462

Tjälve H, Wilander E, Johansson E (1976) Distribution of labelled streptozotocin in mice: uptake and retention in pancreatic islets. J Endocrinol 69(3):455–456

Todd JA (2010) Etiology of type 1 diabetes. Immunity 32(4):457–467

Todd JA, Knip M, Mathieu C (2011) Strategies for the prevention of autoimmune type 1 diabetes. Diabet Med 28(10).1141–1143

Tohgo A, Takasawa S, Noguchi N, Koguma T, Nata K, Sugimoto T, Furuya Y, Yonekura H, Okamoto H (1994) Essential cysteine residues for cyclic ADP-ribose synthesis and hydrolysis by CD38. J Biol Chem 269(46):28555–28557

Tohgo A, Munakata H, Takasawa S, Nata K, Akiyama T, Hayashi N, Okamoto H (1997) Lysine 129 of CD38 (ADP-ribosyl cyclase/cyclic ADP-ribose hydrolase) participates in the binding of ATP to inhibit the cyclic ADP-ribose hydrolase. J Biol Chem 272(7):3879–3882

Toniolo A, Onodera T, Yoon J-W, Notkins AL (1980) Induction of diabetes by cumulative environmental insulitis from viruses and chemicals. Nature 288(5789):383–385

Toyota T, Satoh J, Oya K, Shintani S, Okano T (1986) Streptococcal preparation (OK-432) inhibits development of type I diabetes in NOD mice. Diabetes 35(4):496–499

Tsubouchi S, Suzuki H, Ariyoshi H, Matsuzawa T (1981) Radiation-induced acute necrosis of the pancreatic islet and the diabetic syndrome in the golden hamster (Mesocricetus auratus). Int J Radiat Biol Relat Stud Phys Chem Med 40(1):95–106

Uchigata Y, Yamamoto H, Kawamura A, Okamoto H (1982) Protection by superoxide dismutase, catalase, and poly(ADP-ribose) synthetase inhibitors against alloxan- and streptozotocin-induced islet DNA strand breaks and against the inhibition of proinsulin synthesis. J Biol Chem 257(11):6084–6088

Uchigata Y, Yamamoto H, Nagai H, Okamoto H (1983) Effect of poly(ADP-ribose) synthetase inhibitor administration to rats before and after injection of alloxan and streptozotocin on islet proinsulin synthesis. Diabetes 32(4):316–318

Ueda K, Hayaishi O (1985) ADP-ribosylation. Annu Rev Biochem 54:73–100

Unno M, Nata K, Noguchi N, Narushima Y, Akiyama T, Ikeda T, Nakagawa K, Takasawa S, Okamoto H (2002) Production and characterization of *Reg* knockout mice: reduced proliferation of pancreatic β-cells in *Reg* knockout mice. Diabetes 51(suppl 3):S478–S483

Varadi A, Rutter GA (2002) Dynamic imaging of endoplasmic reticulum Ca^{2+} concentration in insulin-secreting MIN6 Cells using recombinant targeted cameleons: roles of sarco(endo)plasmic reticulum Ca^{2+}-ATPase (SERCA)-2 and ryanodine receptors. Diabetes 51(Suppl 1):S190–S201

Viterbo D, Callender GE, DiMaio T, Mueller CM, Smith-Norowitz T, Zenilman ME, Bluth MH (2009) Administration of anti-Reg I and anti-PAPII antibodies worsens pancreatitis. JOP 10(1):15–23

Von Mering J, Minkowski O (1889) Diabetes mellitus nach Pankreasexstirpation. Zbl Klin Med 10:393–394

Wada R, Yagihashi S (2004) Nitric oxide generation and poly(ADP ribose) polymerase activation precede beta-cell death in rats with a single high-dose injection of streptozotocin. Virchows Arch 444(4):375–382

Wang Y, Kim NS, Haince J-F, Kang HC, David KK, Andrabi SA, Poirier GG, Dawson VL, Dawson TM (2011) Poly(ADP-ribose) (PAR) binding to apoptosis-inducing factor is critical for PAR polymerase-1-dependent cell death (parthanatos). Sci Signal 4(167): ra20

Watanabe T, Yonekura H, Terazono K, Yamamoto H, Okamoto H (1990) Complete nucleotide sequence of human reg gene and its expression in normal and tumoral tissues. The reg protein, pancreatic stone protein, and pancreatic thread protein are one and the same product of the gene. J Biol Chem 265(13):74732–77439

Watanabe T, Yonemura Y, Yonekura H, Suzuki Y, Miyashita H, Sugiyama K, Moriizumi S, Unno M, Tanaka O, Kondo H, Bone AJ, Takasawa S, Okamoto H (1994) Pancreatic beta-cell replication and amelioration of surgical diabetes by Reg protein. Proc Natl Acad Sci USA 91(9):3589–3592

Webb DL, Islam MS, Efanov AM, Brown G, Köhler M, Larsson O, Berggren PO (1996) Insulin exocytosis and glucose-mediated increase in cytoplasmic free Ca^{2+} concentration in the pancreatic β-cell are independent of cyclic ADP-ribose. J Biol Chem 271(32):19074–19079

Wöhler F, Liebig J (1838) Untersuchungen über die Natur der Harnsäure. Ann Pharm 26(3):241–336

Yagui K, Shimada F, Mimura M, Hashimoto N, Suzuki Y, Tokuyama Y, Nata K, Tohgo A, Ikehata F, Takasawa S, Okamoto H, Makino H, Saito Y, Kanatsuka A (1998) A missense mutation in the CD38 gene, a novel factor for insulin secretion: association with Type II diabetes mellitus in Japanese subjects and evidence of abnormal function when expressed in vitro. Diabetologia 41(9):1024–1028

Yamada K, Nonaka K, Hanafusa T, Miyazaki A, Toyoshima H, Tarui S (1982) Preventive and therapeutic effects of large-dose nicotinamide injections on diabetes associated with insulitis. An observation in nonobese diabetic (NOD) mice. Diabetes 31(9):749–753

Yamagami T, Miwa A, Takasawa S, Yamamoto H, Okamoto H (1985) Induction of rat pancreatic B-cell tumors by the combined administration of streptozotocin or alloxan and poly(adenosine diphosphate ribose) synthetase inhibitors. Cancer Res 45(4):1845–1849

Yamamoto H, Uchigata Y, Okamoto H (1981a) Streptozotocin and alloxan induce DNA strand breaks and poly(ADP-ribose) synthetase in pancreatic islets. Nature 294(5838):284–286

Yamamoto H, Uchigata Y, Okamoto H (1981b) DNA strand breaks in pancreatic islets by in vivo administration of alloxan or streptozotocin. Biochem Biophys Res Commun 103(3):1014–1020

Ylipaasto P, Smura T, Gopalacharyulu P, Paananen A, Seppänen-Laakso T, Kaijalainen S, Ahlfors H, Korsgren O, Lakey JR, Lahesmaa R, Piemonti L, Oresic M, Galama J, Roivainen M (2012) Enterovirus-induced gene expression profile is critical for human pancreatic islet destruction. Diabetologia 55(12):3273–3283

Yonemura Y, Takashima T, Miwa K, Miyazaki I, Yamamoto H, Okamoto H (1984) Amelioration of diabetes mellitus in partially depancreatized rats by poly(ADP-ribose) synthetase inhibitors. Evidence of islet B-cell regeneration. Diabetes 33(4):401–404

Yoon KH, Ko SH, Cho JH, Lee JM, Ahn YB, Song KH, Yoo SJ, Kang MI, Cha BY, Lee KW, Son HY, Kang SK, Kim HS, Lee IK, Bonner-Weir S (2003) Selective β-cell loss and α-cell expansion in patients with Type 2 diabetes mellitus in Korea. J Clin Endocrinol Metab 88(5):2300–2308

Zingarelli B, Salzman AL, Szabó C (1998) Genetic disruption of poly (ADP-ribose) synthetase inhibits the expression of P-selectin and intercellular adhesion molecule-1 in myocardial ischemia/reperfusion injury. Circ Res 83(1):85–94

Zingarelli B, Szabó C, Salzman AL (1999) Blockade of Poly(ADP-ribose) synthetase inhibits neutrophil recruitment, oxidant generation, and mucosal injury in murine colitis. Gastroenterology 116(2):335–345

Zóka A, Műzes G, Somogyi A, Varga T, Szémán B, Al-Aissa Z, Hadarits O, Firneisz G (2013) Altered immune regulation in type 1 diabetes. Clin Dev Immunol 2013:254874

Chapter 7
MicroRNAs: A Link Between Type 1 Diabetes and the Environment?

Ammira Al-Shabeeb Akil, Andy Ho, Carah A. Figueroa-Crisostomo, William D. Rawlinson and Maria E. Craig

Introduction

Type 1 diabetes (T1D) is an autoimmune disease characterized by selective destruction of insulin-producing pancreatic β-cells. Although T1D is known to be largely immunological, increasing evidence suggests the importance of genetic and environmental factors (including viruses) in the initiation of islet

A.A.-S. Akil (✉)
Division of Translational Medicine, Sidra Medical and Research Centre,
Education City North Campus, Qatar Foundation, Al Luqta Street, 26999 Doha, Qatar
e-mail: aakil@sidra.org

A.A.-S. Akil · M.E. Craig
Faculty of Medicine, School of Women's and Children's Health,
Health University of New South Wales, Kensington, NSW 2052, Australia

M.E. Craig
The Children's Hospital, Institute of Endocrinology and Diabetes, Westmead, NSW 2145, Australia

A.A.-S. Akil · W.D. Rawlinson · M.E. Craig
Virology Division, South Eastern Area Laboratory Services SEALS,
Prince of Wales Hospital, University of New South Wales, Randwick, NSW 2031, Australia

M.E. Craig
Discipline of Pediatrics and Child Health, Health University of Sydney,
Westmead, NSW 2052, Australia

A. Ho
Faculty of Medicine, School of Biotechnology and Biomolecular Sciences,
University of New South Wales, Kensington, NSW 2052, Australia

C.A. Figueroa-Crisostomo · W.D. Rawlinson
Faculty of Science, School of Biotechnology and Biomolecular Sciences,
University of New South Wales, Kensington, NSW 2052, Australia

A.A. Hardikar (ed.), *Pancreatic Islet Biology*, Stem Cell Biology and Regenerative Medicine, DOI 10.1007/978-3-319-45307-1_7

autoimmunity and/or acceleration to clinical T1D. In particular, epidemiological data as well as substantial data obtained through animal models, cell lines and human studies support a role for enterovirus (EV) infections. EV infections are more frequent in newly diagnosed T1D patients compared to healthy controls. EV-RNA and protein have been detected in intestine, blood and pancreatic biopsies of patients with T1D. EVs contribute to β-cell damage by multiple mechanisms, such as direct cytolysis of infected β-cells, islet inflammation, and molecular mimicry are thought to be involved. MicroRNAs (miRNAs) are endogenous small non-coding RNAs molecules that regulate gene expression mostly through translational repression. They affect cell growth, proliferation, differentiation, development, apoptosis and maintenance of tissue identity. They are implicated in immune system regulation and potentially important in autoimmune destruction of pancreatic β-cells. The relationship between EVs infections, miRNAs, and development of T1D is discussed aiming to describe genes encoding miRNA signatures that may be involved in T1D. Many miRNAs, and potentially other small regulatory RNA molecules discussed through this chapter, are involved in immune regulation and modulation, which may provide novel diagnostic, prognostic and therapeutic alternatives for type 1 diabetes (T1D) in future.

Type 1 Diabetes

Type 1 diabetes (T1D) is an autoimmune disorder, which accounts for 5–10 % of all diagnosed cases of diabetes. It is characterized by severe autoimmune destruction of insulin-producing β-cells (Bluestone et al. 2010). The major autoimmune markers in T1D include (pro)insulin/insulin, L-glutamic acid decarboxylase (GAD65), zinc transporter (ZnT8), phosphotyrosine phosphatase-related proteins (IA2 and IA2β) and islet cell autoantigen 69 (ICA69). Normally, one or more of these autoantibodies can be detected in circulation for most patients at T1D diagnosis (Arvan et al. 2012; Abuhatzira et al. 2015).

Similar to other autoimmune diseases, T1D is considered as a complex genetic disorder with an environmental component. The human leukocyte antigen (HLA) genes cluster on chromosome 6p21 is known to be the pivotal T1D susceptibility region (Bluestone et al. 2010; Storling and Brorsson 2013; Steck and Rewers 2011). In addition to HLA, more than 50 additional loci significantly affect T1D risk, including the insulin gene (chromosome 11p5), the *CTLA-4* gene (chromosome 2q33), *IL2RA* (chromosome 10p15), interferon induced with helicase C domain 1 *(IFIH1)* gene (chromosome2p24), *PTPN22* gene (chromosome 2q24) (Barrett et al. 2009), *ERBB3* (*chromosome* 12q13), *KIAA0350* (*chromosome*16p13.13), *SH2B3* (*chromosome* 12q24), *IL18RAP* (chromosome 2q12), *IL10* (*chromosome* 12q15) and *COBL* gene (*chromosome* 7p12.1) (Storling and Brorsson 2013; Zouidi et al. 2014; Santin and Eizirik 2013). However, non-HLA-risk alleles contribute to risk of developing islet autoantibodies and progression from islet autoimmunity to diabetes. The greatest diabetes discrimination was achieved by the summation of risk alleles for eight genes in children of parents with T1D; these are: *IFIH1, CTLA4, PTPN22, IL18RAP, SH2B3, KIAA0350, COBL* and *ERBB3* genes (Winkler et al. 2012).

There are multiple lines of evidence supporting a strong environmental component in T1D etiology. While genetic susceptibility is important, twin studies of T1D show less than 40 % concordance (Pyke 1989). The European incidence of T1D is increasing by approximately 3 %, each year, which is higher than what can be attributed to genetics alone (Patterson et al. 2009). Furthermore, incidence of T1D varies between European nations (displaying more than tenfold difference), and migration studies have demonstrated that T1D incidence is elevated in people who move from a low-incidence to a high-incidence region (Knip et al. 2005). Together, this indicates environmental causes are involved in the initiation and/or advancement of some β-cell damage leading to T1D (Horn et al. 1988; Jun and Yoon 2003).

Presumptive environmental determinants include diet, vitamin D insufficiency, vaccination, toxins and viruses (Sebastiani et al. 2011; Smyk et al. 2012). There is a large body of evidence that implicates viral pathogens, such as rubella virus, mumps virus, rotavirus, cytomegalovirus, parvoviruses and particularly EV, as major contributors to β-cell destruction (Pak 1988; Filippi and von Herrath 2008; Hober and Sauter 2010; Hyoty et al. 1988; Hyöty and Taylor 2002; Honeyman et al. 1998, 2000). In a recent meta-analysis, we found strong evidence for an association between EVs at the clinical onset of T1D and between EV infection and islet autoimmunity (Yeung et al. 2011).

Enterovirus in the Pathogenesis of T1D

EVs are ubiquitous with more than 150 genotypes that produce self-limited infections, with the exclusion of polioviruses and EV71 which periodically cause severe neurologic complications (Chiang et al. 2012). Transmission of EV infection occurs via fecal–oral transmission as the viruses infect mucosa of the respiratory or gastrointestinal tracts. EVs reach the pancreas through a secondary infection and spread via the lymphatic and hematopoietic systems (Smura et al. 2010).

Serological Studies of EV in T1D

Many patients with T1D demonstrate seroreactivity to islet cells, suggesting an autoantibody-mediated etiology in this disease (Miersch et al. 2013). At the time of T1D diagnosis, islet cell antigens (ICA) are present in 90 % of patients; insulin and pro-insulin at 23 and 34 %; GAD65 in 73 %; and IA-2 in 75 % of patients at diagnosis (Narendran et al. 2005). Studies on the seasonality of EV infections demonstrate infection rates are usually higher during winter and correlate with the increased incidence of autoantibodies (Knip et al. 2005). Furthermore, antibodies against EVs are more frequent in newly diagnosed T1D patients, than in healthy subjects (Jaidane and Hober 2008; Hyoty 2002; Haverkos et al. 2003; Sarmiento et al. 2007; Moya-Suri et al. 2005; Elfaitouri et al. 2007). EV infection may occur in the uterus; IgM antibodies against EV has been detected in 30 % of maternal serum at delivery from non-T1D mothers whose children developed T1D, while it is detected in only 16 % of children born from non-T1D mothers who did not develop T1D (Elfving et al. 2008).

Human coxsackievirus type B (CVB) infections have been associated with onset of T1D. Recently, a subset of immune dominant peptides that are homogenous between CVB4-VP1 protein and β-cell-specific autoantigens have been detected at diagnosis (Bason et al. 2013). These 'anti-CVB' antibodies were capable of inducing pancreatic β-cell apoptosis, which is the ultimate step to development of T1D (van der Werf et al. 2007; Yoon et al. 1978).

Screening EV-neutralizing antibodies can identify EV serotypes, which may be involved in the initiation of T1D in infected individuals. In a longitudinal-prospective birth-cohort study, 183 children tested persistently positive for at least two autoantibodies comparing with 366 autoantibody-negative matched controls. CVB1 serotype was associated with an increased risk of autoimmunity, while CVB3 and CVB6 were associated with reduced risk (Laitinen et al. 2013).

The possible role of CVB1 as a diabetogenic virus was investigated in different European populations including Finland, Sweden, UK, France and Greece. Antibodies against CVB1 were more frequent among diabetic children than controls (odds ratio (OR) 1.7 and 95 % CI 1.0–2.9) with the risk associated with CVB1 infection not linked to HLA genotype (Oikarinen et al. 2013).

Molecular Studies in T1D

A large proportion (30 vs. 4 % of non-diabetic controls) (Craig et al. 2003) of T1D patients have prolonged EV infection, and the EV-RNA can be detected in blood at the time of T1D diagnosis. Children with islet autoantibodies had higher EV-RNA and EV-specific antibody titers, compared to controls (Hyoty 2002). EV-RNA has been detected in peripheral blood mononuclear cells (PBMCs) from 50 % of newly diagnosed T1D children, but in none of the controls. Sequencing of EV strains isolated from these children revealed homology with CVB as well as other EVs (Yin et al. 2002).

In the Finnish type 1 Diabetes Prediction and Prevention (Dippe et al. 1975) study, 5.1 % of T1D patients were positive for EV-RNA, compared with 1.9 % in the control group. The strongest association between the presence of EV-RNA and the autoimmunity occurred within the six months' period following the initial positive autoantibody detection (Oikarinen et al. 2011).

An association between elevated levels of IFN-α and the presence of CVB-RNA has been found in the blood of T1D at different stages of the disease process

(Chehadeh et al. 2000). The Environmental Determinates of Diabetes in the Young (TEDDY) study demonstrated the presence of both anti-EV-IgM and EV-RNA during pregnancy in mothers of babies who developed T1D before the age of 15 (Chehadeh et al. 2000; Dahlquist et al. 1999).

EV-RNA was detected in 11.5 % of blood samples collected from the Norwegian 'Environmental Triggers of Type 1 Diabetes (EIDIA) study' participants. The EV-RNA in the blood was significantly associated with the common *IFIH1-SNP* 'rs1990760' comparing with the other genotypes, a relationship that may lead to autoimmunity or T1D (Cinek et al. 2012).

Enterovirus in the Diabetic Pancreas

Pathological studies of the pancreas in T1D demonstrate that continued loss of pancreatic β-cells takes place before the clinical onset of the disease, and this effect is accelerated with EV infection. Pancreatic autopsy tissue from patients with the recent onset of T1D demonstrated a higher proportion of EV-VP1 capsid protein (61 %) compared to non-diabetic controls (6 %) (Richardson et al. 2009). The VP1 capsid protein was detected in pancreatic tissues from 3/6 T1D organ donors, but not from non-diabetic organ donors (17 ± 7 % and 33 ± 14 % of β cells after 4 and 7 days of co-culture cellular model, respectively) (Dotta et al. 2007).

The impact of CVB4 infection has been identified on pancreatic ductal cells that are known to be involved in the renewal of pancreatic β-cells (Sane et al. 2013). The presence of CVB4E2 infectious particles and the VP1 capsid protein remains detectable in the supernatant fluid for up to 37 weeks post-viral infection of primary pancreas ductal cells and human pancreatic carcinoma, epithelial-like (PANC-1) cells.

Persistent EV infections result in down-regulation of the *Pdx1* gene, which is essential for the development of endocrine pancreas. A substantial proportion of EV-RNA was detected in the small-intestine biopsies from patients with T1D (75 %), when compared with healthy controls (10 %), suggesting the possible role of EV in human T1D initiation (Oikarinen et al. 2008).

MicroRNAs Biogenesis and Mechanisms of Action

MicroRNAs (miRNAs) are small non-coding highly conserved RNA molecules, approximately 19–23 nucleotides (nt) in length (Huntzinger and Izaurralde 2011). They target one or more messenger RNA (mRNA) molecules through complementary target sequences generally located in their 3′UTR ultimately resulting in down-regulation of gene expression. This is achieved through mechanisms including translational inhibition, mRNA destabilization, and deadenylation (Sonenberg 2012). MiRNAs posttranscriptionally regulate protein synthesis by base pairing to partially complementary sequences in the 3′ untranslated regions (UTRs) of target mRNAs (Sonenberg 2012; Bartel 2009; Eulalio et al. 2008). A considerable number of studies have linked impairment of miRNAs with certain diseases, suggesting these as powerful regulators of mRNAs activity (Chang and Mendell 2007). Some of these regulatory mechanisms have been shown to alter miRNAs biogenesis and activity (Krol et al. 2010).

The process of miRNAs biogenesis is a multistage event assembled by distinct RNA polymerases (Kim 2005) (Fig. 1). The miRNA genes are positioned in clusters and initially transcribed under strict tissue-specific regulation as part of a long

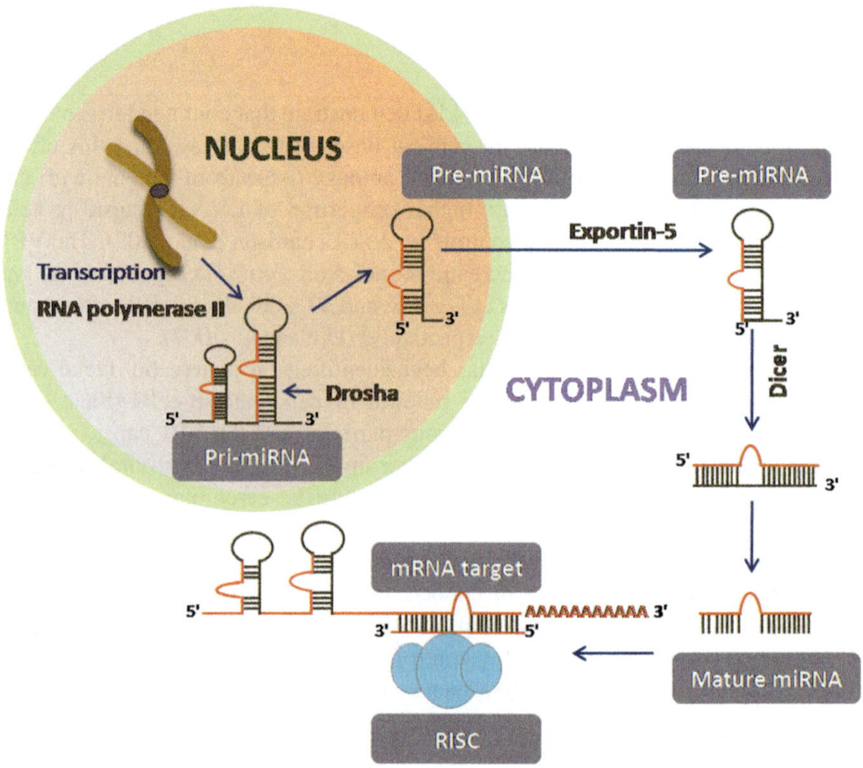

Fig. 1 MiRNA biogenesis and mechanisms of action. MiRNAs are short, double-stranded RNA sequences of 19–23 nucleotide in length. The primary transcripts of miRNAs (pri-miRNAs) are transcribed by RNA ploymerase II enzyme. Subsequently, the pri-miRNAs transcripts are processed by RNase enzyme, Drosha, to precursor miRNA (pre-miRNA) stem-loop of about 60 nucleotides in length. Pre-miRNA transcripts are then exported into the cytoplasm by exportin 5 (a Ran-GTP-dependent nuclear transport receptor. In the cytoplasm, the pre-miRNAs further processed into approximately 23 nucleotide miRNA duplex by type III RNase, Dicer enzyme. The last step of mature miRNA molecule formation is by incorporating the functional (mature) miRNA strand of the miRNA duplex (miRNAAniRNA*) into RNA-induced silencing complex (RISC), to form the miRISC. Within the miRISC, miRNAs base pair to target mRNA and regulate gene expression by translation repression, mRNA cleavage or degradation and deadenylation

primary miRNA (pri-miRNA) transcript, which contains the mature miRNA as part of RNA hairpin (Krol et al. 2010; Setyowati et al. 2012; Zeng et al. 2005).

Transcription of the primary miRNA is controlled mainly by the RNA polymerase II and in some cases by RNA polymerase III (Kim 2005). Pri-miRNAs are then processed by Drosha, a nuclear RNase III enzyme that silences the upper part of the RNA hairpin to release the precursor miRNA (pre-miRNA), which is generally 60–70 nucleotide (nt) long with a 2-nt 3′-overhang end structure (Lee et al. 2002, 2003). The 'end structure' is then recognized by the nuclear transporters exportin 5 and Ran-GTP, and exported into the cytoplasm (Kim 2005; Lund et al. 2004; Yi et al. 2003). During the second cleavage step, the pre-miRNA is further processed by a specific endoribonuclease RNase III enzyme called Dicer. Dicer operates near the hairpin loop to produce a ~23-nt miRNA–miRNA duplex (Shivdasani 2006). The RNA-induced silencing complex (RISC) has the ability to determine the strand with the weak hydrogen bond and then integrate this strand (Khvorova et al. 2003) to form the miRISC complex. In contrast, the other remaining strand denoted by a '*' or the 'passenger strand' will generally be degraded to low level in most cells. However, certain miRNA* are expressed abundantly and can be employed into the silencing complex RISC to regulate gene expression (Zeng et al. 2005; Dp 2004).

Mechanisms of MiRNA Target Gene Silencing

Since their discovery, the role of miRNAs as key posttranscriptional regulators of gene expression is becoming more recognizable. Many scientific reports have been shown evidence of mRNA destabilization, translational repression, degradation and activation of gene expression (Huntzinger and Izaurralde 2011) In fact, translation of mRNA into protein represents the last step in gene expression pathway, which arbitrates the generation of the proteome from genomic information (Sonenberg 2012; Fabian et al. 2010).

The active process of mRNA translation can be divided into three stages: initiation (recruitment of the ribosomal subunits to the mRNA initiation codon), elongation (elongation of polypeptide chain) and termination (release of the mature protein molecule). Since the elongation and termination stages are facilitated by a restricted group of exclusive factors such as transcription factor IIF, elongins, 5,6-dichloro-1-beta-D-ribofuranosylbenzimidazole-sensitivity-inducing factor DSIF and negative elongation factor NELF (Filipowicz et al. 2008), the initiation of translation is a highly delicate critical event that is aided by large number of polypeptides (Huntzinger and Izaurralde 2011; Filipowicz et al. 2008; Preiss and Hentze 2003). Efficient translation initiation occurs by recognition of the 5′ Cap or the 3′ UTR tail of the mature mRNAs by a number of core translation factors, such as elongation initiation factor 2 eIF2, eIF3, eIF1 and eIF4F (Jackson et al. 2010; Talsky and Collins 2010).

MiRNAs function as guides for the RISC by forming miRISC complex to bind mRNA and to repress their translation and/or decrease their stability (Huntzinger and Izaurralde 2011; Pillai et al. 2005; Gebauer and Hentze 2004). Binding the miRISC complex to the 3′UTR target sequence through partial base pairing can end up with removal of the poly(A) tail by certain deadenylation factors and make the mRNA accessible to 'exonucleolytic' degradation (Bartel 2009). Consequently, each miRNA can potentially interact with multiple target mRNAs. Furthermore, miRNAs may reduce protein (but not mRNA) levels, proposing that translational repression is mediated by miRISC (Chekulaeva et al. 2011; Braun et al. 2011; Wu et al. 2006). In contrast, there is evidence for translation initiation, elongation inhibition or direct proteolysis of the synthesized peptide from targeted miRISC (Pillai et al. 2005; Wu et al. 2006; Petersen et al. 2006; Maroney et al. 2006).

Role of MiRNAs in the Pancreatic Environment

Pancreatic β-cells play a crucial role in controlling normal glycaemia via regulation of insulin synthesis and production. There is growing evidence that miRNAs play an important role in the control of β-cell function within normal and pathophysiological conditions (Table 7.1) (Roggli et al. 2012; Ventriglia et al. 2015).

In a comparison of miRNAs fraction profile of primary human islets and enriched β-cells, 366 miRNAs were expressed in human islets, 346 in β-cells, and 328 were shared. In addition, 40 miRNAs were predominantly expressed in islets when compared with 15 other human tissues (van de Bunt et al. 2013). Correspondingly, global profiling of 667 miRNAs identified 141 differentially expressed miRNAs in both α and β-cells, 134 were expressed in β-cells, while only seven in α-cells. Most of the recognized target genes of 'β-cells miRNAs' were T1D associated (Klein et al. 2013).

The role of miRNAs in the insulin pathway begins at the early stages of pancreatic islet expansion during embryonic development. Multiple miRNAs are involved in this process, such as miRNA-503, miRNA-541, miRNA-214 (van de Bunt et al. 2013; Lynn et al. 2007), miRNA-15a/b, miRNA-16, miRNA-195 (Joglekar et al. 2007) and miRNA-124a (Baroukh et al. 2007), miRNA-376, miRNA-9, miRNA-7 (which act as a brake for β-cell proliferation) and miRNA-375 (Joglekar et al. 2007, 2009, 2011) and miRNA-17/92, miRNA-143–145, miRNA-130, let-7, miR-221/222, miRNA-200 and miRNA-29 (Deiuliis 2015). An interlinked network of certain miRNAs and their mRNA target genes cooperate with each other during human fetal pancreas development. MiRNA-218, miRNA-484, miRNA-107, miRNA-30d, miRNA-21, miRNA-34a and miRNA-96 demonstrated different expression levels throughout this process, indicating the potential of two-way level of gene regulation through miRNAs targeting multiple mRNA genes and target mRNAs regulated by other miRNAs in parallel (Rosero et al. 2010).

Table 7.1 miRNAs involved in β-cells biology and dysregulation

miRNA	Validated target(s)	Function	Reference
miRNA-7a	Ribosomal protein S6 kinase and eukaryotic translation initiation factor 4E	Inhibits cellular proliferation and replication by repressing downstream proteins of the mTOR pathway.	Wang et al. (2013)
miRNA-9	Onecut-2	Increases granuphilin expression, a GTPase effector associated with β-cell secretory granules. Leads to inhibition of insulin exocytosis.	Plaisance et al. (2006)
miRNA-15a	Uncoupling protein-2	Up-regulates biosynthesis of insulin in glucose-rich environment.	Sun et al. (2011)
miRNA-15b, miRNA-16, miRNA-195	Neurogenin-3	Reduces insulin secreting cells in developing pancreas and inhibits neurogenin-3-mediated regeneration.	Joglekar et al. (2007)
miRNA-21	Programmed cell death 4 and Piccolo	Inhibits pro-inflammatory cytokine-induced cellular apoptosis, thereby promoting the survival of pancreatic islet β-cells. Also impairs cAMP insulin secretion.	Ruan et al. (2011), Bravo-Egana et al. (2012)
miRNA-24, miRNA-26, miRNA-148, miRNA-182	Basic helix-loop-helix family member e22 and SOX6	Up-regulates insulin biosynthesis by suppressing translation of proteins involved in repressing insulin activators.	Melkman-Zehavi et al. (2011)
miRNA-29a/b/c	Induced myeloid leukemia cell differentiation protein.	Mediates cytokine-triggered cellular death by repressing anti-apoptotic protein.	Roggli et al. (2012)
miRNA-30a-5p/b/c/e-5p	Vimentin and Snail1	Prevents β-cell proliferation by inhibiting epithelial-to-mesenchymal phenotype transition.	Joglekar et al. (2009)
miRNA-34a, miRNA-146a	Not verified	Mediates cytokine-triggered β-cell dysfunction, reducing glucose-induced insulin secretion and making them prone to apoptosis.	Roggli et al. (2010)
miRNA-218, miRNA-495	Hepatocyte nuclear factor 6 and Onecut-2	Modulates early pancreas morphogenesis.	Simion et al. (2010)
miRNA-375	Myotrophin, Caveolin1, and Inhibitor of DNA binding 3	Multi-functional miRNA: known to promote morphological development of pancreatic islets; down-regulates negative growth regulators, promoting cellular growth and proliferation.	Poy et al. (2004, 2009), Kloosterman et al. (2007)

cAMP = cyclic adenosine monophosphate; mTOR = mammalian target of rapamycin

Several in vitro studies confirmed the early involvement of certain miRNAs in the insulin pathway. Using a *pdx-1-Cre* mouse model, deletion of Dicer-1, in the early stages of pancreatic cell development, resulted in the suppression of the entire miRNA pathway, dramatically reducing the number of insulin-producing β-cells and pancreas agenesis (Lynn et al. 2007). Dicer-1 is necessary for maintaining the adult pancreas. In mice with Dicer-1-deficient β-cells, substantial decline in insulin mRNA and protein levels was observed, effectively rendering them diabetic (Melkman-Zehavi et al. 2011). Similarly, Dicer-1-hypomorphic-adult mice were histologically normal except for the pancreas which was typically normal at the fetal and neonatal stages (Morita et al. 2009).

In pancreatic β-cells, insulin expression is strictly controlled by a set of transcriptional activators and inhibitors that stimulate the insulin gene in response to increasing plasma glucose level. Knocking down miRNA-24, miRNA-26, miRNA-182 or miRNA-148 in cultured β-cells or primary islets led to a reduction in insulin promoter function, lowered insulin mRNA levels and up-regulation of transcriptional repressors. This suggests that these miRNAs regulate insulin expression by down-regulating transcriptional repressors (Melkman-Zehavi et al. 2011). Thus, the modulation of specific insulin-regulating miRNAs may be a promising therapeutic target (Melkman-Zehavi et al. 2011; Plaisance et al. 2006). In human β-cell, PdX1 and MafA are transcriptional factors that are involved in insulin secretion and are regulated by certain miRNAs. MiRNA-30d induces insulin production and increases the expression of *MafA* gene (Babu et al. 2008). Overexpression of miRNA-30d protected cells from tumor necrosis factor (TNF)-α-mediated suppression of insulin secretion and transcription by down-regulating the mitogen-activated protein 4 kinase 4, a TNF-α-activated kinase, demonstrating the multiple roles of miRNA-30d in activating insulin production and protecting the cell from pro-inflammatory cytokines (Zhao et al. 2012).

The islet-specific miRNA-375 is the most well-studied and characterized miRNA. It is implicated in regulating insulin expression and controlling glucose homeostasis (Joglekar et al. 2007, 2007, 2009, 2011; Poy et al. 2009). Overexpression of miRNA-375 suppressed glucose-induced insulin secretion, and at the same time, knocking down miRNA-375 function enhanced insulin secretion (Poy et al. 2004). The genetic deletion of miRNA-375 reveals weakened glycemic control as a result of decreased β-cell and increased α-cell mass and function in a mice model (Latreille et al. 2015).

In addition to miRNA-375, studies have identified other potential key miRNAs implicated in the network of regulation of insulin secretion. A panel of 10 miRNAs (miRNA-27a, miRNA-130a, miRNA-192, miRNA-200a, miRNA-320, miRNA-337, miRNA-369-5p, miRNA-379, miRNA-410 and miRNA-532) were down-regulated in glucose non-responsive MIN6 cells compared to glucose-responsive cells. Knocking down miRNA-200a, miRNA-130 and miRNA-410 may enhance the levels of glucose-stimulated insulin secretion (Hennessy et al. 2010).

In a recent study, the adipose tissue from women with polycystic ovary syndrome as well as control subjects with insulin resistance has been investigated. MiRNA-93 and miRNA-223 were both reported to be expressed in adipose tissues and influencing the glucose metabolism (Chen et al. 2013). The overexpression of miRNA-223 in human differentiated adipocytes was correlated with the decreased level of glucose transporter isoform 4 (GLUT4) protein and insulin-stimulated glucose intake, suggesting the potential therapeutic role of miRNA-223 for the insulin resistance-related disorders (Deiuliis 2015; Chuang et al. 2015).

Similarly, studies have further confirmed the presence of another group of glucose-regulated miRNAs in MIN6 cells. A group of 61 miRNAs were identified as glucose-regulated miRNAs out of 108 detected. Most of these miRNAs, including miRNA-124a, miRNA-107 and miRNA-30d, were overexpressed in the presence of high glucose level, while miRNA-296, miRNA-484 and miRNA-690 were substantially down-expressed by high glucose treatment (Tang et al. 2009). Cytokine-mediated β-cell dysfunction represented by IL-1 beta stimulates the expression level of miRNA-101a and miRNA-30b in MIN-6 cells. These two miRNAs are typically involved in beta-cell dysfunction such as decline insulin content, elevated in β-cell death and also gene expression (Zheng et al. 2015).

Cell-Associated MiRNAs and T1D

MiRNAs are expressed in a tissue-specific pattern and are implicated in regulation of various cellular activities, such as cell cycle regulation, cellular homeostasis, inflammatory pathways and apoptosis (Table 7.2). Pancreatic β-cells exposed

Table 7.2 miRNAs that modulate apoptotic, proliferative and inflammatory pathways

Signaling pathway	miRNA	References
JAK/STAT	Let-7a, miRNA-18a, miRNA-19a, miRNA-30d, miRNA-155	Iliopoulos et al. (2009), Wu et al. 2013), Collins et al. (2013, Kobayashi et al. (2012), Su et al. (2011)
NF-κB	miRNA-9, miRNA-22, miRNA-27b, miRNA-125a/b, miRNA-218, miRNA-301a, miRNA-362	Guo et al. (2009), Polioudakis et al. (2013), Rastogi et al. (2013), Kim et al. (2012), Gao et al. (2010), Xia et al. (2013), Huang et al. (2013), Lu et al. (2011), Xia et al. (2014)
Cytokine production	miRNA-27b, miRNA-29, miRNA-31, miRNA-146a, miRNA-155, miRNA-181a, miRNA-187	Su et al. (2011), Rastogi et al. (2013, Ma et al. (2011), Steiner et al. 2011), Fan et al. 2012), Lu et al. (2010), Tang et al. (2009), Zhao et al. (2012), Rossato et al. (2012)
Cellular proliferation	miRNA-30d, miRNA-93, miRNA-106b, miRNA-181d, miRNA-195, miRNA-205	Kobayashi et al. (2012), Fang et al. (2011), Ivanovska et al. (2008), Wang et al. (2012), Hui et al. (2013), Yue et al. (2012)
Cellular apoptosis	miRNA-34a, miRNA-106a, miRNA-221, miRNA-222, miRNA-451	Wang et al. (2013), Li et al. (2009), Chen et al. (2012a), Tian et al. 2012)

JAK/STAT = Janus kinase/signal transducer and activator of transcription; NF-κB = nuclear factor kappa-light-chain-enhancer of activated B-cells to pro-inflammatory cytokines demonstrate alterations in gene expression leading to inefficient insulin secretion and apoptosis (Roggli et al. 2010). MiRNA transcriptome alterations of rodent pancreatic α (αTC1–6) and β (βTC1) cells pre- and post-treatment with IL-1β, IFN-γ and TNF-α demonstrated that miRNA-296-3p and miRNA-298-5p were highly expressed in αTC1-6, but not in βTC1 cells at the steady state. Both miRNAs were significantly down-regulated after treatment with IL-1β, IFN-γ and TNF-α. Specific downregulation of miRNA-296-3p and miRNA-298-5p coupled to up-regulation of their targets as IGF1Rβ and TNF-α are a major factor for apoptosis resistance induced by cytokines (Barbagallo et al. 2013).

Responsiveness of the pancreatic β-cells to cytokine treatment demonstrated significant effects on global and specific miRNAs expression level. The effect of cytokine treatment on the global miRNA profiling of pancreatic β-cells demonstrated a substantial increase in miRNA-21 and miRNA-34 levels post-cytokine exposure (Ruan et al. 2011). MiRNA-34 up-regulation has been linked to p53 activation, which leads to sensitization to apoptosis. MiRNA-21 targets the programmed cell death 4 (PDCD4) protein, a tumor suppressor, and its upstream transcriptional activator nuclear factor κB (NF-κB).

c-Real and p65 are two transcriptional factors, from the NF-κB family, that up-regulate the miRNA-21 gene promoter and increase miRNA-21 expression levels. Consequently, down-regulation of PDCD4, which is able to induce apoptosis through the Bcl-2-associated X protein (Abdool Karim 2010) family, by miRNA-21, renders pancreatic β-cells resistant to death (Ruan et al. 2011). Reduced expression of miRNA-21 and miRNA-34, by antisense treatment, has been observed to attenuate apoptosis and improve glucose-simulated insulin release (Roggli et al. 2010; Lovis et al. 2008). MiRNA-21 and miRNA-34 could also be involved in T-cell regulation and apoptosis signaling pathways alterations as well as maintaining the pro-inflammatory environment (Salas-Perez et al. 2013).

Since miRNAs are firmly implicated in the regulation and pathogenesis of autoimmune diseases including T1D, miRNA analysis of selected cell populations might assist in defining at-risk individuals. In a comparison of human T1D associated loci and the chromosomal locations of 530 miRNAs, 27 miRNAs were located in 9 human genetic loci linked with T1D (Zhou and He 2008). Surprisingly, autoimmune and β-cell-related genes, such as T-cell co-stimulator and CD28, are predicted targets for these miRNAs (miRNA-16-2), autoimmune regulators (miRNA-877), insulin secretion regulator (miRNA-375), *IFNγ* and *FasL* (miRNA-551b) (Zhou and He 2008), β-cells mass expansion and insulin sensitivity (miRNA-338-3p) (Jacovetti et al. 2012).

Specific miRNAs have different expression signatures in PBMCs from T1D patients compared to controls, implying their involvement in T-cell regulation (Yang et al. 2015). Salas-Perez et al. showed that the expression of both miRNA-21a and miRNA-93 was down-regulated in T1D samples, compared with controls. However, after exposing PBMCs to 11 and 25 mM glucose, the level of expression of miRNA-21a decreased in controls and increased in T1D group, while the expression level of miRNA-93 did not change in the T1D group compared with the controls, suggesting that decreased levels of miRNA-21a might be an indicator of persistent hyperglycemia (Salas-Perez et al. 2013).

Moreover, specific miRNAs may represent fundamental biomarkers of diabetes, whereas shared miRNAs may distinguish diabetes as a metabolic-inflammatory disorder. To get an overall picture of the miRNA and mRNA expression and use this 'signature' as a tool to categorize the different types of diabetes mellitus, PBMCs were collected from T1D, T2D and gestational diabetes patients and total RNA was extracted for analysis. Twenty miRNAs were specific for T1D, 14 for T2D, and 19 miRNAs for the gestational diabetes group (Salas-Perez et al. 2013; Takahashi et al. 2014). Nine miRNAs (miRNA-126, miRNA-1307, miRNA-142-3p, miRNA-142-5p, miRNA-144, miRNA-199a-5p, miRNA-27a, miRNA-29b, and miRNA-342-3p) were shared between the three types of diabetes mellitus. Receiver operating characteristic (ROC) curve analysis was used to search for potential miRNAs which targeted specific mRNAs associated with diabetes pathogenesis and were found to be unique to each type of diabetes (Griner et al. 1981). Accordingly, miRNA-1274a, miRNA-1274b, miRNA-let-7f presented a good score for T1D; miRNA-222, miRNA-30e and miRNA-140-3p for T2D; and miRNA-181a and miRNA-1268 for gestational diabetes. These findings highlight the usefulness of PBMCs as reporter cells to characterize the miRNA signature associated with diverse diabetes mellitus manifestations (Collares et al. 2013).

Regulatory T-cells (T-regs) are critical to the maintenance of immune cell homeostasis via effective negative regulation of other immune cells. Specific miRNAs are differentially expressed and occasionally influence the function of T-regs (Sebastiani et al. 2011; Gilad et al. 2008). In T-regs and conventional T-cells populations isolated from T1D patients, the level of miRNA-510 was significantly up-regulated and the level of miRNA-342 was down-regulated in T-regs compared with the control group (Garzon et al. 2007). The level of miRNA-146a was higher in T-Regs compared to conventional T-cells, and the level of miRNA-20b, miRNA-31, miRNA-99a, miRNA-125b, miRNA-151, miRNA-335 and miRNA-365 was lower in T1D patients. This information suggests that these miRNAs have a role in the autoimmune destruction of β-cells (Hezova et al. 2010).

Numerous investigations have demonstrated an altered miRNA signature in lymphocytes of patients with several autoimmune disorders including T1D. For example, a group of T1D patients with ongoing islet autoimmunity (positive for GAD65 or IA-2A autoantibodies) were compared to patients who were negative for both. The level of miRNA-326 was increased in peripheral blood lymphocytes isolated from the former group, compared with the latter (Sebastiani et al. 2011).

The miRNA-326 was previously found to be altered in Th-17 cell subset during autoimmune multiple sclerosis, in which the levels of this miRNA greatly correlated with disease severity. Interestingly, vitamin D receptor and erythroblastosis virus E26 oncogene homolog 1 are predicted target genes for miRNA-326, suggesting the involvement of this miRNA in the immune regulation process in T1D (Sebastiani et al. 2011; Du et al. 2009).

Circulating MiRNAs and T1D

It has been indicated that extracellular miRNAs can be detected in the circulation and act as biomarkers of various human diseases (Table 7.4). In cancer research, blood-based miRNAs, either cell associated or cell free, arise as a novel, repetitive, non-invasive and real time fingerprint of great potential for cancer detection and treatment (Qi et al. 2013; Ji and Wang 2009). Stable expression in serum, plasma, saliva, urine and other body fluids is the key advantage of circulating miRNAs (Zen and Zhang 2012; Zhu et al. 2009). Circulating miRNAs are present in a remarkably stable form and protected from endogenous RNase activity. There is minimum to no effect on the levels of miRNA-15b, miRNA-16, or miRNA-24 when incubating plasma samples at room temperature for up to 24 h or exposing the samples to multiple freeze-thawing cycles (Mitchell et al. 2008). Serum miRNA stability was also confirmed under other harsh conditions such as low/high pH and boiling without any serious differences comparing to non-treated serum samples (Chen et al. 2008).

The measurement of miRNAs in serum or plasma represents a valuable approach for the blood-based detection of various human diseases including autoimmune T1D. In 2008, patients with prostate cancer were readily distinguished from healthy controls by measuring the levels of circulating miRNA-141 in their serum (Mitchell et al. 2008). The first evidence for the presence of circulating miRNAs in plasma of patients with diabetes was published in 2010 (Zampetaki et al. 2010). A deficiency of endothelial miR-126 and low plasma levels of miRNA-20b, miRNA-21, miRNA-24, miRNA-15a, miRNA-126, miRNA-191, miRNA-197, miRNA-223, miRNA-320 and miRNA-486 may explain the impaired peripheral angiogenic signaling in diabetic patients. In addition, low expression of miRNA-15a, miRNA-29b, miRNA-126, miRNA-223 and high expression of miRNA-28-3p was detected in 80 participants with either pre-diabetes or T2DM, suggesting their dysregulation may accelerate the disease manifestation (Roivainen et al. 2002).

Identifying a new blood-based biomarker from children with newly diagnosed T1D might anticipate the status of the residual β-cell function. This is important since children with T1D are predicted to lose approximately 80–90 % of their insulin-producing β-cell function at the time of diagnosis (Aly and Gottlieb 2009). Remission phase of T1D represents a unique intervention treatment opportunity; during this phase, an increase in the endogenous insulin production is followed by reduced need of exogenous insulin. The administration of β-cell growth factors has the potential to improve glycemic control and reduce T1D complications (Aly and Gottlieb 2009; Eldor et al. 2009).

Circulating miRNAs are useful predictive biomarkers for tissue physiopathology. Unique circulating miRNA signatures, which predict ongoing destruction or regeneration of endogenous residual β-cell, have been identified (Nielsen et al. 2012). In two T1D cohorts and one control group, miRNA-152, miRNA-30a-5p, miRNA-181a, miRNA-24, miRNA-148a, miRNA-210, miRNA-27a, miRNA-29a, miRNA-26a, miRNA-27b, miRNA-25 and miRNA-200a were linked to apoptosis and β-cell networks. Interestingly, this study demonstrated that miRNA-25 might be a 'tissue-specific' miRNA due to its positive association with glycemic control (glycated hemoglobin, HbA1c) few months after the new onset of T1D(Nielsen et al. 2012).

Global miRNA profiling of 20 newly diagnosed T1D patients and healthy controls revealed a specific miRNA signature. A group of 206 miRNAs were detected in both groups, 64 of which were specifically differentially expressed in T1D patients (Sebastiani et al. 2012). Importantly, several of these miRNAs function as immune cells regulators (miRNA-31, miRNA-155, miRNA-147a, miRNA 181a and miRNA-199a) (Guay and Regazzi 2013).

The importance of miRNA-375 in β-cell function and as a blood marker to detect β-cell death and predict T1D has been described (El Ouaamari et al. 2008; Farr et al. 2013). The level of miRNA-375, in plasma samples of non-obese diabetic mice (NOD) and in C57BL/6 mice treated with streptozotocin, was increased dramatically 2 weeks before the onset of autoimmunity. Additionally, inducing cell death in pancreatic islets treated with cytokines and streptozotocin generated a marked elevation of extracellular miRNA-375 which was reduced by cell death inhibitors (Erener et al. 2013). These promising findings described in mice require further validation in humans as the reduction of β-cell mass takes longer than in NOD mice (Guay and Regazzi 2013; Erener et al. 2013).

Microalbuminuria gives the ultimate clinical marker of diabetic nephropathy among patients with Type 1 diabetes (Argyropoulos et al. 2015). The expression levels of 723 unique miRNAs in the normoalbuminuric urine of patients who did not develop nephropathy relative to patients who afterward developed microalbuminuria. Results demonstrated the association of microalbuminuria development with several miRNAs (miRNA-105-3p, miRNA-1972, miRNA-28-3p, miRNA-30b-3p, miRNA-363-3p, miRNA-424-5p, miRNA-486-5p, miRNA-495, miRNA-548o-3p), while another set of miRNAs were expressed gender-related (female-related) differences in their levels of expression such as miRNA-192-5p and miRNA-720. Interestingly, these predicted target genes and biological pathways are known to be key players involved in the pathogenesis of diabetic renal diseases (Argyropoulos et al. 2015).

HNF1A is a transcriptional activator that monitors the tissue-specific expression of several genes, particularly in pancreatic islet cells and in liver (Delvecchio et al. 2014). Abnormalities in this gene are a cause of maturity onset diabetes of the young type 3 (MODY3) and also in the development of hepatic adenomas (Dominguez-Lopez et al. 2005). In a recent study, the level of circulating miRNA-224 and miRNA-103 was determined in the urine of individuals with T1D, T2D and also in carriers of HNF1A mutation. MiRNA-224 was extremely higher in the urine of HNF1A mutation carriers and individuals with T1D. However, miRNA-103 was significantly expressed in urine in all diabetes cohorts when compared to the healthy controls (Bacon et al. 2015).

Role of MiRNAs in Host–Virus Interactions

RNA viruses encode a few proteins, yet still create complex interactions and networks with the host cell elements to accomplish replication and spreading. MiRNAs, as unique regulators of viral replication and gene expression, play a fundamental role in this synergy. Considering their small size, lack of immunogenicity, and extraordinary functional flexibility, miRNAs are appealing targets as virally encoded regulators of both host and viral gene expression (Sullivan and Ganem 2005; Berkhout and Jeang 2007).

Viruses may utilize both viral and host miRNAs to influence their own gene expression, via modulating the host immune-physiology components, or directly through modifying their own-life cycle and replication, respectively (Cai et al. 2005; Wang et al. 2013). Despite the mechanism(s) viruses might use to employ miRNAs to their benefit, expression of certain cellular miRNAs might be unfavorable to the virus because of the interplay with the viral mRNAs or due to specific miRNA function(s) (Wang et al. 2013).

Several studies have revealed a list of cellular miRNAs that are differentially expressed during viral infections in both DNA and RNA viruses (Table 7.3). Furthermore, several viruses, such as the Epstein–Barr virus (EBV) and herpes simplex virus (HSV), can regulate both host and viral gene expression by virally encoded miRNAs (Pfeffer et al. 2004; Feederle et al. 2011). Some intracellular miRNAs can bind to the viral mRNAs and exhibit antiviral properties, for example, miRNA-32-1 targets the primate foamy virus (PFV) genome and effectively reduces viral replication (Muller and Imler 2007) (Lecellier et al. 2005). Similarly, miRNA-125-5P and miRNA-122 can both suppress the replication of hepatitis B virus (HBV) by targeting the S protein (Potenza et al. 2011) and binding to conserved region of the virus genome, respectively (Chen et al. 2011). MiRNA-29 is reported to inhibit the replication of human immunodeficiency virus (HIV) (Shivdasani 2006) by binding to the UTR of the viral genome (Nathans et al. 2009). HIV-1 is suppressed by some cellular miRNAs but in turn might affect miRNAs abundance by producing viral proteins such as 'Tat' protein (Lecellier et al. 2005). The Tat protein of HIV-1, and Tas protein of PFV, can inactivate the cellular RNA-defense mechanism by destroying the 'Dicer function' in processing the precursor double-stranded RNAs into small interference RNAs (siRNAs) (Lecellier et al. 2005; Bennasser et al. 2005) (Tables 7.4, 7.5).

Similar to HIV-1 Tat protein, influenza A virus protein NS1, Ebola virus protein VP35 and vaccinia virus protein E3L all function as commonly powerful inhibitors of RNA silencing (Haasnoot et al. 2007). MiRNA-448, miRNA-431, miRNA-296, miRNA-196 and miRNA-296 are known as interferon (IFN)-induced miRNAs and have an interesting inhibitory effect on the replication of hepatitis C virus (HCV) in hepatocellular carcinoma (Huh7) cell line in vitro (Pedersen et al. 2007). Lastly, the vesicular stomatitis virus (VSV) replication is inhibited by miRNA-24 via viral L protein-encoding gene and by miRNA-93, through targeting the P protein-encoding gene (Otsuka et al. 2007).

Table 7.3 miRNAs that modulate apoptotic, proliferative and cytokine signaling pathways

miRNA	Validated targets	Function	Reference
Let-7a	Interleukin-6	Reduces JAK/STAT pathway activation and inflammation.	Iliopoulos et al. (2009).
miRNA-9	Nuclear factor κ-B1	Down-regulates NF-κB pathway to inhibit cellular proliferation.	Guo et al. (2009).
miRNA-18a	Protein inhibitor of activated STAT3	Indirectly induces STAT3 signaling pathways by alleviating upstream suppressors. Leads to an anti-apoptotic cellular state and increased proliferation.	Wu et al. (2013).
miRNA-19a	Suppressor of cytokine signaling 3	Enhances JAK/STAT signaling and STAT3 phosphorylation in response to pro-inflammatory cytokines.	Collins et al. (2013).
miRNA-22	High-mobility group Box-1 and interferon regulatory factor-5	Suppresses the interferon signaling pathways by inhibiting the activation of NK-κB pathways. Results in sequential proliferation induction.	Polioudakis et al. (2013).
miRNA-25, miRNA-93, and miRNA-106b	Mothers against decapentaplegic homology	Enhances the TGF-β signaling pathway.	Smith et al. (2012).
miRNA-27b	Peroxisome proliferator-activated receptor γ	Suppresses NF-κB pathway activity, thus reducing expression of pro-inflammatory cytokines such as interleukin-1b, interleukin-6.	Rastogi et al. (2013).
miRNA-29	IFN-γ and T-box transcription factor	Reduces IFN-γ production.	Ma et al. (2011), Steiner et al. (2011)
miRNA-30d	Suppressor of cytokine signaling 1	Promotes Proliferation by alleviating suppression of cytokine-mediated JAK/STAT pathways.	Kobayashi et al. (2012).
miRNA-31	Ras homolog gene family, member A	Increases IL-2 production by removing the protein suppressing the IL-2 promoter.	Fan et al. (2012).
miRNA-34a	Notch 1 and 2	Inhibits cellular proliferation and induces apoptosis.	Li et al. (2009)
miRNA-93	Integrin-β8	Enhances cellular survival and growth by removing negative regulators of cellular proliferation.	Fang et al. (2011).
miRNA-106a	FAS	Inhibits the FAS-mediated extrinsic cell death pathway.	Wang et al. (2013).
miRNA-106b	Cyclin-dependent kinase inhibitor 1	Loss of cell cycle checkpoint protein, cyclin-dependent kinase inhibitor 1, results in abnormal cellular phenotype and dysfunction.	Ivanovska et al. (2008).

(continued)

Table 7.3 (continued)

miRNA	Validated targets	Function	Reference
miRNA-125a/b	TNF-α-induced protein 3	Loss of antagonistic action on proteins downstream of TNF-receptor engagement, resulting in constitutive activation of the NF-κB pathway.	Kim et al. (2012).
miRNA-145	TNF-α convertase	Activates NF-κB pathway and increases production of TNF-α. Also prevents proteolytic release of membrane-bound TNF-α.	Lorente-Cebrian et al. (2014).
miRNA-146a	STAT1 and Interferon regulatory factor 5	Reduces induction of Type 1 IFN and represses proteins downstream of the Type 1 IFN signaling pathway. Also prevents production of IFN-γ in regulatory T-cells	Lu et al. (2010), Tang et al. (2009)
miRNA-155	Suppressor of cytokine signaling 1	Indirectly activates the JAK/STAT pathway to increase proliferation, promotes type 1 IFN signaling and up-regulates expression of IFN-inducible genes.	Su et al. (2011)
miRNA-181a	p300/Creb Binding Protein-associated factor	Initiates a negative feedback mechanism in TNF-α-induced transcription of pro-inflammatory genes.	Zhao et al. (2012)
miRNA-181d	K-ras and B-cell lymphoma 2	Arrest G1 Cell Cycle, resulting in reduced proliferation and induced apoptosis.	Wang et al. (2012)
miRNA-187	TNF-α and nuclear factor kappa-B inhibitor ζ	Mediates IL-10 suppression of pro-inflammatory cytokine synthesis (TNF-α, IL-6, and IL-12) in activated monocytes.	Rossato et al. (2012)
miRNA-195	Cyclin D1 and E1	Reduces cellular proliferation by degrading key cell cycle proteins.	Hui et al. (2013)
miRNA-205	Vascular endothelial growth factor A	Cell cycle arrest at G0/G1 phase, resulting in inhibited proliferation.	Yue et al. (2012)
miRNA-218	Epidermal growth factor receptor-co-amplified and over-expressed protein	Down-regulation of key NF-κB regulators results in inhibition of NF-κB pathway. Leads to a pro-apoptotic state.	Gao et al. (2010), Xia et al. (2013)
miRNA-221, miRNA-222	p53 up-regulated modulator of apoptosis protein		

(continued)

Table 7.3 (continued)

miRNA	Validated targets	Function	Reference
	Reduces apoptosis by regulating the mitochondrial apoptosis pathway. Also interferes with IFN-α signaling pathway, by indirectly modulating STAT1 and 2, thereby promoting cellular proliferation.		Chen et al. (2012a), Zhang et al. (2010).
miRNA-301a	Nuclear factor-κB repressing factor	Enhances NF-κB activation, resulting in increased pro-inflammatory cytokine production.	Huang et al. (2013), Lu et al. (2011).
miRNA-330	Endophilin-A1.	Induced cellular proliferation and anti-apoptotic state due to loss of regulatory endophilin-A1 complex.	Qu et al. (2012).
miRNA-362	Ubiquitin carboxyl-terminal hydrolase	Loss of negative regulatory check on the NF-κB signaling pathway, resulting in proliferation and resistance to apoptosis.	Xia et al. (2014).
miRNA-451	Calcium-binding protein 39	Reduces proliferation by down-regulating upstream regulators of PI3 K/AKT pathway.	Tian et al. (2012).
miRNA-939	Human inducible nitric oxide synthase	Inhibits cytokine-induced human inducible nitric oxide synthase and decreases nitric oxide synthesis.	Guo et al. (2012).

Table 7.4 Examples of viral regulation of cellular miRNAs and their effects

Virus species	Diseases caused	miRNA affected	Effect	Reference
Epstein–Barr virus	Burkitt's lymphoma, nasopharyngeal carcinoma	Up-regulation of miRNA-146a.	Suppression of interferon-mediated antiviral response.	Cameron et al. (2008)
Hepatitis C virus	Hepatocellular carcinoma, hepatitis	Down-regulation of miRNA-152.	Increased WnT1 production, resulting in cellular proliferation.	Huang et al. (2014)
		Down-regulation of miRNA-107 and miRNA-449a.	Up-regulation of CCL-2, resulting in chronic inflammation.	Sarma et al. (2014)
Human cytomegalovirus	Mononucleosis	Down-regulation of miRNA-100 and miRNA-101.	Promotion of mTOR signaling pathway, resulting in increased viral replication.	Wang et al. (2008)
Human immunodeficiency virus-1	Acquired immunodeficiency syndrome	Down-regulation of miRNA-17/92 cluster.	Up-regulation of co-factors necessary for HIV-1 protein function.	Triboulet et al. (2007)
Human papillomavirus	Cervical carcinoma	Down-regulation of miRNA-34a.	Increased cellular proliferation.	Wang et al. (2009)
Influenza A Virus	Influenza	Down-regulation of miRNA-548an.	Induces anti-apoptotic cellular state, promoting viral persistence.	Othumpangat et al. (2013)

CCL-2 = chemokine ligand 2; miR = miRNA; mTOR = mammalian target of rapamycin; HIV = human immunodeficiency virus

The role of miRNAs in EVs infection, particularly in CVB infections, and how they lead to autoimmunity and development of T1D have yet to be investigated. Few studies have focused on the role of certain miRNAs during viral-induced disease. Some documented examples include CVB3-induced myocarditis and EV71-induced hand, foot and mouth disease (HFMD). Novel regulatory mechanisms of host–virus interaction where miRNA-126 controls CVB3-induced infectious heart disease have been explored. ERK1/2 and Wnt/β-catenin signaling pathways are regulated by miRNA-126, and the coordination between SPRED1, LRP6, and WRCH1 target genes and the identified regulatory pathways is essential for the life cycle and cytopathogenicity of CVB3 (Ye et al. 2013).

A novel molecular link between miRNA-203 and zinc finger protein-148 (ZFP-148) has been revealed. Ectopic expression of miRNA-203 regulates CVB3 viral myocarditis via activation of protein kinase C/transcription factor AP-1 pathway and ZFP-148 transcription factor as a novel target gene. Up-regulation of miRNA-203 leads to down-regulation of ZFP-148 and enhanced cell viability and,

Table 7.5 Circulating miRNAs in blood associated with various human diseases

Disease	miRNAs differentially expressed[a]	Selected references
Type 1 diabetes	miRNA-24, miRNA-25, miRNA-26a, miRNA-27a, miRNA-27b, miRNA-29a; miRNA-30a-5p, miRNA-148a, miRNA-152, miRNA-181a, miRNA-200a and miRNA-210; miRNA-9, miRNA-31, miRNA-34a, miRNA-146a, miRNA-155, miRNA-181a and miRNA-199a; miRNA-21a and miRNA-93; miRNA-326	Sebastiani et al. (2011), Nielsen et al. (2012), Salas-Perez et al. (2013)
Non-small cell lung cancer	miRNA-20, miRNA-24, miRNA-25, miRNA-145, miRNA-152, miRNA-199a5p, miRNA-221, miRNA-222, miRNA-223 and miRNA-320; let-7a; miRNA-21, miRNA-126, miRNA-210 and miRNA-486-5p	Jeong et al. (2011), Shen et al. (2011), Chen et al. (2012b)
Breast cancer	Let-7c, miRNA-589 (Caucasian American) and miRNA-let-7d*; miRNA-425* (African American); miRNA-195; miRNA-16, miRNA-25, miRNA-222, miRNA-324-3p	Heneghan et al. (2010), Zhao et al. (2010), Hu et al. (2012)
Renal cell cancer	miRNA-378, miRNA-451	Redova et al. (2012)
Hepatocellular carcinoma	miRNA-23b, miRNA-423, miRNA-375, miRNA-23a; miRNA-342-3p (HBV +); miRNA-10a, miRNA-125b; miRNA-15b, miRNA-130b, miRNA-122, miRNA-192, miRNA-21, miRNA-223, miRNA-26a, miRNA-27a, miRNA-801	Li et al. (2010), Zhou et al. (2011), Liu et al. (2012)
Gastric cancer	miRNA-1, miRNA-20a, miRNA-27a, miRNA-34, miRNA-423-5p; miRNA-106a/let-7a, miRNA-106b	Tsujiura et al. (2010), Liu et al. (2011)
Pancreatic cancer	miRNA-16, miRNA-196a, CA19-9; miRNA-18a	Morimura et al. (2011), Liu et al. (2012)
Colorectal cancer	miRNA-92; miRNA-29a, miRNA-92a	Ng et al. (2009), Huang et al. (2010)
Acute myocardial infarction	miRNA-208b; miRNA-208a; miRNA-499; miRNA-663b, miRNA-1291; miRNA-133, miRNA-328; miR-1	Ai et al. (2010), Wang et al. (2010), Corsten et al. (2010), Gidlof et al. (2011), Meder et al. (2011), Wang et al. (2011)
Alzheimer's disease	let-7d-5p, let-7 g-5p, miRNA-15b-5p, miRNA-142-3p, miRNA-191-5p; miRNA-301a-3p, miRNA-545	Kumar et al. (2013)
Multiple sclerosis, relapsing-remitting	hsa-miRNA-145	Keller et al. (2009)
Viral hepatitis	miRNA-375, miRNA-10a, miRNA-223; miRNA-92a, miRNA-423	Li et al. (2010)
Hand, foot and mouth disease	miRNA-148a, miRNA-143, miRNA-324-3p, miRNA-628-3p, miRNA-140-5p, miRNA-362-3p	Cui et al. (2011)

[a]In serum, plasma and/or PBMCs. HBV = hepatitis B virus. (;) is used to separate miRNAs described by different studies

subsequently, increased CVB3 replication (Hemida et al. 2013). Several miRNAs regulate the pathogenesis of viral myocarditis, such as miRNA-21, miRNA-146b and miRNA-451. MiRNA-21 and miRNA-146b were up-regulated in a murine viral myocarditis model, while miRNA-451 was down-regulated. Differential expression of these miRNAs might ameliorate myocarditis by decreasing levels of Th17 and RORγt pathways (Liu et al. 2013).

miRNA-21 could protect myocardial apoptosis by inhibiting the PDCD4 expression (He et al. 2013). Moreover, miRNA-146a, miRNA-374, miRNA-21, miRNA-29a* and miRNA-23a are involved in controlling several important innate and antiviral pathways like the Toll-like receptor pathway, RIG-I-like receptor pathway, NOD-like receptor, cytokine–cytokine receptor pathway, MAPK signaling pathway, JAK–STAT signaling pathway and natural killer cell-mediated cytotoxicity, suggesting these miRNAs may possibly be a novel therapeutic target for treatment of CVB3-induced myocarditis (Zhang et al. 2013)

Potential miRNA-342-5p targets in the CVB3 genome have been identified based on sequence comparison. MiRNA-324 can inhibit CVB3 biogenesis process via targeting its 2C-coding sequence, and therefore it might be a potential therapeutic agent in the treatment of CVB3 infection (Wang et al. 2012). MiRNA-10a*, which was detected in the cardiac tissues of Balb/c mice, significantly improved the CVB3 biosynthesis during the cardiac infection, indicating that miRNA-10a* might be involved in CVB3 pathogenesis (Tong et al. 2013). Enterovirus 71 (EV71) is an aggressive neurotropic EV known to cause HFMD in young children and the development of severe neurologic diseases such as meningitis and cerebellar encephalitis. Down-regulation of miRNA-23b is linked to increased EV71 replication in the virus-infected rhabdomyosarcoma (RD) cells. Thus, the up-regulation of this miRNA may be used to prevent EV71 replication and treat EV71-induced HFMD (Wen et al. 2013).

Overexpression of miRNA-296-5p in EV71 (strain BrCr)-infected cells appear to inhibit EV71 replication, while inhibition of endogenous miRNA-296-5p facilitated EV71 infection. Interestingly, introduction of an equivalent mutation into the EV71 strains BrCr genome by site-directed mutagenesis damaged the antiviral inhibitory effects of miRNA-296-5p and facilitated mutant virus infection (Zheng et al. 2013). At the same time, introducing a compensatory mutation in matching miRNA-296-5p target sequence in EV71 strain HeN restored the inhibitory effect of this miRNA, indicating that miRNA-296-5p participate in host antiviral defense mechanism and the mutated virus avoids suppression by cellular miRNAs (Zheng et al. 2013). Picornaviruses including EVs can utilize their own viral proteases to shut off the host Cap-dependent protein translation via cleavage of the host translation initiation factors eIF4GI and eIF4GII. EV71- induced miRNA-141 targets the cap-dependent translation initiation factor, eIF4E, to shut off host protein synthesis. Overexpression of miRNA-141 promotes the switch from cap-dependent into cap-independent translation via EGR1 transcriptional factor during EV71 infection, suggesting that miRNA-141 can facilitate viral replication and expansion by accelerating the translational shift (Ho et al. 2011).

Recent high-throughput sequencing technologies have allowed deep sequencing of large libraries of miRNAs. By using deep sequencing, EV71-infected and non-infected Hep2 cells, a group of 569 miRNAs were detected in EV71-infected cells, while 540 miRNAs were detected in non-infected cells (WHO 2001). Of these, 64 miRNA were differentially expressed between infected and non-infected cells, 42 were up-regulated. Interestingly, the motif of miRNAs observed upon EV71 infection (more up-regulated than down-regulated) is similar in some way to what has been previously observed with cells infected with HCV(Liu et al. 2010) but different with those with EBV and cytomegalovirus (CMV) infection where more down-regulated miRNAs have been reported (Godshalk et al. 2008; Wang et al. 2008).

Similar to EV71, coxsackievirus A16 (CVA16) is known as a major causative agent of HFMD. Serum miRNAs profile might be useful tool for subtyping EV HFMD infections. Using two-step screening approach and applying (ROC) curve analysis, five miRNAs (miRNA-148a, miRNA-143, miRNA-324-3p, miRNA-545 and miRNA-140-5p) were highly up-regulated in CVA16 patient's sera versus those with EV71. Moreover, six serum circulating miRNAs (miRNA-148a, miRNA-143, miRNA-324-3p, miRNA-628-3p, miRNA-140-5p and miRNA-362-3p) were significantly able to discriminate between EV infections from other microbial infections with a sensitivity of 97.1 % and specificity of 92.7 % (Cui et al. 2011).

Of the six, miRNA-148a has been the best studied within a viral context, but studies have yielded contrasting findings. It has been reported the HBV X protein both up- and down-regulates miR-148a, to inhibit and promote cellular proliferation, respectively (Xu et al. 2013; Yuan et al. 2012). Results from cancer research demonstrated that miRNA-148a directly targets the hematopoietic pre-B-cell leukemia transcription factor-interacting protein (HPIP). This protein can promote cell growth through action of the AKT/mTOR and ERK/mTOR pathway (Li et al. 2013).

MiRNA-140-5p suppresses cellular proliferation by inhibiting the transforming growth factor beta (TGF-beta) and mitogen-activated protein kinase/extracellular signal-regulated kinase (MAPK/ERK) signaling pathway, which occurs by arresting cells at S phase. Restoration of miRNA-140-5- targets, TGFB receptor 1 (TGFBR1) and fibroblast growth factor 9 (FGF9) restored cells to their pathological phenotype (Yang et al. 2013). MiRNA-324-3p is able to promote *RelA* gene expression by directly binding to its promoter in DNA and transfection of pre-miRNA-324-3p elevates level of cleaved Caspase-3. Furthermore, Ago2 siRNA significantly prevented levels of cleaved Caspase-3. This is most likely due to miRNA-324-3p's up-regulation of the RelA protein, a known enhancer of apoptotic pathway (Dharap et al. 2013). MiRNA-362-3p is down-regulated in the colon mucosa of patients with recurrent colorectal cancers. MiRNA-362-3p targets E2F1, USF2, and PTPN1 and significantly increases the number of cells arrest at the G1 phase. Specific knockdown of these proteins resulted in a similar phenotype, except for USF2, which had no impact on cellular proliferation (Christensen et al. 2013). Overexpression of miRNA-143 significantly inhibited HeLa cell proliferation and promoted apoptosis. This is due to miRNA-143 directly inhibiting Bcl-2 expression, an anti-apoptotic protein (Liu et al. 2012).

Conclusion and Future Directions

There is accumulating evidence that miRNAs are master regulators of gene expression both in the normal physiological milieu and in the context of disease. Assessing the miRNAs expression profiling is acquiring widespread attention because miRNAs, as fundamental regulators of gene expression process, can control multiple biological processes and demonstrate promising role in multifactorial and complex diseases such as diabetes. EVs, which are among the most important of human pathogens, are frequently associated with T1D pathogenesis. Growing evidence supports the significant role of miRNAs in EVs replication and in the manipulation of host–viral interactions. MicroRNAs may regulate the pathogenesis of T1D: Research in this area has identified miRNAs that are pivotal in insulin secretion and signaling, designating these with specific roles in autoimmunity to pancreatic β-cells. Future studies should be directed at inspecting the possible interplay between miRNAs, EVs infections and T1D pathogenesis. Additional investigations into the role of specific miRNAs fingerprint in CVBs infection to pancreatic β-cells may shed greater light on the viral–host interaction and ultimately provide a greater understanding of EV-induced T1D.

References

Abdool Karim Q (2010) Effectiveness and safety of tenofovir gel, an antiretroviral microbicide, for the prevention of HIV infection in women. Science 329(5996):1168–1174

Abuhatzira L et al (2015) Multiple microRNAs within the 14q32 cluster target the mRNAs of major type 1 diabetes autoantigens IA-2, IA-2beta, and GAD65. FASEB J 29(10):4374–4383

Ai J et al (2010) Circulating microRNA-1 as a potential novel biomarker for acute myocardial infarction. Biochem Biophys Res Commun 391(1):73–77

Aly H, Gottlieb P (2009) The honeymoon phase: intersection of metabolism and immunology. Curr Opin Endocrinol Diabetes Obes 16(4):286–292

Argyropoulos C et al (2015) Urinary microrna profiling predicts the development of microalbuminuria in patients with Type 1 diabetes. J Clin Med 4(7):1498–1517

Arvan P et al (2012) Islet autoantigens: structure, function, localization, and regulation. Cold Spring Harb Perspect Med 2(8)

Babu DA et al (2008) Pdx1 and BETA2/NeuroD1 participate in a transcriptional complex that mediates short-range DNA looping at the insulin gene. J Biol Chem 283(13):8164–8172

Bacon S et al (2015) MicroRNA-224 is readily detectable in urine of individuals with diabetes mellitus and is a potential indicator of beta-cell demise. Genes (Basel) 6(2):399–416

Barbagallo D et al (2013) miR-296-3p, miR-298-5p and their downstream networks are causally involved in the higher resistance of mammalian pancreatic alpha cells to cytokine-induced apoptosis as compared to beta cells. BMC Genom 14:62

Baroukh N, Ravier MA et al (2007) MicroRNA-124a regulates Foxa2 expression and intracellular signaling in pancreatic beta-cell lines. J Biol Chem 282:19575–19588

Barrett JC, Clayton DG, Concannon P, Akolkar B, Cooper JD, Erlich HA, Julier C, Morahan G, Nerup J, Nierras C, Plagnol V, Pociot F, Schuilenburg H, Smyth DJ, Stevens H, Todd JA, Walker NM, Rich SS, The Type 1 Diabetes Genetics Consortium (2009) Genome-wide association study and meta-analysis find that over 40 loci affect risk of type 1 diabetes. Nat Genet 41(6):703–707

Bartel DP (2009a) MicroRNAs: target recognition and regulatory functions. Cell 136:215–233

Bartel DP (2009b) MicroRNAs: target recognition and regulatory functions. Cell 136(2):215–233

Bason C et al (2013) In type 1 diabetes a subset of anti-coxsackievirus b4 antibodies recognize autoantigens and induce apoptosis of pancreatic Beta cells. PLoS ONE 8(2):e57729

Bennasser Y et al (2005) Evidence that HIV-1 encodes an siRNA and a suppressor of RNA silencing. Immunity 22(5):607–619

Berkhout B, Jeang KT (2007) RISCy business: microRNAs, pathogenesis, and viruses. J Biol Chem 282(37):26641–26645

Bluestone JA, Herold K, Eisenbarth G (2010) Genetics, pathogenesis and clinical interventions in type 1 diabetes. Nature 464(7293):1293–1300

Braun JE et al (2011) GW182 proteins directly recruit cytoplasmic deadenylase complexes to miRNA targets. Mol Cell 44(1):120–133

Bravo-Egana V et al (2012) Inflammation-mediated regulation of MicroRNA expression in transplanted pancreatic islets. J Transplant 2012:723614

Cai X et al (2005) Kaposi's sarcoma-associated herpesvirus expresses an array of viral microRNAs in latently infected cells. Proc Natl Acad Sci USA 102(15):5570–5575

Cameron JE et al (2008) Epstein-Barr virus latent membrane protein 1 induces cellular MicroRNA miR-146a, a modulator of lymphocyte signaling pathways. J Virol 82(4):1946–1958

Chang TC, Mendell JT (2007) microRNAs in vertebrate physiology and human disease. Annu Rev Genom Hum Genet 8:215–239

Chehadeh W et al (2000) Increased level of interferon-alpha in blood of patients with insulin-dependent diabetes mellitus: relationship with coxsackievirus B infection. J Infect Dis 181(6):1929–1939

Chekulaeva M et al (2011) miRNA repression involves GW182-mediated recruitment of CCR4-NOT through conserved W-containing motifs. Nat Struct Mol Biol 18(11):1218–1226

Chen X et al (2008) Characterization of microRNAs in serum: a novel class of biomarkers for diagnosis of cancer and other diseases. Cell Res 18(10):997–1006

Chen Y et al (2011) A liver-specific microRNA binds to a highly conserved RNA sequence of hepatitis B virus and negatively regulates viral gene expression and replication. FASEB J 25(12):4511–4521

Chen L et al (2012a) Downregulation of miR-221/222 sensitizes glioma cells to temozolomide by regulating apoptosis independently of p53 status. Oncol Rep 27(3):854–860

Chen X et al (2012b) Identification of ten serum microRNAs from a genome-wide serum microRNA expression profile as novel noninvasive biomarkers for nonsmall cell lung cancer diagnosis. Int J Cancer 130(7):1620–1628

Chen YH et al (2013) miRNA-93 inhibits GLUT4 and is overexpressed in adipose tissue of polycystic ovary syndrome patients and women with insulin resistance. Diabetes 62(7):2278–2286

Chiang PS, Huang ML, Luo ST, Lin TY, Tsao KC, Lee MS (2012) Comparing molecular methods for early detection and serotyping of enteroviruses in throat swabs of pediatric patients. PLoS ONE 7(10):e48269

Christensen LL et al (2013) MiRNA-362-3p induces cell cycle arrest through targeting of E2F1, USF2 and PTPN1 and is associated with recurrence of colorectal cancer. Int J Cancer 133(1):67–78

Chuang TY et al (2015) MicroRNA-223 expression is upregulated in insulin resistant human adipose tissue. J Diabetes Res 2015:943659

Cinek O, Tapia G, Witsø E, Kramna L, Holkova K, Rasmussen T, Stene LC, Rønningen KS (2012) Enterovirus RNA in peripheral blood may be associated with the variants of

rs1990760, a common type 1 diabetes associated polymorphism in IFIH1. PLoS ONE 7(11):e48409
Collares CV et al (2013) Identifying common and specific microRNAs expressed in peripheral blood mononuclear cell of type 1, type 2, and gestational diabetes mellitus patients. BMC Res Notes 6:491
Collins AS et al (2013) miR-19a: an effective regulator of SOCS3 and enhancer of JAK-STAT signalling. PLoS ONE 8(7):e69090
Corsten MF et al (2010) Circulating MicroRNA-208b and MicroRNA-499 reflect myocardial damage in cardiovascular disease. Circ Cardiovasc Genet 3(6):499–506
Craig ME et al (2003) Reduced frequency of HLA DRB1*03-DQB1*02 in children with type 1 diabetes associated with enterovirus RNA. J Infect Dis 187(10):1562–1570
Cui L et al (2011) Serum microRNA expression profile distinguishes enterovirus 71 and coxsackievirus 16 infections in patients with hand-foot-and-mouth disease. PLoS ONE 6(11):e27071
Dahlquist GG, Boman JE, Juto P (1999) Enteroviral RNA and IgM antibodies in early pregnancy and risk for childhood-onset IDDM in offspring. Diabetes Care 22(2):364–365
Deiuliis JA (2015) MicroRNAs as regulators of metabolic disease: pathophysiologic significance and emerging role as biomarkers and therapeutics. Int J Obes (Lond) 40:88–101
Delvecchio M et al (2014) Low prevalence of HNF1A mutations after molecular screening of multiple MODY genes in 58 Italian families recruited in the pediatric or adult diabetes clinic from a single Italian hospital. Diabetes Care 37(12):e258–e260
Dharap A et al (2013) MicroRNA miR-324-3p induces promoter-mediated expression of RelA gene. PLoS ONE 8(11):e79467
Dippe SE et al (1975) Lack of causal association between Coxsackie B4 virus infection and diabetes. Lancet 1(7920):1314–1317
Dominguez-Lopez A et al (2005) Mutations in MODY genes are not common cause of early-onset type 2 diabetes in Mexican families. Jop 6(3):238–245
Dotta F et al (2007) Coxsackie B4 virus infection of beta cells and natural killer cell insulitis in recent-onset type 1 diabetic patients. Proc Natl Acad Sci USA 104(12):5115–5120
Dp B (2004) MicroRNAs: genomics, biogenesis, mechanism, and function. Cell 116(2):281–297
Du C et al (2009) MicroRNA miR-326 regulates TH-17 differentiation and is associated with the pathogenesis of multiple sclerosis. Nat Immunol 10(12):1252–1259
El Ouaamari A et al (2008) miR-375 targets 3′-phosphoinositide-dependent protein kinase-1 and regulates glucose-induced biological responses in pancreatic beta-cells. Diabetes 57(10):2708–2717
Eldor R, Kassem S, Raz I (2009) Immune modulation in type 1 diabetes mellitus using DiaPep277: a short review and update of recent clinical trial results. Diabetes Metab Res Rev 25(4):316–320
Elfaitouri A et al (2007) Recent enterovirus infection in type 1 diabetes: evidence with a novel IgM method. J Med Virol 79(12):1861–1867
Elfving M et al (2008) Maternal enterovirus infection during pregnancy as a risk factor in offspring diagnosed with type 1 diabetes between 15 and 30 years of age. Exp Diabetes Res 2008:271958
Erener S et al (2013) Circulating miR-375 as a biomarker of beta-cell death and diabetes in mice. Endocrinology 154(2):603–608
Eulalio A, Huntzinger E, Izaurralde E (2008) Getting to the root of miRNA-mediated gene silencing. Cell 132:9–14
Fabian MR, Sonenberg N, Filipowicz W (2010) Regulation of mRNA translation and stability by microRNAs. Annu Rev Biochem 79:351–379
Fan W et al (2012) Identification of microRNA-31 as a novel regulator contributing to impaired interleukin-2 production in T cells from patients with systemic lupus erythematosus. Arthritis Rheum 64(11):3715–3725
Fang L et al (2011) MicroRNA miR-93 promotes tumor growth and angiogenesis by targeting integrin-beta8. Oncogene 30(7):806–821

Farr RJ et al (2013) Circulating non-coding RNAs as biomarkers of beta cell death in diabetes. Pediatr Endocrinol Rev 11(1):14–20
Feederle R et al (2011) A viral microRNA cluster strongly potentiates the transforming properties of a human herpesvirus. PLoS Pathog 7(2):e1001294
Filipowicz W, Bhattacharyya SN, Sonenberg N (2008) Mechanisms of post-transcriptional regulation by microRNAs: are the answers in sight? Nat Rev Genet 9(2):102–114
Filippi CM, von Herrath MG (2008) Viral trigger for type 1 diabetes: pros and cons. Diabetes 57(11):2863–2871
Gao C et al (2010) Reduced microRNA-218 expression is associated with high nuclear factor kappa B activation in gastric cancer. Cancer 116(1):41–49
Garzon R, Pichiorri F, Palumbo T et al (2007) MicroRNA gene expression during retinoic acid-induced differentiation of human acute promyelocytic leukemia. Oncogene 26:4148–4157
Gebauer F, Hentze MW (2004) Molecular mechanisms of translational control. Nat Rev Mol Cell Biol 5(10):827–835
Gidlof O et al (2011) Cardiospecific microRNA plasma levels correlate with troponin and cardiac function in patients with ST elevation myocardial infarction, are selectively dependent on renal elimination, and can be detected in urine samples. Cardiology 118(4):217–226
Gilad S et al (2008) Serum microRNAs are promising novel biomarkers. PLoS ONE 3(9):e3148
Godshalk SE, Bhaduri-McIntosh S, Slack FJ (2008) Epstein-Barr virus-mediated dysregulation of human microRNA expression. Cell Cycle 7(22):3595–3600
Griner PF et al (1981) Selection and interpretation of diagnostic tests and procedures. Principles and applications. Ann Intern Med 94(4 Pt 2):557–592
Guay C, Regazzi R (2013) Circulating microRNAs as novel biomarkers for diabetes mellitus. Nat Rev Endocrinol 9(9):513–521
Guo LM et al (2009) MicroRNA-9 inhibits ovarian cancer cell growth through regulation of NF-kappaB1. FEBS J 276(19):5537–5546
Guo Z et al (2012) miRNA-939 regulates human inducible nitric oxide synthase posttranscriptional gene expression in human hepatocytes. Proc Natl Acad Sci USA 109(15):5826–5831
Haasnoot J et al (2007) The Ebola virus VP35 protein is a suppressor of RNA silencing. PLoS Pathog 3(6):e86
Haverkos HW et al (2003) Enteroviruses and type 1 diabetes mellitus. Biomed Pharmacother 57(9):379–385
He J et al (2013) MiR-21 confers resistance against CVB3-induced myocarditis by inhibiting PDCD4-mediated apoptosis. Clin Invest Med 36(2):E103–E111
Hemida MG et al (2013) MicroRNA-203 enhances coxsackievirus B3 replication through targeting zinc finger protein-148. Cell Mol Life Sci 70(2):277–291
Heneghan HM et al (2010) Systemic miRNA-195 differentiates breast cancer from other malignancies and is a potential biomarker for detecting noninvasive and early stage disease. Oncologist 15(7):673–682
Hennessy E et al (2010) Identification of microRNAs with a role in glucose stimulated insulin secretion by expression profiling of MIN6 cells. Biochem Biophys Res Commun 396(2):457–462
Hezova R, Slaby O, Faltejskova P et al (2010) microRNA-342, microRNA-191 and microRNA-510 are differentially expressed in T regulatory cells of type 1 diabetic patients. Cell Immunol 260:70–74
Ho BC et al (2011) Enterovirus-induced miR-141 contributes to shutoff of host protein translation by targeting the translation initiation factor eIF4E. Cell Host Microbe 9(1):58–69
Hober D, Sauter P (2010) Pathogenesis of type 1 diabetes mellitus: interplay between enterovirus and host. Nat Rev Endocrinol 6(5):279–289
Honeyman MC, Coulson BS, Stone NL, Gellert SA, Goldwater PN, Steele CE, Couper JJ, Tait BD, Colman PG, Harrison LC (2000) Association between rotavirus infection and pancreatic islet autoimmunity in children at risk of developing type 1 diabetes. Diabetes 49:1319–1324

Honeyman MC, Stone NL, Harrison LC (1998) T-cell epitopes in type 1 diabetes autoantigen tyrosine phosphatase IA-2: potential for mimicry with rotavirus and other environmental agents. Mol Med 4:231–239

Horn G, Long C, Erlich H (1988) Allelic sequence of the HLA-DQ loci: relationship to serology and to insulin-dependent diabetes susceptibility. Proc Natl Acad Sci USA 85:6012–6016

Hu Z et al (2012) Serum microRNA profiling and breast cancer risk: the use of miR-484/191 as endogenous controls. Carcinogenesis 33(4):828–834

Huang Z et al (2010) Plasma microRNAs are promising novel biomarkers for early detection of colorectal cancer. Int J Cancer 127(1):118–126

Huang L et al (2013) Down-regulation of miR-301a suppresses pro-inflammatory cytokines in Toll-like receptor-triggered macrophages. Immunology 140(3):314–322

Huang S et al (2014) HCV Core Protein-Induced Down-Regulation of microRNA-152 Promoted Aberrant Proliferation by Regulating Wnt1 in HepG2 Cells. PLoS ONE 9(1):e81730

Hui W et al (2013) MicroRNA-195 inhibits the proliferation of human glioma cells by directly targeting cyclin D1 and cyclin E1. PLoS ONE 8(1):e54932

Huntzinger E, Izaurralde E (2011) Gene silencing by microRNAs: contributions of translational repression and mRNA decay. Nat Rev Genet 12(2):99–110

Hyoty H (2002) Enterovirus infections and type 1 diabetes. Ann Med 34(3):138–147

Hyoty H, Reunanen A, Ilonen J, Surcel HM, Rilva A, Kaar ML, Huupponen T, Makela AL et al (1988) Mumps infections in the etiology of type 1 (insulin-dependent) diabetes. Diabetes Res 9:111–116

Hyöty H, Taylor K (2002) The role of viruses in human diabetes. Diabetologia 45:1353–1361

Iliopoulos D, Hirsch HA, Struhl K (2009) An epigenetic switch involving NF-kappaB, Lin28, Let-7 MicroRNA, and IL6 links inflammation to cell transformation. Cell 139(4):693–706

Ivanovska I et al (2008) MicroRNAs in the miR-106b family regulate p21/CDKN1A and promote cell cycle progression. Mol Cell Biol 28(7):2167–2174

Jackson RJ, Hellen CU, Pestova TV (2010) The mechanism of eukaryotic translation initiation and principles of its regulation. Nat Rev Mol Cell Biol 11(2):113–127

Jacovetti C, Abderrahmani A, Parnaud G, Jonas JC, Peyot ML, Cornu M, Laybutt R, Meugnier E, Rome S, Thorens B, Prentki M, Bosco D, Regazzi R (2012) MicroRNAs contribute to compensatory β cell expansion during pregnancy and obesity. J Clin Invest 122(10):3541–3551

Jaidane H, Hober D (2008) Role of coxsackievirus B4 in the pathogenesis of type 1 diabetes. Diabetes Metab 34(6 Pt 1):537–548

Jeong HC et al (2011) Aberrant expression of let-7a miRNA in the blood of non-small cell lung cancer patients. Mol Med Rep 4(2):383–387

Ji J, Wang XW (2009) New kids on the block: diagnostic and prognostic microRNAs in hepatocellular carcinoma. Cancer Biol Ther 8(18):1686–1693

Joglekar MV et al (2007b) MicroRNA profiling of developing and regenerating pancreas reveal post-transcriptional regulation of neurogenin3. Dev Biol 311(2):603–612

Joglekar MV et al (2009b) The miR-30 family microRNAs confer epithelial phenotype to human pancreatic cells. Islets 1(2):137–147

Joglekar MV, Joglekar VM, Hardikar AA (2009a) Expression of islet-specific microRNAs during human pancreatic development. Gene Expr Patterns 9:109–113

Joglekar MV, Parekh VS, Hardikar AA (2011) Islet-specific microRNAs in pancreas development, regeneration and diabetes. Indian J Exp Biol 49(6):401–408

Joglekar MV, Parekh VS, Mehta S, Bhonde RR, Hardikar AA (2007a) MicroRNAprofiling of developing and regenerating pancreas reveal post-transcriptional regulation of neurogenin3. Dev Biol 311:603–612

Jun H-S, Yoon J-W (2003) A new look at viruses in type 1 diabetes. Diabetes Metab Res Rev 19:8–31

Keller A et al (2009) Multiple sclerosis: microRNA expression profiles accurately differentiate patients with relapsing-remitting disease from healthy controls. PLoS ONE 4(10):e7440

Khvorova A, Reynolds A, Jayasena SD (2003) Functional siRNAs and miRNAs exhibit strand bias. Cell 115:209–216

Kim VN (2005) MicroRNA biogenesis: coordinated cropping and dicing. Nat Rev Mol Cell Biol 6(5):376–385

Kim SW et al (2012) MicroRNAs miR-125a and miR-125b constitutively activate the NF-kappaB pathway by targeting the tumor necrosis factor alpha-induced protein 3 (TNFAIP3, A20). Proc Natl Acad Sci USA 109(20):7865–7870

Klein D, Misawa R, Bravo-Egana V, Vargas N, Rosero S et al (2013) MicroRNA expression in alpha and beta cells of human pancreatic islets. PLoS ONE 8(1):e55064. doi:10.1371/journal.pone.0055064

Kloosterman WP et al (2007) Targeted inhibition of miRNA maturation with morpholinos reveals a role for miR-375 in pancreatic islet development. PLoS Biol 5(8):e203

Knip M et al (2005) Environmental triggers and determinants of type 1 diabetes. Diabetes 54(Suppl 2):S125–S136

Kobayashi N et al (2012) Identification of miR-30d as a novel prognostic maker of prostate cancer. Oncotarget 3(11):1455–1471

Krol J, Loedige I, Filipowicz W (2010) The widespread regulation of microRNA biogenesis, function *and decay*. Nat Rev Genet 11:597–610

Kumar P et al (2013) Circulating miRNA biomarkers for Alzheimer's disease. PLoS ONE 8(7):e69807

Laitinen OH et al (2013) Coxsackievirus B1 Is associated with induction of beta-cell autoimmunity that portends type 1 diabetes. Diabetes

Latreille M et al (2015) miR-375 gene dosage in pancreatic beta-cells: implications for regulation of beta-cell mass and biomarker development. J Mol Med (Berl) 93:1159–1169

Lecellier CH et al (2005) A cellular microRNA mediates antiviral defense in human cells. Science 308(5721):557–560

Lee Y, Ahn C, Han J, Choi H, Kim J, Yim J, Lee J, Provost P, Rådmark O, Kim S, Kim VN (2003) The nuclear RNase III Drosha initiates microRNA processing. Nature 425:415–419

Lee Y, Jeon K, Lee JT, Kim S, Kim VN (2002) MicroRNA maturation: stepwise processing and subcellular localization. EMBO J 21:4663–4670

Li Y et al (2009) MicroRNA-34a inhibits glioblastoma growth by targeting multiple oncogenes. Cancer Res 69(19):7569–7576

Li LM et al (2010) Serum microRNA profiles serve as novel biomarkers for HBV infection and diagnosis of HBV-positive hepatocarcinoma. Cancer Res 70(23):9798–9807

Li Y et al (2013) MicroRNA-548 down-regulates host antiviral response via direct targeting of IFN-lambda1. Protein Cell 4(2):130–141

Liu X et al (2010) Systematic identification of microRNA and messenger RNA profiles in hepatitis C virus-infected human hepatoma cells. Virology 398(1):57–67

Liu R et al (2011) A five-microRNA signature identified from genome-wide serum microRNA expression profiling serves as a fingerprint for gastric cancer diagnosis. Eur J Cancer 47(5):784–791

Liu L et al (2012a) miR-143 is downregulated in cervical cancer and promotes apoptosis and inhibits tumor formation by targeting Bcl-2. Mol Med Rep 5(3):753–760

Liu AM et al (2012b) Circulating miR-15b and miR-130b in serum as potential markers for detecting hepatocellular carcinoma: a retrospective cohort study. BMJ Open 2(2):e000825

Liu J et al (2012c) Combination of plasma microRNAs with serum CA19-9 for early detection of pancreatic cancer. Int J Cancer 131(3):683–691

Liu YL et al (2013) MicroRNA-21 and -146b are involved in the pathogenesis of murine viral myocarditis by regulating TH-17 differentiation. Arch Virol 158(9):1953–1963

Lorente-Cebrian S et al (2014) MicroRNAs regulate human adipocyte lipolysis: effects of miR-145 are linked to TNF-alpha. PLoS ONE 9(1):e86800

Lovis P et al (2008) Alterations in microRNA expression contribute to fatty acid-induced pancreatic beta-cell dysfunction. Diabetes 57(10):2728–2736

Lu LF et al (2010) Function of miR-146a in controlling Treg cell-mediated regulation of Th1 responses. Cell 142(6):914–929

Lu Z et al (2011) miR-301a as an NF-kappaB activator in pancreatic cancer cells. EMBO J 30(1):57–67

Lund E, Güttinger S, Calado A, Dahlberg JE, Kutay U (2004) Nuclear export of microRNA precursors. Science 303:95–98

Lynn FC et al (2007b) MicroRNA expression is required for pancreatic islet cell genesis in the mouse. Diabetes 56(12):2938–2945

Lynn FC, Skewes-Cox P, Kosaka Y, McManus MT, Harfe BD, German MS (2007a) MicroRNA expression is required for pancreatic islet cell genesis in the mouse. Diabetes 56:2938–2945

Ma F et al (2011) The microRNA miR-29 controls innate and adaptive immune responses to intracellular bacterial infection by targeting interferon-gamma. Nat Immunol 12(9):861–869

Maroney PA et al (2006) Evidence that microRNAs are associated with translating messenger RNAs in human cells. Nat Struct Mol Biol 13(12):1102–1107

Meder B et al (2011) MicroRNA signatures in total peripheral blood as novel biomarkers for acute myocardial infarction. Basic Res Cardiol 106(1):13–23

Melkman-Zehavi T et al (2011) miRNAs control insulin content in pancreatic beta-cells via downregulation of transcriptional repressors. EMBO J 30(5):835–845

Miersch S et al (2013) Serological autoantibody profiling of type 1 diabetes by protein arrays. J Proteomics 94:486–496

Mitchell PS et al (2008) Circulating microRNAs as stable blood-based markers for cancer detection. Proc Natl Acad Sci USA 105(30):10513–10518

Morimura R et al (2011) Novel diagnostic value of circulating miR-18a in plasma of patients with pancreatic cancer. Br J Cancer 105(11):1733–1740

Morita S et al (2009) Dicer is required for maintaining adult pancreas. PLoS ONE 4(1):e4212

Moya-Suri V et al (2005) Enterovirus RNA sequences in sera of schoolchildren in the general population and their association with type 1-diabetes-associated autoantibodies. J Med Microbiol 54(Pt 9):879–883

Muller S, Imler JL (2007) Dicing with viruses: microRNAs as antiviral factors. Immunity 27(1):1–3

Narendran P, Estella E, Fourlanos S (2005) Immunology of type 1 diabetes. QJM 98(8):547–556

Nathans R et al (2009) Cellular microRNA and P bodies modulate host-HIV-1 interactions. Mol Cell 34(6):696–709

Ng EK et al (2009) Differential expression of microRNAs in plasma of patients with colorectal cancer: a potential marker for colorectal cancer screening. Gut 58(10):1375–1381

Nielsen LB et al (2012) Circulating levels of microRNA from children with newly diagnosed type 1 diabetes and healthy controls: evidence that miR-25 associates to residual beta-cell function and glycaemic control during disease progression. Exp Diabetes Res 2012:896362

Oikarinen M et al (2008) Detection of enteroviruses in the intestine of type 1 diabetic patients. Clin Exp Immunol 151(1):71–75

Oikarinen S, Martiskainen M, Tauriainen S, Huhtala H, Ilonen J, Veijola R, Simell O, Knip M, Hyöty H (2011) Enterovirus RNA in blood is linked to the development of type 1 diabetes. Diabetes 60(1):276–279

Oikarinen S et al (2013) Virus antibody survey in different European populations indicates risk association between coxsackievirus B1 and type 1 diabetes. Diabetes

Othumpangat S et al (2013) Expression of non-structural-1A binding protein in lung epithelial cells is modulated by miRNA-548an on exposure to influenza A virus. Virology 447(1–2):84–94

Otsuka M et al (2007) Hypersusceptibility to vesicular stomatitis virus infection in Dicer1-deficient mice is due to impaired miR24 and miR93 expression. Immunity 27(1):123–134

Pak CY (1988) E.H., McArthur RG, Yoon JW: Association of cytomegalovirus infection with autoimmune type 1 diabetes. Lancet 2:1–4

Patterson CC et al (2009) Incidence trends for childhood type 1 diabetes in Europe during 1989-2003 and predicted new cases 2005–20: a multicentre prospective registration study. Lancet 373(9680):2027–2033

Pedersen IM et al (2007) Interferon modulation of cellular microRNAs as an antiviral mechanism. Nature 449(7164):919–922

Petersen CP et al (2006) Short RNAs repress translation after initiation in mammalian cells. Mol Cell 21(4):533–542

Pfeffer S, Zavolan M, Grasser FA, Chien M, Russo JJ, Ju J, John B, Marks D, Enright AJ, Sander C, Tuschl T (2004) Identification of virus-encoded microRNAs. Science 304:734–736

Pillai RS et al (2005) Inhibition of translational initiation by Let-7 MicroRNA in human cells. Science 309(5740):1573–1576

Plaisance V et al (2006) MicroRNA-9 controls the expression of Granuphilin/Slp4 and the secretory response of insulin-producing cells. J Biol Chem 281(37):26932–26942

Polioudakis D et al (2013) A Myc-microRNA network promotes exit from quiescence by suppressing the interferon response and cell-cycle arrest genes. Nucleic Acids Res 41(4):2239–2254

Potenza N et al (2011) Human microRNA hsa-miR-125a-5p interferes with expression of hepatitis B virus surface antigen. Nucleic Acids Res 39(12):5157–5163

Poy MN et al (2004) A pancreatic islet-specific microRNA regulates insulin secretion. Nature 432(7014):226–230

Poy MN et al (2009) miR-375 maintains normal pancreatic alpha- and beta-cell mass. Proc Natl Acad Sci USA 106(14):5813–5818

Preiss T, Hentze M (2003) Starting the protein synthesis machine: eukaryotic translation initiation. BioEssays 25(12):1201–1211

Pyke DA () *The genetic perspective-putting research into practice.* In DiabetesLarkins RG, Zimmet PZ, Chisholm DJ (eds). Exerpta Medica: Amsterdam, 1989; 1227-1230., 1988

Qi J et al (2013) Circulating microRNAs (cmiRNAs) as novel potential biomarkers for hepatocellular carcinoma. Neoplasma 60(2):135–142

Qu S et al (2012) MicroRNA-330 is an oncogenic factor in glioblastoma cells by regulating SH3GL2 gene. PLoS ONE 7(9):e46010

Rastogi N et al (2013) (6)-Gingerolinduced myeloid leukemia cell death is initiated by reactive oxygen species and activation of miR-27b expression. Free Radic Biol Med 68C:288–301

Redova M et al (2012) Circulating miR-378 and miR-451 in serum are potential biomarkers for renal cell carcinoma. J Transl Med 10:55

Richardson SJ et al (2009) The prevalence of enteroviral capsid protein vp1 immunostaining in pancreatic islets in human type 1 diabetes. Diabetologia 52(6):1143–1151

Roggli E et al (2010) Involvement of microRNAs in the cytotoxic effects exerted by proinflammatory cytokines on pancreatic beta-cells. Diabetes 59(4):978–986

Roggli E et al (2012) Changes in microRNA expression contribute to pancreatic beta-cell dysfunction in prediabetic NOD mice. Diabetes 61(7):1742–1751

Roivainen M, Ylipaasto P, Savolainen C, Galama J, Hovi T, Otonkokski T (2002) Functional impairment and killing of human beta cells by enteroviruses: the capacity is shared by a wide range of serotypes, but the extent is characteristic of individual virus strains. Diabetologia 45:693–702

Rosero S et al (2010) MicroRNA signature of the human developing pancreas. BMC Genom 11:509

Rossato M et al (2012) IL-10-induced microRNA-187 negatively regulates TNF-alpha, IL-6, and IL-12p40 production in TLR4-stimulated monocytes. Proc Natl Acad Sci USA 109(45):E3101–E3110

Ruan Q et al (2011b) The microRNA-21-PDCD4 axis prevents type 1 diabetes by blocking pancreatic beta cell death. Proc Natl Acad Sci USA 108(29):12030–12035

Ruan Q, Wang T, Kameswaran V, Wei Q, Johnson DS, Matschinsky F, Shi W (2011a) Chen YH (2011) The microRNA-21 – PDCD4 axis prevents type 1 diabetes by blocking pancreatic β cell death. PNAS 108(29):10230–10235

Salas-Perez F et al (2013) MicroRNAs miR-21a and miR-93 are down regulated in peripheral blood mononuclear cells (PBMCs) from patients with type 1 diabetes. Immunobiology 218(5):733–737

Sane F et al (2013) Coxsackievirus B4 can infect human pancreas ductal cells and persist in ductal-like cell cultures which results in inhibition of Pdx1 expression and disturbed formation of islet-like cell aggregates. Cell Mol Life Sci 70(21):4169–4180

Santin I, Eizirik DL (2013) Candidate genes for type 1 diabetes modulate pancreatic islet inflammation and beta-cell apoptosis. Diabetes Obes Metab 15(Suppl 3):71–81

Sarma NJ et al (2014) Hepatitis C virus induced changes in miRNA-107 and miRNA-449a modulate CCL2 by targeting IL6 receptor complex in hepatitis. J Virol

Sarmiento L et al (2007) Occurrence of enterovirus RNA in serum of children with newly diagnosed type 1 diabetes and islet cell autoantibody-positive subjects in a population with a low incidence of type 1 diabetes. Autoimmunity 40(7):540–545

Sebastiani G et al (2011a) Increased expression of microRNA miR-326 in type 1 diabetic patients with ongoing islet autoimmunity. Diabetes Metab Res Rev 27(8):862–866

Sebastiani G et al (2012) MicroRNA expression fingerprint in serum of type 1 diabetic patients. Diabetologia 55:S48

Sebastiani G, Vendrame F, Dotta F (2011b) MicroRNAs as new tools for exploring type 1 diabetes: relevance for immunomodulation and transplantation therapy. Transplant Proc 43(1):2

Setyowati D et al (2012) miRNAs and diabetes mellitus. Expert Rev Endocrinol Metab 7(3):281–300

Shen J et al (2011) Plasma microRNAs as potential biomarkers for non-small-cell lung cancer. Lab Invest 91(4):579–587

Shivdasani RA (2006) MicroRNAs: regulators of gene expression and cell differentiation. Blood 108(12):3646–3653

Simion A et al (2010) MiR-495 and miR-218 regulate the expression of the Onecut transcription factors HNF-6 and OC-2. Biochem Biophys Res Commun 391(1):293–298

Smith AL et al (2012) The miR-106b-25 cluster targets Smad7, activates TGF-beta signaling, and induces EMT and tumor initiating cell characteristics downstream of Six1 in human breast cancer. Oncogene 31(50):5162–5171

Smura T et al (2010) Cellular tropism of human enterovirus D species serotypes EV-94, EV-70, and EV-68 in vitro: implications for pathogenesis. J Med Virol 82(11):1940–1949

Smyk D et al (2012) Autoimmunity and environment: am I at risk? Clin Rev Allergy Immunol 42(2):199–212

Sonenberg MRFN (2012) The mechanics of miRNA-mediated gene silencing: a look under the hood of miRISC. Nat Struct Mol Biol 19(6)

Steck AK, Rewers MJ (2011) Genetics of type 1 diabetes. Clin Chem 57(2):176–185

Steiner DF et al (2011) MicroRNA-29 regulates T-box transcription factors and interferon-gamma production in helper T cells. Immunity 35(2):169–181

Storling J, Brorsson CA (2013) Candidate genes expressed in human islets and their role in the pathogenesis of type 1 diabetes. Curr Diab Rep 13(5):633–641

Su C et al (2011) Ectopic expression of microRNA-155 enhances innate antiviral immunity against HBV infection in human hepatoma cells. Virol J 8:354

Sullivan CS, Ganem D (2005) MicroRNAs and viral infection. Mol Cell 20(1):3–7

Sun LL et al (2011) MicroRNA-15a positively regulates insulin synthesis by inhibiting uncoupling protein-2 expression. Diabetes Res Clin Pract 91(1):94–100

Takahashi P et al (2014) MicroRNA expression profiling and functional annotation analysis of their targets in patients with type 1 diabetes mellitus. Gene 539(2):213–223

Talsky KB, Collins K (2010) Initiation by a eukaryotic RNA-dependent RNA polymerase requires looping of the template end and is influenced by the template-tailing activity of an associated uridyltransferase. J Biol Chem 285(36):27614–27623

Tang X et al (2009a) Identification of glucose-regulated miRNAs from pancreatic beta cells reveals a role for miR-30d in insulin transcription. RNA 15(2):287–293

Tang Y et al (2009b) MicroRNA-146A contributes to abnormal activation of the type I interferon pathway in human lupus by targeting the key signaling proteins. Arthritis Rheum 60(4):1065–1075

Tian Y et al (2012) MicroRNA miR-451 downregulates the PI3 K/AKT pathway through CAB39 in human glioma. Int J Oncol 40(4):1105–1112

Tong L et al (2013) MiR-10a* up-regulates coxsackievirus B3 biosynthesis by targeting the 3D-coding sequence. Nucleic Acids Res 41(6):3760–3771

Triboulet R et al (2007) Suppression of microRNA-silencing pathway by HIV-1 during virus replication. Science 315(5818):1579–1582
Tsujiura M et al (2010) Circulating microRNAs in plasma of patients with gastric cancers. Br J Cancer 102(7):1174–1179
van de Bunt M, Gaulton KJ, Parts L, Moran I, Johnson PR, Lindgren CM, Ferrer J, Gloyn AL, McCarthy MI (2013) The miRNA profile of human pancreatic islets and beta-cells and relationship to type 2 diabetes pathogenesis. PLoS ONE 8(1):e55272. doi:10.1371/journal.pone.0055272
van der Werf N et al (2007) Viral infections as potential triggers of type 1 diabetes. Diabetes Metab Res Rev 23(3):169–183
Ventriglia G et al (2015) MicroRNAs: novel players in the dialogue between pancreatic islets and immune system in autoimmune diabetes. Biomed Res Int 2015:749734
WHO (2001) Manual for the virological investigation of polio WHO/EPI/GEN/97.01. World Health Organization, Geneva
Wang FZ et al (2008) Human cytomegalovirus infection alters the expression of cellular microRNA species that affect its replication. J Virol 82(18):9065–9074
Wang X et al (2009) Oncogenic HPV infection interrupts the expression of tumor-suppressive miR-34a through viral oncoprotein E6. RNA 15(4):637–647
Wang GK et al (2010) Circulating microRNA: a novel potential biomarker for early diagnosis of acute myocardial infarction in humans. Eur Heart J 31(6):659–666
Wang R et al (2011) Circulating microRNAs are promising novel biomarkers of acute myocardial infarction. Intern Med 50(17):1789–1795
Wang L et al (2012a) MiR-342-5p suppresses coxsackievirus B3 biosynthesis by targeting the 2C-coding region. Antiviral Res 93(2):270–279
Wang XF et al (2012b) MiR-181d acts as a tumor suppressor in glioma by targeting K-ras and Bcl-2. J Cancer Res Clin Oncol 138(4):573–584
Wang Z et al (2013a) miR-106a is frequently upregulated in gastric cancer and inhibits the extrinsic apoptotic pathway by targeting FAS. Mol Carcinog 52(8):634–646
Wang Y et al (2013b) MicroRNA-7 regulates the mTOR pathway and proliferation in adult pancreatic beta-cells. Diabetes 62(3):887–895
Wen BP et al (2013) MicroRNA-23b inhibits enterovirus 71 replication through downregulation of EV71 VPl protein. Intervirology 56(3):195–200
Winkler C et al (2012) A strategy for combining minor genetic susceptibility genes to improve prediction of disease in type 1 diabetes. Genes Immun 13(7):549–555
Wu W et al (2013) MicroRNA-18a modulates STAT3 activity through negative regulation of PIAS3 during gastric adenocarcinogenesis. Br J Cancer 108(3):653–661
Wu L, Fan J, Belasco JG (2006) MicroRNAs direct rapid deadenylation of mRNA. Proc Natl Acad Sci USA 103(11):4034–4039
Xia H et al (2013) MiR-218 sensitizes glioma cells to apoptosis and inhibits tumorigenicity by regulating ECOP-mediated suppression of NF-kappaB activity. Neuro Oncol 15(4):413–422
Xia JT et al (2014) MicroRNA-362 induces cell proliferation and apoptosis resistance in gastric cancer by activation of NF-kappaB signaling. J Transl Med 12(1):33
Xu X et al (2013) Hepatitis B virus X protein represses miRNA-148a to enhance tumorigenesis. J Clin Invest 123(2):630–645
Yang H et al (2013) MicroRNA-140-5p suppresses tumor growth and metastasis by targeting transforming growth factor beta receptor 1 and fibroblast growth factor 9 in hepatocellular carcinoma. Hepatology 58(1):205–217
Yang M et al (2015) Decreased miR-146 expression in peripheral blood mononuclear cells is correlated with ongoing islet autoimmunity in type 1 diabetes patients 1miR-146. J Diabetes 7(2):158–165
Ye X et al (2013) MiR-126 promotes coxsackievirus replication by mediating cross-talk of ERK1/2 and Wnt/beta-catenin signal pathways. Cell Mol Life Sci 70(23):4631–4644
Yeung W-CG, Rawlinson WD, Craig ME (2011) Enterovirus infection and type 1 diabetes mellitus: systematic review and meta-analysis of observational molecular studies. BMJ 342:d35

Yi R, Qin Y, Macara IG, Cullen BR (2003) Exportin-5 mediates the nuclear export of pre-microRNAs and short hairpin RNAs. Genes Dev 17:3011–3016

Yin H et al (2002) Enterovirus RNA is found in peripheral blood mononuclear cells in a majority of type 1 diabetic children at onset. Diabetes 51(6):1964–1971

Yoon JW, Onodera T, Notkins AL (1978) Virus-induced diabetes mellitus. XV. Beta cell damage and insulin-dependent hyperglycemia in mice infected with coxsackie virus B4. J Exp Med 148(4):1068–1080

Yuan K et al (2012) Role of miR-148a in hepatitis B associated hepatocellular carcinoma. PLoS ONE 7(4):e35331

Yue X et al (2012) MicroRNA-205 functions as a tumor suppressor in human glioblastoma cells by targeting VEGF-A. Oncol Rep 27(4):1200–1206

Zampetaki A et al (2010) Plasma microRNA profiling reveals loss of endothelial miR-126 and other microRNAs in type 2 diabetes. Circ Res 107(6):810–817

Zen K, Zhang CY (2012) Circulating microRNAs: a novel class of biomarkers to diagnose and monitor human cancers. Med Res Rev 32(2):326–348

Zeng Y, Yi R, Cullen BR (2005) Recognition and cleavage of primary microRNA precursors by the nuclear processing enzyme Drosha. EMBO J 24:138–148

Zhang C et al (2010) Global changes of mRNA expression reveals an increased activity of the interferon-induced signal transducer and activator of transcription (STAT) pathway by repression of miR-221/222 in glioblastoma U251 cells. Int J Oncol 36(6):1503–1512

Zhang Q et al (2013) MicroRNAs regulate the pathogenesis of CVB3-induced viral myocarditis. Intervirology 56(2):104–113

Zhao H et al (2010) A pilot study of circulating miRNAs as potential biomarkers of early stage breast cancer. PLoS ONE 5(10):e13735

Zhao X et al (2012a) MicroRNA-30d induces insulin transcription factor MafA and insulin production by targeting mitogen-activated protein 4 kinase 4 (MAP4K4) in pancreatic beta-cells. J Biol Chem 287(37):31155–31164

Zhao J et al (2012b) Downregulation of PCAF by miR-181a/b provides feedback regulation to TNF-alpha-induced transcription of proinflammatory genes in liver epithelial cells. J Immunol 188(3):1266–1274

Zheng Z et al (2013) Human microRNA hsa-miR-296-5p suppresses enterovirus 71 replication by targeting the viral genome. J Virol 87(10):5645–5656

Zheng Y et al (2015) miR-101a and miR-30b contribute to inflammatory cytokine-mediated beta-cell dysfunction. Lab Invest 95:1387–1397

Zhou J et al (2011) Plasma microRNA panel to diagnose hepatitis B virus-related hepatocellular carcinoma. J Clin Oncol 29(36):4781–4788

Zhou L, He H (2008) MicroRNA genes. Ann N Y Acad Sci 1150:72

Zhu W et al (2009) Circulating microRNAs in breast cancer and healthy subjects. BMC Res Notes 2:89

Zouidi F et al (2014) Contribution of PTPN22, CD28, CTLA-4 and ZAP-70 variants to the risk of type 1 diabetes in Tunisians. Gene 533(1):420–426

Chapter 8
Transdifferentiation of Extra-Pancreatic Tissues for Cell Replacement Therapy for Diabetes

Irit Meivar-Levy, Hila Barash and Sarah Ferber

Abbreviations

2 DOG	2 Deoxy glucose
AAV	Adeno-associated virus
Ad	Adenovirus (recombinant, replication deficient)
AFP	Alpha fetoprotein
ARX	Aristaless-related homeobox
CAD	Cyclophosphamide-accelerated diabetes
C/EBP	CCAAT/enhancer binding protein
EGF	Epidermal growth factor
EMT	Epithelial to mesenchymal transition
ESC	Embryonic stem cells
FGAD	First generation recombinant adenovirus
FGF	Fibroblast growth factor
FoxA	Forkhead box A
GATA1	Globin transcription factor 1
iPS	Induced pluripotent stem cells
MafA	V-maf avian musculoaponeurotic fibrosarcoma oncogene homolog
MSX1	Msh homeobox 1

I. Meivar-Levy · H. Barash · S. Ferber (✉)
Sheba Regenerative Medicine, Stem Cells and Tissue Engineering Center,
Sheba Medical Center, 52621 Tel-Hashomer, Israel
e-mail: sferber@sheba.health.gov.il

S. Ferber
Department of Human Genetics and Molecular Medicine,
Sackler School of Medicine, Tel-Aviv University, 69978 Tel Aviv, Israel

A.A. Hardikar (ed.), *Pancreatic Islet Biology*, Stem Cell Biology
and Regenerative Medicine, DOI 10.1007/978-3-319-45307-1_8

MyoD1	Myogenic differentiation 1
NIC	Nicotinamide
NOD	Non-obese diabetic
NeuroD1	Neurogenic differentiation factor 1
Ngn3	Neurogenin 3
Nkx6.1	NKX homeobox 6.1
Pax4	Paired homeobox Pax 4
Pax6	Paired homeobox Pax 6
Pdx1	Pancreatic and duodenum homeobox gene 1
SCID	Severe combined immunodeficiency
STZ	Streptozotocin
T1DM	Type 1 diabetes mellitus
TD	Transdifferentiation
pTFs	Pancreatic transcription factors
TFs	Transcription factors
Vp16	Viral protein 16

What Is Transdifferentiation?

Cells are the ultimate structural unit of an animal or plant body. There are about 210 kinds of cell in the human body (Alberts et al. 2002). Most of the cells are differentiated cells expressing unique set of genes and displaying a specific function and appearance. It has long been thought that animal cells, once committed to a specific lineage, generally, can no longer change their fate and thus are terminally differentiated (Grafi 2004; Slack and Tosh 2001). In the last decade, increasing evidence has accumulated demonstrating the remarkable ability of differentiated cells' ability to convert into a completely different phenotype (Slack and Tosh 2001). This occurs via a process called developmental redirection, nuclear reprogramming or transdifferentiation (Meivar-Levy and Ferber 2003, 2006; Pomerantz and Blau 2004; Slack and Tosh 2001). A robust example of such a process is the induction of pluripotency in mice and human cells, which was awarded with the Nobel Prize (Yamanaka 2008, 2009).

Adult cell reprogramming into committed cells (transdifferentiation, TD) is the direct conversion of adult cells into cells with a different phenotype and function. During transdifferentiation, the pattern of the expressed genetic information is altered, and active genes are being silenced while the expression of many silent genes is activated to produce new proteins and novel cellular function.

Classical gene therapy utilizes ectopic expression of relevant genetic information for a desired function, which persists as long as the ectopic gene is transcribed. By contrast, transdifferentiation utilizes ectopic expression of one or very few genes (transcription factors) to induce alterations in the expression of hundreds of genes, allowing adult cells to gain alternate developmental options and therefore functions (Meivar-Levy and Ferber 2003, 2006). Transdifferentiation and the newly acquired profile of the activated genes persist long after the ectopic gene expression declines (Ber et al. 2003; Sapir et al. 2005; Zhou et al. 2008).

Since the TFs serve as a short-term trigger and the generated TD cells are post-mitotic, many of the TD protocols introduce the transcription factors (TFs) using replication deficient recombinant adenoviruses (Capasso et al. 2014; Meivar-Levy and Ferber 2006, 2008). Recombinant adenoviruses are non-enveloped DNA viruses carrying linear double stranded DNA of about 35 kb. About 50 distinct serotypes of distinguishable viruses were found, but the most commonly used in gene therapy are the serotypes 2 and 5. Recombinant adenoviruses efficiently infect nondividing cells and are considered relatively safe, since they do not integrate into the host genome and are quickly cleared from the cells. By contrast, most (but not all) reprogramming protocols that are used for generating induced pluripotent cells use integrating genetic information for activating the process (Ieda et al. 2010; Vierbuchen et al. 2010; Yamanaka 2008). A role for microRNA ectopic expression in mediating transdifferentiation has been also suggested (Yoo et al. 2011).

The first example of experimentally induced transdifferentiation was the ability of the myogenic factor MyoD to reprogram a variety of primary and differentiated cell lines to a myogenic phenotype (Weintraub et al. 1989). Several other examples have been described, including the conversion of pancreatic exocrine cells to hepatocytes (Shen et al. 2000), B lymphocytes into macrophages (Xie et al. 2004), pancreatic exocrine cells to endocrine cells (Zhou et al. 2008), fibroblasts to neurons (Vierbuchen et al. 2010), to cardiomyocytes (Ieda et al. 2010) and to hepatocytes (Huang et al. 2014). The direct conversion of somatic cells into alternative committed lineages in one step (transdifferentiation), without reverting to pluripotency that may pose a cancer threat to the patient, has opened up tremendous opportunities for regenerative medicine.

The Roles of TFs in Transdifferentiation

One or more tissue-specific TFs need to be overexpressed to achieve adult cell reprogramming. Typically the ones that are most effective are those that are involved in the normal embryonic development of the cell type in question (Berneman-Zeitouni et al. 2014; Ieda et al. 2010; Sapir et al. 2005; Shen et al. 2000; Vierbuchen et al. 2010; Xie et al. 2004; Zhou et al. 2008). Their function is not simply to activate direct target genes, but to shift the cell into a new stable profile of gene expression. An important mechanistic question of transdifferentiation is how target sites on silent genes become bound de novo by ectopic TFs, thereby initiating regulatory events in the chromatin (Iwafuchi-Doi and Zaret 2014; Zaret and Carroll 2011). It was suggested that some but not all of the transcription factors that control organogenesis in the embryo may possess the ability to bind to silent genes or to heterochromatin. These factors also called pioneer transcription factors have the structural capacity of binding to the heterochromatin or the core histone proteins and are characterized by their nucleosome-binding properties that actively help initiate the reorganization of regulatory factors on the DNA (Iwafuchi-Doi and Zaret 2014; Zaret and Carroll 2011).

Developmental and genetic studies have identified many transcription factors that are necessary for the specification of different cell types and that are likely to function as pioneer transcription factors (Iwafuchi-Doi and Zaret 2014; Zaret and Carroll 2011). For example, the transcription factor MyoD, which initiates the myogenic program when introduced into fibroblasts, is one such pioneer transcription factor (Weintraub et al. 1989). Other examples include GATA1, whose ectopic expression in monocytes induces the erythroid and megakaryocyte programs (Nerlov et al. 2000; Visvader et al. 1995) and CCAAT/enhancer binding protein α and β (C/EBPα and C/EBPβ), whose expression in B cells activates the macrophage program (Xie et al. 2004). Perhaps the most dramatic example is the Pax6 homolog, whose ectopic expression in imaginal disc primordia can cause the appearance of additional eyes growing out of a fly wings, legs and antennae (Halder et al. 1995a, b). These examples of transdifferentiation are paradigms for direct reprogramming approaches to create cells of clinical interest, which among other organs include creating liver hepatocytes and pancreatic β cells (Jarikji et al. 2007; Sapir et al. 2005; Shen et al. 2000; Yechoor et al. 2009; Zhou et al. 2008).

Liver to Pancreas Transdifferentiation

Liver is the largest organ in the body with a high level of functional redundancy (Desmet 2001). Unlike β cells, the liver regenerates efficiently, mainly by the proliferation of mature hepatocytes (Thorgeirsson 1996). Human liver cells can be propagated in vitro for months, and the numbers of cells can be expanded substantially ex vivo (Meivar-Levy et al. 2011).

The liver and the pancreas are related developmentally. Both are derived from appendages of the upper primitive foregut endoderm (Fukuda and Kikuchi 2005; Ober et al. 2003). It has been suggested that the late separation of liver and pancreas during organogenesis in the primitive ventral endoderm, might have left both tissues with pluripotent cells that are capable of giving rise to both hepatic and pancreatic lineages (Deutsch et al. 2001). Thus, trans-conversion between the liver and the pancreas is conceivable. In addition, the two organs share many characteristics, including responsiveness to glucose and a large group of specific transcription factors mutually expressed in both tissues (Otsuka et al. 2003). Trans-conversion between pancreatic acinar cells and hepatocytes in both rodents and humans has been reported under experimental, pathological and malignant conditions (Shen et al. 2003).

Comparing the development of liver and pancreas in other species further emphasizes the close relationship between these two organs. In lower organisms, such as worms and eels, there is no spatial separation between the two organs

(Kito et al. 1982; Yang et al. 1999) and a single organ, the "hepato-pancreas," functions as both liver and pancreas (Deutsch et al. 2001).

The Role of Pdx-1 in Transdifferentiation

The transcription factor pancreatic and duodenal homeobox gene 1 (Pdx-1) has a central role in regulating both pancreas organogenesis, as well as adult β cell function. Pdx-1 is involved in regulating the expression of multiple β cell-specific genes and has a key role in pancreatic morphogenesis in mice and humans (Offield et al. 1996; Stoffers et al. 1997).

Pdx-1 was the first pancreatic transcription factor (pTF) analyzed for its role in transdifferentiation of extra-pancreatic tissues (Ferber et al. 2000). Transcription factors with the highest reprogramming activity often have the special ability to engage their target sites on nucleosomal DNA, thus behaving as "pioneer factors" to initiate events in closed chromatin (Iwafuchi-Doi and Zaret 2014). There is no direct proof that Pdx-1 functions as a pioneer factor for pancreas organogenesis and in transdifferentiation of adult extra-pancreatic tissues. However, numerous studies suggest its crucial and obligatory roles in both processes. It is used in most transdifferentiation protocols from extra-pancreatic tissues (Aviv et al. 2009; Berneman-Zeitouni et al. 2014; Cao et al. 2004; Gefen-Halevi et al. 2010; Li et al. 2005; Motoyama et al. 2009; Nakajima-Nagata et al. 2004; Sapir et al. 2005; Yang et al. 2013).

Pdx-1 alone exhibits the capacity to activate the pancreatic lineage both in vivo and in vitro in mice and humans, respectively. It induces robust alterations in the liver, inducing the expression of hundreds of genes, including the activation of many specific pTFs, let alone its own endogenous expression (Ber et al. 2003; Sapir et al. 2005). The newly generated expression persists long after the ectopic gene expression diminishes (Ber et al. 2003). Pdx-1 modulation by VP-16 increases its transdifferentiation capacity, suggesting that Pdx-1 cooperation with other TFs may be synergistic in activating the pancreatic lineage in the liver (Horb et al. 2003; Kaneto et al. 2005b). Its obligatory roles in pancreas organogenesis as well as in liver to pancreas transdifferentiation could be attributed to its potential capacity of cooperating with the pioneer factor FOXA2 that preexists in liver cells (Hoffman et al. 2010; Xu and Zaret 2012), or by its capacity to recruit p300 and other chromatin structure modulators (Francis et al. 2006, 2005).

Liver to Pancreas Transdifferentiation, In Vivo

The first publication describing the ability of Pdx-1 to induce a pancreatic phenotype and function in the liver was performed in mice liver in vivo (Ferber et al. 2000). In that study, transient ectopic expression of Pdx-1 in the liver, delivered by the systemic administration of first-generation E1-deleted recombinant adenovirus (FGAD), induced a wide repertoire of pancreatic gene expression. Surprisingly the short-term expression of ectopic Pdx-1 led to a long-lasting production and secretion of processed and biologically active insulin in the liver (Ber et al. 2003). Pdx-1 induced its own expression in the liver (auto-induction), which, in turn, explains the irreversible nature of the "liver to pancreas" developmental redirection process (Ber et al. 2003).

Hepatic insulin production triggered by Pdx-1 administration using FGAD was functional. Not only did it restore euglycemia in streptozotocin (STZ)-induced diabetic mice (Ferber et al. 2000), but it also prevented STZ-induced hyperglycemia, even eight months after the initial FGAD treatment (Ber et al. 2003). These data indicate the irreversibility of the process and that the surrogate liver insulin-producing cells resist β cell-specific toxins. Moreover, since liver has an important role in neutralizing toxins, in contrast to pancreatic β cells, and has high levels of catalase and superoxide dismutase activities (Desmet 2001), the liver surrogate β cells may be more resistant to cellular assaults. Similar approaches were used by different research groups that confirmed the original observation and increased the understanding and efficiency of the process (Banga et al. 2012; Cim et al. 2012; Imai et al. 2005; Kaneto et al. 2005a, b; Kojima et al. 2003).

In vivo Pdx-1-induced liver to pancreas transdifferentiation by FGAD occurred in less than 1 % of the liver cells that specifically surround the central veins, albeit the initial random expression of Pdx-1 in about 50 % of the cells in liver. Utilizing a different reprogramming approach resulted in SOX9+ cells being the majority of the insulin-positive cells in the liver (Banga et al. 2012), or around the portal vein and liver capsule (Kojima et al. 2003). Such a variety could be caused by the heterogeneity of cells in the liver (Jungermann 1987). Different cell populations in the liver could respond distinctly to the pancreatic transgene as a result of ultrastructural modifications in chromatin compaction, the presence of silencing effects or lack of complementing transcription factors that may work in concert with the ectopic gene (Chakrabarti et al. 2003; Chakrabarti and Mirmira 2003; Hashimshony et al. 2003; Mosley and Ozcan 2003).

Liver to Pancreas Transdifferentiation-In Vitro

During the 15 years since our initial publication, the potential for converting liver cells into pancreas in vitro has been demonstrated by many groups in rodents (Cao et al. 2004; Li et al. 2005; Motoyama et al. 2009; Nakajima-Nagata et al. 2004; Yang et al.) and in human tissues (Sapir et al. 2005; Zalzman et al. 2005, 2003). The capacity of inducing functional transdifferentiation of mature *human* liver cells carries a substantial therapeutic significance, since it may allow autologous cell replacement therapy for diabetes patients. Indeed, our studies demonstrated the potential use of primary cultures of adult human liver cells as pancreatic progenitors (Aviv et al. 2009; Berneman-Zeitouni et al. 2014; Gefen-Halevi et al. 2010; Sapir et al. 2005). Cells isolated from >75 different adult human liver donors were enzymatically dissociated and cultured in conditions which do not preserve the hepatocytes state of differentiation (Meivar-Levy et al. 2011; Sapir et al. 2005). The cells were propagated in vitro for many passages, and the propagated cells could be cryopreserved for >15 years in liquid-N_2. Upon addition of Pdx-1 and soluble factors (EGF, nicotinamide and Exendin-4), up to 50 % of the Pdx-1-expressing cells activated an ectopic insulin promoter. The acquisition of pancreatic β cell-like activity was analyzed by the capacity of the transdifferentiated cells to secrete insulin in a physiologically relevant, glucose-regulated manner. The produced hormone was processed, stored in induced secretory granules and secreted upon glucose challenge (Aviv et al. 2009; Berneman-Zeitouni et al. 2014; Gefen-Halevi et al. 2010; Sapir et al. 2005). Moreover, Pdx-1-treated adult human liver cells expressed numerous pTFs, including the endogenous human Pdx-1(Sapir et al. 2005). The most important clinical function of these cells is their capacity to ameliorate hyperglycemia upon implantation in diabetic immunodeficient mice in vivo. Human c-peptide secretion and the parallel amelioration of diabetes persisted for the whole duration of the experiment (60 days) (Sapir et al. 2005), with no episodes of hypoglycemia even after prolonged fasting (Fig. 8.1).

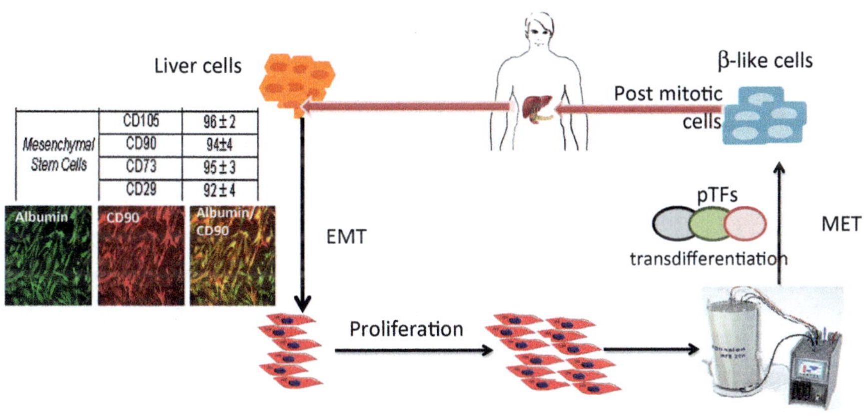

Fig. 8.1 Schematic presentation of generating insulin-producing cells by pTFs-induced liver cells transdifferentiation: Primary culture is generated from 1 to 2 g human liver biopsy. In cell culture, adult human liver cells undergo EMT (Epithelial to Mesenchymal Transition) and efficiently propagate in vitro (Meivar-Levy et al. 2011). The cells are propagated to the final number of 2 billion cells in special bioreactors and the transdifferentiation cocktail (pTFs and SF) is being applied. The insulin producing, transdifferentiated cells can be implanted into the same diabetic patient within 5 days post-pTFs administration

The Role of Dedifferentiation in the Transdifferentiation Process

One of the mechanisms associated with natural regeneration is dedifferentiation, which involves a terminally differentiated cell regression to a less differentiated stage from within its own lineage. This process allows the cell to proliferate before redifferentiating, leading to the replacement of those cells that have been lost (Jopling et al. 2011).

In nonmammalian vertebrates, there are several examples of dedifferentiation that occurs during regeneration; one of the best-characterized examples is limb regeneration in urodeles (salamander) (Fior 2014; Mescher 1996). Upon amputation, the underlying stump cells including muscle, cartilage, bone, tendons, various types of fibroblasts and Schwann cells receive signals that induce dedifferentiation of the cells and the formation of a blastema (Tamura et al. 2010). The regenerative blastema is a mass of proliferating mesenchymal-like cells of mixed origins that are capable of forming all cell types required for the formation of the new limb. The molecular mechanisms of the dedifferentiation process are not fully understood, but involve the suppression of the differentiated functional genes and the activation of genes that characterize earlier developmental stages such as Msx1 (Simon et al. 1995; Yang et al. 2014).

In mammals, dedifferentiation is mainly associated with cancer formation and in deterioration of functional cells due to insult, as in β cell dedifferentiation in diabetes. There are only a few known cases of dedifferentiation which are induced during regeneration that are naturally occurring in mammalian such as during Schwann cells regeneration after nerve injury (Jopling et al. 2011; Mirsky et al. 2008).

Dedifferentiation displays an obligatory role during Pdx-1-induced liver to endocrine pancreas transdifferentiation (Meivar-Levy et al. 2007). Pdx-1, but none of the other pTFs that promote the activation of the pancreatic lineage, suppresses the expression of numerous mature hepatic-specific markers, while activating the immature hepatic marker, AFP (Meivar-Levy et al. 2007). Evidence indicating that Pdx-1 triggers hepatic dedifferentiation by repressing the key hepatic transcription factor CCAAT/enhancer binding protein CEBPβ, have been presented (Meivar-Levy et al. 2007). A similar transcription factor's cross-antagonism, by repressing alternative cell fate options, has been suggested between GATA-1 and PU.1 during myeloid cells fate specification (Arinobu et al. 2007).

Hepatic dedifferentiation is necessary, though not sufficient, for the activation of the pancreatic repertoire in the liver. Inhibition of CEBPβ (without introducing ectopic pTFs) results in hepatic dedifferentiation that is associated with an increased expression of the endocrine pancreatic marker Ngn3, but not by an increase in pancreatic hormone expression (Meivar-Levy et al. 2007). Thus, Pdx-1 plays a dual role in liver to pancreas transdifferentiation by inducing both hepatic dedifferentiation and activating the pancreatic lineage.

Similar observations were reported for reprogramming acinar pancreas and skin cells toward endocrine pancreatic fate (Mauda-Havakuk et al. 2011; Zhou et al. 2008). Dedifferentiation and loss of the host expression repertoire are also suggested in direct reprogramming of fibroblasts to dopaminergic neurons (Kim et al. 2011) and in transdifferentiation between the different blood cells lineages (Graf 2011; Graf and Enver 2009).

Although the mechanisms of the dedifferentiation and the involvement of epigenetic modification are not yet fully understood, it is clear that the cells do not regress into a "stem-like" developmental fate. The cells do not express "stemness" markers and do not exhibit uncontrolled proliferation upon implantation in immune deficient rodents even over extended periods of time (Meivar-Levy et al. 2011).

The identification of dedifferentiation signals may promote the capacity to endow mature tissues in mammals with the plasticity needed for acquiring novel developmental fates and functions to be implemented in the field of regenerative medicine.

Fine Tuning of Liver to Pancreas Transdifferentiation

Additional Pancreatic Transcription Factors Induce Pancreatic Lineage in Liver

Several pTFs have been demonstrated to promote the effect of Pdx-1 on the developmental redirection process improving β cell maturation and amelioration of hyperglycemia. All of these factors are known pTFs that participate in embryonic pancreatic differentiation (Bernardo et al. 2008; Wilson et al. 2003). In most reported studies, the pTFs; NeuroD-1, Nkx6.1, Pax-4, MafA or Ngn-3 were incapable of individually inducing a functional β cell expression profile in the liver although they significantly promote the effect of Pdx-1 in the process (Banga et al. 2012; Berneman-Zeitouni et al. 2014; Gefen-Halevi et al. 2010; Kaneto et al. 2005a, b). In contrast, Kojima et al. reported that NeuroD-1 expression moderately activates the pancreatic repertoire and function (Kojima et al. 2003). This basic helix-loop-helix transcription factor is required for morphogenesis of pancreatic islets and NeuroD-1-deficient mice die from severe diabetes (Naya et al. 1997). Interestingly, ectopic NeuroD-1 expression in the liver induced both downstream and upstream transcription factors, which are part of the pancreatic transcriptional network, including Pdx-1. Transdifferentiation induced by NeuroD-1 differs from that induced by Pdx-1. When liver transdifferentiation as is induced with Pdx-1, glucagon and insulin are produced in distinct cells as occurs in the pancreas (Ber et al. 2003); however, both pancreatic hormones were coproduced within the same cell in the liver of mice treated with NeuroD-1 (Kojima et al. 2003). Moreover, co-expression of Nkx6.1 together with Pdx-1 results in the activation of insulin expression associated with the suppression of glucagon expression (Gefen-Halevi et al. 2010).

The appearance of distinct types of pancreatic hormone-producing cells may suggest that the developmental shift upon Pdx-1 ectopic expression occurs in distinct populations of liver cells (Ber et al. 2003). It is not clear what dictates the distinct characteristics of the cells induced: the individual TFs, different affected host cells in the liver, or possibly both. One of the most important indications of the roles of pTFs in independently inducing true transdifferentiation would be their capacity to repress the host repertoire of genes and in activating a robust irreversible process that does not rely on their continuous ectopic expression.

Small Molecules Promoting Liver to Pancreas Transdifferentiation

The use of ectopic expression of foreign genetic information in regenerative medicine may raise some safety issues. Thus, the ability to reprogram adult extra-pancreatic tissue toward insulin-producing cells using only small molecules or peptides, without the need of ectopic expression of pluripotency reprogramming factors or pTFs, is highly applicable. Indeed, attempts to mimic direct reprogramming using the protocols that promote pancreatic differentiation in embryonic stem cells and induced pluripotent cells (Alipio et al. 2010; D'Amour et al. 2005, 2006) were reported using fibroblasts (Pennarossa et al. 2013; Pereyra-Bonnet et al. 2014).

The ability to reprogram liver or other mature tissue with just small molecules or peptides has not yet been reported; however, a role in promoting TFs-induced liver to pancreas transdifferentiation has been suggested. Early studies explored the promoting effects of growth factors, known to be active in β cells' differentiation (Aviv et al. 2009; Kojima et al. 2003; Sapir et al. 2005; Zalzman et al. 2005). Betacellulin [β cell-stimulating hormone, (Kojima et al. 2003)], EGF (Aviv et al. 2009; Gefen-Halevi et al. 2010; Sapir et al. 2005), Activin-A (Zalzman et al. 2005) and Exendin-4 (Aviv et al. 2009) have been found to have promoting effects on insulin production in reprogrammed liver cells. Nicotinamide is a potent inducer of endocrine differentiation and nowadays is added to most reprogramming protocols (Aviv et al. 2009; Gefen-Halevi et al. 2010; Sapir et al. 2005). Induction of hepatic regeneration using 70 % hepatectomy (Miyatsuka et al. 2003) or by using small molecules for promoting liver regeneration (Banga et al. 2014) was also demonstrated to improve the effect of pancreatic transcription factors in accelerating the transdifferentiation process along the pancreatic lineage. WY14643, also known as pirinixic acid, is an agonist of both peroxisome-proliferator-activated receptors α and γ and is known to cause liver hyperplasia. Co-treatment of diabetic mice with pancreatic transcription factors and WY14643 resulted in insulin positive in the liver that were stable and were able to relieve diabetes in the long term (Banga et al. 2014).

As reprogramming cell fate is an epigenetic process, modifying cellular gene expression patterns would involve the regulation of histone deacetylases (HDACs), histone methyltransferases (HMTs) and DNA methyl transferases (DNMTs) that can modulate genome-wide DNA and histone modifications. In several studies, the reprogramming efficiency and generation of induced pluripotent stem cells could be enhanced using small molecules to inhibit these enzymes (Huangfu et al. 2008; Selvaraj et al. 2010). Chromatin modifying agents were suggested to increase the reprogramming efficiency of the extra-pancreatic tissues toward pancreas. Romidepsin (Romi; an inhibitor of enzymatic activity of histone deacetylases (HDACs)) and 5-Azacytidine (5-AzC, an inhibitor of DNA methyltransferase causing hypomethylation of DNA,) significantly increased the reprogramming of human skin fibroblast into insulin-producing cells (Katz et al. 2013). The use of BIX-01294 (BIX, a diazepin-quinazolin-amine), a histone lysine methyltransferase inhibitor, substantially increased the number of insulin-positive cells in liver (Akinci et al. 2013).

Notch signaling is a highly conserved pathway which plays a central role in pancreas development (Afelik and Jensen 2013; Bar and Efrat 2014); inhibiting the Notch signaling in dedifferentiated β cells following ex vivo expansion has been shown to promote restoration of the β cell phenotype (Bar et al. 2012). Similarly, inhibiting Notch signaling using DAPT, a Notch inhibitor, has a positive impact on pTFs-induced transdifferentiation in liver (Akinci et al. 2013).

Transdifferentiation is a Consecutive and Hierarchical Process

Most transdifferentiation protocols use the concerted expression of several tissue-specific TFs (Banga et al. 2012; Berneman-Zeitouni et al. 2014; Kaneto et al. 2005a, b). To understand the consequences of such an approach, it is essential to analyze whether transdifferentiation is a single-step process or rather is a gradual and sequential process, resembling embryonic organogenesis. Pancreas specification initiates by the homeobox transcription factor Pdx-1, and then the endocrine differentiation is mediated by the basic helix–loop–helix factor Ngn3 (Gradwohl et al. 2000), followed by NeuroD1 (Glick et al. 2000; Huang et al. 2002). The paired homeobox factors Pax-6 and Pax4 and Arx have been implicated as key factors in the segregation of the different endocrine cell types (Brun and Gauthier 2008; Collombat et al. 2003). The final maturation along the β cell lineage and function is attributed to selective expression of MafA in β cells in the adult pancreas (Kataoka et al. 2002). Targeted disruption or temporal mis-expression of pancreatic transcription factors during pancreas organogenesis hampers pancreas development, islet cells' differentiation and function, as well as the segregation between the different endocrine pancreatic lineages (Nishimura et al. 2009).

A recent study suggests that transdifferentiation of human liver cells along the pancreatic lineage is a gradual and consecutive process (Berneman-Zeitouni et al. 2014). Only the sequential ectopic expression of three pTFs led by Pdx-1 in a direct hierarchical manner resulted in increased transdifferentiation of *mature* β-like cells. Transdifferentiation induced by either a concerted pTFs expression or by a sequential nonhierarchical mode resulted in the generation of *immature* pancreatic endocrine progenitors. These progenitors-like-cells were characterized by multi-hormone coproducing cells, impaired insulin processing and ablated glucose-regulated hormone secretion (Berneman-Zeitouni et al. 2014). A hierarchical role of TFs has been suggested in hematopoietic stem cell specification along the different hematopoietic lineages (Iwasaki et al. 2006). However, while stem cell specification has been already suggested to be a stepwise, gradual process (D'Amour et al. 2005; Kroon et al. 2008), this is the first demonstration that transdifferentiation may be also a consecutive process. It will be important to analyze whether a similar sequential and hierarchical process occurs in the recently described neuronal and cardiac transdifferentiation systems (Ambasudhan et al. 2011; Ieda et al. 2010; Pang et al. 2011; Vierbuchen et al. 2010).

Thus, this study suggests a crucial role for temporally and hierarchically controlling the transdifferentiation process in order to optimize the maturation of the newly generated cells along the desired lineages.

Transdifferentiation Crosses the Barrier of Distinct Developmental Layers

The first example of transdifferentiation toward insulin-producing cells was demonstrated in liver cells and later in the exocrine pancreas (Zhou et al. 2008; Zhou and Melton 2008). The success in activating the endocrine pancreatic lineage in liver or exocrine pancreas could be attributed to the close developmental relationships between the transconverting tissues that both originate in the primitive foregut endoderm (Deutsch et al. 2001; Gershengorn et al. 2005; Kahn 2000). However, the current understanding of stem cell biology and induced pluripotency favors the notion that every adult tissue may have the potential to undergo reprogramming to a pluripotency or to alternate committed lineages.

During the past 5 years, many studies indicate the ability to generate functional β-like cells not only from liver but also from ectodermal cells [skin keratinocytes (Mauda-Havakuk et al. 2011), gall bladder (Hickey et al. 2013), neuroendocrine cells in thyroid (Thule et al. 2014)] and from mesodermal cells [bone marrow-derived mesenchymal stromal cells (Karnieli et al. 2007) and skin fibroblasts (Katz et al. 2013)].

In particular, the skin attracts attention as a cell source for regenerative medicine due to its accessibility. The skin is formed from two main layers, the epidermal and dermal-derived tissues. Both keratinocytes and fibroblasts were reported to be induced toward insulin-producing cells by ectopic expression of pancreatic TFs (Katz et al. 2013; Mauda-Havakuk et al. 2011).

Transdifferentiation of Keratinocytes

Pdx-1 treated keratinocytes acquire "endocrine-like" characteristics and functions, including insulin production, processing and secretion in response to elevated glucose concentration (Mauda-Havakuk et al. 2011). Moreover, like in pancreatic β cells and in reprogrammed liver cells (German et al. 1990; Sapir et al. 2005), insulin secretion could only be stimulated by glucose. 2-DOG, the non-metabolizable glucose analog, has no effect on insulin secretion from reprogrammed human keratinocytes. However, although the keratinocytes secreted insulin in a glucose-regulated manner, the concentration of the glucose required for efficient secretion was 25 mM, higher than required for β cells or transdifferentiated liver cells (Mauda-Havakuk et al. 2011) and above the normal physiological range.

The ability of the keratinocytes to directly reprogram into pancreatic β-like cells prove the notion that transdifferentiation can cross the barrier of the different developmental layers. The use of keratinocytes in regenerative medicine per se may be limited, as the proliferative capacity of these cells is limited as is their ability to function as proper β cells.

Transdifferentiation of Fibroblasts

Skin fibroblasts are the major cellular source for iPS cells and for direct reprogramming in regenerative medicine. It was reported that ectopic expression of specific TFs can directly reprogram fibroblast to neurons, cardiomyoblasts, multilineage blood progenitors, hepatocytes and pancreatic β cells (Eguizabal et al. 2013; Nizzardo et al. 2013). Human skin fibroblasts treated with ectopic expression of Pdx-1 with two epigenetic modifying compounds; romidepsin (Romi, a HDACi) and 5-Azacytidine (5-AzC), a cytidine analog that cannot be methylated, results in activation of insulin and glucagon gene expression and reprogramming of the fibroblasts to the endocrine pancreas (Katz et al. 2013). Although, the majority of the cells stain are positive for glucagon rather than insulin (80 % GCG+ vs 2–4 % INS+), insulin secretion is stimulated only by glucose, and not by the glucose analog, 2-DOG. The transplanted reprogrammed fibroblasts, secret human c-peptide and slightly ameliorate hyperglycemia in diabetic mice (Katz et al. 2013).

It is clear that the protocol suggested by Katz et al. (2013) may not be the optimal protocol for generating functional β cells from skin fibroblasts. Additional combinations of pancreatic transcription factors, soluble factors and miRNAs may promote this process in the mesodermal derived tissue. Additional studies activated the pancreatic lineage and function in adult human and pig skin fibroblasts without ectopic transgenes expression. Human fibroblasts were exposed to DNA methyltransferase inhibitor 5-Azacytidine, followed by a three-step protocol for the induction of endocrine pancreatic differentiation that lasted 36 days. At the end of this treatment, 35 ± 8.9 % fibroblasts produced insulin and released the hormone in response to glucose (Pennarossa et al. 2013). This study suggests that it is possible to convert adult fibroblasts into insulin-secreting cells, avoiding both a stable pluripotent stage and any transgenic modification, in a process which resembles the step-wise induction of the pancreatic differentiation in ESC and IPS cells (D'Amour et al. 2005, 2006; Rezania et al. 2013; Schulz et al. 2012).

Are Transdifferentiated Liver Cells Sustainable to Recurrent Autoimmune Attack?

Recurrence of pancreatic autoantibodies after kidney–pancreas transplantation or islet cells transplantation is a disturbing finding. It is estimated that half of the immunological losses of pancreas grafts may be due to autoimmunity. It seems that patients with positive pancreatic autoantibodies, compared to negative patients, could be more likely to present higher HbA1c and lower c-peptide levels (Martins 2014).

It is therefore important to determine whether transdifferentiated cells express diabetogenic antigens and thus are a target for autoimmune attack.

Using a diabetic NOD mouse model, Shternhall-Ron et al. (2007) demonstrated that 43 % of the overtly diabetic CAD-NOD mice treated with Pdx-1 become normoglycemic and maintain stable body weight. The amelioration of hyperglycemia, in Pdx-1 treated diabetic mice is associated with an immune modulation manifested by a Th1 to Th2 immune cytokine shift. Thus, liver to pancreas transdifferentiation ameliorates T1DM in a process that is associated with a concomitant modulation of the autoimmune attack.

However, Pdx-1 therapy alone does not completely eliminate potentially diabetogenic autoimmune T cells detectable upon adoptive transfer in vivo; however, the induction of hyperglycemia is substantially and significantly delayed (Shternhall-Ron et al. 2007). On the other hand, adoptive transfer of activated lymphocytes from diabetic NOD mice into Pdx-1 pretreated SCID mice differentially affected pancreatic β cells compared to transdifferentiated liver cells. While the insulin-producing cells in the liver survived the diabetogenic lymphocytes, the insulin-producing cells in the pancreas were completely destroyed (KS and SF unpublished data).

Under a different transdifferentiation protocol, Tang et al. (2013) suggested that insulin-producing cells generated by transdifferentiation of the liver using ectopic Pdx-1 and Ngn3 expression become a target to autoimmunity. The study demonstrated that the glucose-responsive liver—derived insulin-producing cells were susceptible to autoimmune destruction by diabetogenic splenocytes, as indicated by progressive elevation in blood glucose levels as well as mixed T, and B lymphocytic infiltrates surrounding the cells, 2–3 weeks following the transfer of diabeto-genic splenocytes into NOD/SCID mice (Han et al. 2013; Tang et al. 2013). The difference between their results and ours may be due to their use of a different vector (AAV vs adenovirus), different genes (Pdx-1 and Ngn3) and/or a different model (adoptive transfer of immune cells).

One must consider that the present accumulated data generated in the mouse model of TIDM may not reflect the diabetic state in humans, which should be directly analyzed. This may be performed by injecting human T and B lymphocytes from diabetic patients into SCID/NOD mice preimplanted with human liver transdifferentiated cells.

However, even if there was a consensus regarding the potential immune status of the transdifferentiated insulin-producing cells in the rodent TIDM model, similar studies in man should be replicated. If needed, an immune-modulatory approach, directed at preserving the implanted cells, should be developed.

Translating Extra-pancreatic Transdifferentiation into a Potential Cell Replacement Therapy for Diabetics

It is estimated that in order to replace the β cell mass in a diabetic patient, 10,000 islet equivalents per kilogram of body weight are required (Poradzka et al. 2013; Shapiro 2011). Based on that up to a billion insulin-producing cells are expected to suffice to replace the function of the entire β cell mass in a 75 kg diabetic patient. Insulin-producing cells generated by transdifferentiation are post-mitotic. Therefore, the transdifferentiation protocol should be applied to the final number of desired source cells. Thus, to translate cell transdifferentiation into a cell replacement therapy for diabetes, suitable cell expansion protocols are required and only cells with a high proliferative capacity should be selected. To increase autologous cell replacement therapy safety, cells should be derived from a minimal biopsy from the diabetic patient.

Indeed, we have shown that 1–2 g of liver tissue is sufficient to result in the generation of 2 billion cells within 6–8 weeks (Meivar-Levy et al. 2011). In our procedure, liver tissue is enzymatically dissociated and cells are cultured in conditions that do not preserve the hepatocyte state of differentiation (Meivar-Levy et al. 2011; Sapir et al. 2005). The liver cells undergo epithelial to mesenchymal transition (EMT) and efficiently proliferate in vitro (Meivar-Levy et al. 2011). The expression of many hepatic markers decreases and 90–97 % of the proliferating cells express mesenchymal membrane markers. The proliferating cells display a dedifferentiated hepatic phenotype that is further repressed during pTFs-induced transdifferentiation (Meivar-Levy et al. 2011). Importantly, lineage tracing analyses reveal that insulin-producing cells are indeed, generated in cells that originally expressed albumin (Meivar-Levy et al. 2011).

To be able to generate the desired number of cells needed to control blood glucose levels in adults, cell growth is being adjusted in closed loop bioreactors (Pall Corporation, http://www.pall.com/). These disposable bioreactors are temperature and nutrient-controlled and are suitable to grow about 2 billion of these adherent cells. The transdifferentiation protocol is applied within the same bioreactor. Cells are washed from the excess of virus, and a few days later are collected and implanted after determination of the predesigned release criteria. This approach brings β cell replenishment closer to its clinical implementation (http://www.orgenesis.com).

Summary and Concluding Remarks

Reconstruction of β cell function in extra-pancreatic tissues may constitute a substantial advance toward the development of cell replacement therapy for diabetic patients. The usage of adult extra-pancreatic cells for generating functional insulin-producing tissue may pave the way to autologous implantations, thus allowing the diabetic patient to be the donor of his or her own insulin-producing tissue. This approach may circumvent the shortage in tissue availability, the need for immunosuppressive drugs, and the ethical and safety issues associated with the use of fetal or embryonic stem cells for this purpose.

T1DM is a multi-decade disease, if these cells will not last long enough they could be replenished from the diabetic patient's own biopsy derived cells which could be bio-banked for decades. Could these cells be delivered to the liver, as in the currently used Edmonton protocol for islet cell implantation in humans? Several clinical trials are being presently performed in children with hepatic metabolic disorders (Nussler et al. 2006; Sokal 2014). Insulin-producing cells generated by transdifferentiation produce relatively lower quantities of insulin compared to human islets (Berneman-Zeitouni et al. 2014; Sapir et al. 2005). It is reasonable to assume that such cells upon three-dimensional culture or upon in vivo implantation will further mature to β cell; however, this should be further analyzed. Lastly, whether transdifferentiated cells become a target for autoimmune attack that characterizes type 1 diabetes should be further analyzed in clinical trials, since parallel studies in rodents may have restricted significance.

Acknowledgments We thank the members of our laboratory who contributed to the studies discussed in this review: Keren Shternhall, Helit Cohen, Adi Har-Zahav, Vered Aviv, Alexander Osnis and Moran Fatal, Elad Chernichovsky, Kfir Molakanov and Michal Mauda Havakuk. We also thank Stephan Wolpe for fruitful discussions and for critically reviewing the manuscript. Funding supporting part of the presented studies include JDRF, ISF, D-Cure, BioDisc and EFSD.

References

Afelik S, Jensen J (2013) Notch signaling in the pancreas: patterning and cell fate specification. Wiley Interdiscip Rev Dev Biol 2:531–544

Akinci E, Banga A, Tungatt K, Segal J, Eberhard D, Dutton JR, Slack JM (2013) Reprogramming of various cell types to a beta-like state by Pdx1, Ngn3 and MafA. PLoS One 8:e82424

Alberts B, Johnson A, Lewis J, Raff M, Roberts K, Walter P (2002) Molecular biology of the cell, 4th edn. Garland Science, New York

Alipio Z, Liao W, Roemer EJ, Waner M, Fink LM, Ward DC, Ma Y (2010) Reversal of hyperglycemia in diabetic mouse models using induced-pluripotent stem (iPS)-derived pancreatic beta-like cells. Proc Natl Acad Sci USA 107:13426–13431

Ambasudhan R, Talantova M, Coleman R, Yuan X, Zhu S, Lipton SA, Ding S (2011) Direct reprogramming of adult human fibroblasts to functional neurons under defined conditions. Cell Stem Cell 9:113–118

Arinobu Y, Mizuno S, Chong Y, Shigematsu H, Iino T, Iwasaki H, Graf T, Mayfield R, Chan S, Kastner P et al (2007) Reciprocal activation of GATA-1 and PU.1 marks initial specification of hematopoietic stem cells into myeloerythroid and myelolymphoid lineages. Cell Stem Cell 1:416–427

Aviv V, Meivar-Levy I, Rachmut IH, Rubinek T, Mor E, Ferber S (2009) Exendin-4 promotes liver cell proliferation and enhances PDX-1-induced liver to pancreas transdifferentiation. J Biol Chem 284:33509–33520

Baeyens L, De Breuck S, Lardon J, Mfopou JK, Rooman I, Bouwens L (2005) In vitro generation of insulin-producing beta cells from adult exocrine pancreatic cells. Diabetologia 48:49–57

Banga A, Akinci E, Greder LV, Dutton JR, Slack JM (2012) In vivo reprogramming of Sox9+ cells in the liver to insulin-secreting ducts. Proc Natl Acad Sci USA 109:15336–15341

Banga A, Greder LV, Dutton JR, Slack JM (2014) Stable insulin-secreting ducts formed by reprogramming of cells in the liver using a three-gene cocktail and a PPAR agonist. Gene Ther 21:19–27

Bar Y, Efrat S (2014) The NOTCH pathway in beta-cell growth and differentiation. Vitam Horm 95:391–405

Bar Y, Russ HA, Sintov E, Anker-Kitai L, Knoller S, Efrat S (2012) Redifferentiation of expanded human pancreatic beta-cell-derived cells by inhibition of the NOTCH pathway. J Biol Chem 287:17269–17280

Ber I, Shternhall K, Perl S, Ohanuna Z, Goldberg I, Barshack I, Benvenisti-Zarum L, Meivar-Levy I, Ferber S (2003) Functional, persistent, and extended liver to pancreas transdifferentiation. J Biol Chem 278:31950–31957

Bernardo AS, Hay CW, Docherty K (2008) Pancreatic transcription factors and their role in the birth, life and survival of the pancreatic beta cell. Mol Cell Endocrinol 294:1–9

Berneman-Zeitouni D, Molakandov K, Elgart M, Mor E, Fornoni A, Dominguez MR, Kerr-Conte J, Ott M, Meivar-Levy I, Ferber S (2014) The temporal and hierarchical control of transcription factors-induced liver to pancreas transdifferentiation. PLoS One 9:e87812

Bonner-Weir S, Taneja M, Weir GC, Tatarkiewicz K, Song KH, Sharma A, O'Neil JJ (2000) In vitro cultivation of human islets from expanded ductal tissue. Proc Natl Acad Sci USA 97:7999–8004

Brun T, Gauthier BR (2008) A focus on the role of Pax4 in mature pancreatic islet beta-cell expansion and survival in health and disease. J Mol Endocrinol 40:37–45

Cao LZ, Tang DQ, Horb ME, Li SW, Yang LJ (2004) High glucose is necessary for complete maturation of Pdx1-VP16-expressing hepatic cells into functional insulin-producing cells. Diabetes 53:3168–3178

Capasso C, Garofalo M, Hirvinen M, Cerullo V (2014) The evolution of adenoviral vectors through genetic and chemical surface modifications. Viruses 6:832–855

Chakrabarti SK, Francis J, Ziesmann SM, Garmey JC, Mirmira RG (2003) Covalent histone modifications underlie the developmental regulation of insulin gene transcription in pancreatic beta cells. J Biol Chem 278:23617–23623

Chakrabarti SK, Mirmira RG (2003) Transcription factors direct the development and function of pancreatic b-cells. Trends Endocrinol Metabo TEM 14:78–84

Cim A, Sawyer GJ, Zhang X, Su H, Collins L, Jones P, Antoniou, M, Reynes JP, Lipps HJ, Fabre JW (2012) In vivo studies on non-viral transdifferentiation of liver cells towards pancreatic beta cells. J Endocrinol 214(3):277–288

Collombat P, Mansouri A, Hecksher-Sorensen J, Serup P, Krull J, Gradwohl G, Gruss P (2003) Opposing actions of Arx and Pax4 in endocrine pancreas development. Genes Dev 17:2591–2603

Courtney M, Pfeifer A, Al-Hasani K, Gjernes E, Vieira A, Ben-Othman N, Collombat P (2011) In vivo conversion of adult alpha-cells into beta-like cells: a new research avenue in the context of type 1 diabetes. Diabetes Obes Metab 13:47–52

Cozar-Castellano I, Stewart AF (2005) Molecular engineering human hepatocytes into pancreatic beta cells for diabetes therapy. Proc Natl Acad Sci USA 102:7781–7782

D'Amour KA, Agulnick AD, Eliazer S, Kelly OG, Kroon E, Baetge EE (2005) Efficient differentiation of human embryonic stem cells to definitive endoderm. Nat Biotechnol 23:1534–1541

D'Amour KA, Bang AG, Eliazer S, Kelly OG, Agulnick AD, Smart NG, Moorman MA, Kroon E, Carpenter MK, Baetge EE (2006) Production of pancreatic hormone-expressing endocrine cells from human embryonic stem cells. Nat Biotechnol 24:1392–1401

Desmet VJ (2001) Organization principles. In: Arias IM, Boyer JL, Chisari FV, Fausto N, Schachter D, Shafritz DA (eds) The Liver; biology and pathobiology. Lippincott Williams and Wilkins, Philadelphia, pp 3–15

Deutsch G, Jung J, Zheng M, Lora J, Zaret KS (2001) A bipotential precursor population for pancreas and liver within the embryonic endoderm. Development 128:871–881

Efrat S, Russ HA (2012) Making beta cells from adult tissues. Trends Endocrinol Metab TEM 23:278–285

Eguizabal C, Montserrat N, Veiga A, Izpisua Belmonte JC (2013) Dedifferentiation, transdifferentiation, and reprogramming: future directions in regenerative medicine. Semin Reprod Med 31:82–94

Ferber S, Halkin A, Cohen H, Ber I, Einav Y, Goldberg I, Barshack I, Seijffers R, Kopolovic J, Kaiser N et al (2000) Pancreatic and duodenal homeobox gene 1 induces expression of insulin genes in liver and ameliorates streptozotocin-induced hyperglycemia. Nat Med 6:568–572

Fior J (2014) Salamander regeneration as a model for developing novel regenerative and anticancer therapies. J Cancer 5:715–719

Francis J, Babu DA, Deering TG, Chakrabarti SK, Garmey JC, Evans-Molina C, Taylor DG, Mirmira RG (2006) Role of chromatin accessibility in the occupancy and transcription of the insulin gene by the pancreatic transcription factor Pdx-1. Mol Endocrinol 10:10

Francis J, Chakrabarti SK, Garmey JC, Mirmira RG (2005) Pdx-1 links histone H3-Lys-4 methylation to RNA polymerase II elongation during activation of insulin transcription. J Biol Chem 280:36244–36253

Fukuda K, Kikuchi Y (2005) Endoderm development in vertebrates: fate mapping, induction and regional specification. Dev Growth Differ 47:343–355

Gefen-Halevi S, Rachmut IH, Molakandov K, Berneman D, Mor E, Meivar-Levy I, Ferber S (2010) NKX6.1 promotes PDX-1-induced liver to pancreatic beta-cells reprogramming. Cell Reprogram 12:655–664

German MS, Moss LG, Rutter WJ (1990) Regulation of insulin gene expression by glucose and calcium in transfected primary islet cultures. J Biol Chem 265:22063–22066

Gershengorn MC, Geras-Raaka E, Hardikar AA, Raaka BM (2005) Are better islet cell precursors generated by epithelial-to-mesenchymal transition? Cell Cycle 4:380–382

Glick E, Leshkowitz D, Walker MD (2000) Transcription factor BETA2 acts cooperatively with E2A and PDX1 to activate the insulin gene promoter. J Biol Chem 275:2199–2204

Gradwohl G, Dierich A, LeMeur M, Guillemot F (2000) neurogenin3 is required for the development of the four endocrine cell lineages of the pancreas. Proc Natl Acad Sci USA 97:1607–1611

Graf T (2011) Historical origins of transdifferentiation and reprogramming. Cell Stem Cell 9:504–516

Graf T, Enver T (2009) Forcing cells to change lineages. Nature 462:587–594

Grafi G (2004) How cells dedifferentiate: a lesson from plants. Dev Biol 268:1–6

Halder G, Callaerts P, Gehring WJ (1995a) Induction of ectopic eyes by targeted expression of the eyeless gene in Drosophila. Science (New York, NY) 267:1788–1792

Halder G, Callaerts P, Gehring WJ (1995b) New perspectives on eye evolution. Curr Opin Genet Dev 5:602–609

Han S, Donelan W, Wang H, Reeves W, Yang LJ (2013) Novel autoantigens in type 1 diabetes. Am J Transl Res 5:379–392

Hashimshony T, Zhang J, Keshet I, Bustin M, Cedar H (2003) The role of DNA methylation in setting up chromatin structure during development. Nat Genet 34:187–192

Hickey RD, Galivo F, Schug J, Brehm MA, Haft A, Wang Y, Benedetti E, Gu G, Magnuson MA, Shultz LD et al (2013) Generation of islet-like cells from mouse gall bladder by direct ex vivo reprogramming. Stem Cell Res 11:503–515

Hoffman BG, Robertson G, Zavaglia B, Beach M, Cullum R, Lee S, Soukhatcheva G, Li L, Wederell ED, Thiessen N et al (2010) Locus co-occupancy, nucleosome positioning, and H3K4me1 regulate the functionality of FOXA2-, HNF4A-, and PDX1-bound loci in islets and liver. Genome Res 20:1037–1051

Horb ME, Shen CN, Tosh D, Slack JM (2003) Experimental conversion of liver to pancreas. Curr Biol 13:105–115

Huang HP, Chu K, Nemoz-Gaillard E, Elberg D, Tsai MJ (2002) Neogenesis of beta-cells in adult BETA2/NeuroD-deficient mice. Mol Endocrinol 16:541–551

Huang P, Zhang L, Gao Y, He Z, Yao D, Wu Z, Cen J, Chen X, Liu C, Hu Y et al (2014) Direct reprogramming of human fibroblasts to functional and expandable hepatocytes. Cell Stem Cell 14:370–384

Huangfu D, Maehr R, Guo W, Eijkelenboom A, Snitow M, Chen AE, Melton DA (2008) Induction of pluripotent stem cells by defined factors is greatly improved by small-molecule compounds. Nat Biotechnol 26:795–797

Ieda M, Fu JD, Delgado-Olguin P, Vedatham V, Hayashi Y, Bruneau BG, Sirvastava D (2010) Direct reprogramming of fibroblasts into functional cardiomyocytes by defined factors. Cell 142:375–386

Imai J, Katagiri H, Yamada T, Ishigaki Y, Ogihara T, Uno K, Hasegawa Y, Gao J, Ishihara H, Sasano H et al (2005) Constitutively active PDX1 induced efficient insulin production in adult murine liver. Biochem Biophys Res Commun 326:402–409

Iwafuchi-Doi M, Zaret KS (2014) Pioneer transcription factors in cell reprogramming. Genes Dev 28:2679–2692

Iwasaki H, Mizuno S, Arinobu Y, Ozawa H, Mori Y, Shigematsu H, Takatsu K, Tenen DG, Akashi K (2006) The order of expression of transcription factors directs hierarchical specification of hematopoietic lineages. Genes Dev 20:3010–3021

Jarikji ZH, Vanamala S, Beck CW, Wright CV, Leach SD, Horb ME (2007) Differential ability of Ptf1a and Ptf1a-VP16 to convert stomach, duodenum and liver to pancreas. Dev Biol 304:786–799

Jopling C, Boue S, Izpisua Belmonte JC (2011) Dedifferentiation, transdifferentiation and reprogramming: three routes to regeneration. Nat Rev Mol Cell Biol 12:79–89

Jungermann K (1987) Metabolic zonation of liver parenchyma: significance for the regulation of glycogen metabolism, gluconeogenesis, and glycolysis. Diabetes Metab Rev 3:269–293

Kahn A (2000) Converting hepatocytes to beta-cells—a new approach for diabetes? Nat Med 6:505–506

Kaneto H, Matsuoka TA, Nakatani Y, Miyatsuka T, Matsuhisa M, Hori M, Yamasaki Y (2005a) A crucial role of MafA as a novel therapeutic target for diabetes. J Biol Chem 280:15047–15052

Kaneto H, Nakatani Y, Miyatsuka T, Matsuoka TA, Matsuhisa M, Hori M, Yamasaki Y (2005b) PDX-1/VP16 fusion protein, together with NeuroD or Ngn3, markedly induces insulin gene transcription and ameliorates glucose tolerance. Diabetes 54:1009–1022

Karnieli O, Izhar-Prato Y, Bulvik S, Efrat S (2007) Generation of insulin-producing cells from human bone marrow mesenchymal stem cells by genetic manipulation. Stem Cells (Dayton, Ohio) 5:5

Kataoka K, Han SI, Shioda S, Hirai M, Nishizawa M, Handa H (2002) MafA is a glucose-regulated and pancreatic beta-cell-specific transcriptional activator for the insulin gene. J Biol Chem 277:49903–49910

Katz LS, Geras-Raaka E, Gershengorn MC (2013) Reprogramming adult human dermal fibroblasts to islet-like cells by epigenetic modification coupled to transcription factor modulation. Stem Cells Dev 22:2551–2560

Kim J, Su SC, Wang H, Cheng AW, Cassady JP, Lodato MA, Lengner CJ, Chung CY, Dawlaty MM, Tsai LH et al (2011) Functional integration of dopaminergic neurons directly converted from mouse fibroblasts. Cell Stem Cell 9:413–419

Kito H, Ose Y, Mizuhira V, Sato T, Ishikawa T, Tazawa T (1982) Separation and purification of (Cd, Cu, Zn)-metallothionein in carp hepato-pancreas. Comp Biochem Physiol C 73:121–127

Kojima H, Fujimiya M, Matsumura K, Younan P, Imaeda H, Maeda M, Chan L (2003) NeuroD-betacellulin gene therapy induces islet neogenesis in the liver and reverses diabetes in mice. Nat Med 9:596–603

Kroon E, Martinson LA, Kadoya K, Bang AG, Kelly OG, Eliazer S, Young H, Richardson M, Smart NG, Cunningham J et al (2008) Pancreatic endoderm derived from human embryonic stem cells generates glucose-responsive insulin-secreting cells in vivo. Nat Biotechnol 26:443–452

Li WC, Horb ME, Tosh D, Slack JM (2005) In vitro transdifferentiation of hepatoma cells into functional pancreatic cells. Mech Dev 122:835–847

Martins LS (2014) Autoimmune diabetes recurrence should be routinely monitored after pancreas transplantation. World J Transpl 4:183–187

Mauda-Havakuk M, Litichever N, Chernichovski E, Nakar O, Winkler E, Mazkereth R, Orenstein A, Bar-Meir E, Ravassard P, Meivar-Levy I et al (2011) Ectopic PDX-1 expression directly reprograms human keratinocytes along pancreatic insulin-producing cells fate. PLoS One 6:e26298

Meivar-Levy I, Ferber S (2003) New organs from our own tissues: liver-to-pancreas transdifferentiation. Trends Endocrinol Metab TEM 14:460–466

Meivar-Levy I, Ferber S (2006) Regenerative medicine: using liver to generate pancreas for treating diabetes. Isr Med Assoc J 8:430–434

Meivar-Levy I, Ferber S (2008) Use of extra-pancreatic tissues for cell replacement therapy for diabetes. In: Seino S, Bell G (eds) Pancreatic beta cell in health and disease. Springer, Tokyo, pp 285–312

Meivar-Levy I, Sapir T, Berneman D, Weissbach T, Polak-Charcon S, Ravassard P, Tzakis AG, Mor E, Ricordi C, Ferber S (2011) Human liver cells expressing albumin and mesenchymal characteristics give rise to insulin-producing cells. Int J Transplant 2011. Article ID 252387

Meivar-Levy I, Sapir T, Gefen-Halevi S, Aviv V, Barshack I, Onaca N, Mor E, Ferber S (2007) Pancreatic and duodenal homeobox gene 1 induces hepatic dedifferentiation by suppressing the expression of CCAAT/enhancer-binding protein beta. Hepatology 46:898–905

Mescher AL (1996) The cellular basis of limb regeneration in urodeles. Int J Dev Biol 40:785–795

Mirsky R, Woodhoo A, Parkinson DB, Arthur-Farraj P, Bhaskaran A, Jessen KR (2008) Novel signals controlling embryonic Schwann cell development, myelination and dedifferentiation. J Peripher Nerv Syst JPNS 13:122–135

Miyatsuka T, Kaneto H, Kajimoto Y, Hirota S, Arakawa Y, Fujitani Y, Umayahara Y, Watada H, Yamasaki Y, Magnuson MA et al (2003) Ectopically expressed PDX-1 in liver initiates endocrine and exocrine pancreas differentiation but causes dysmorphogenesis. Biochem Biophys Res Commun 310:1017–1025

Mosley AL, Ozcan S (2003) Glucose regulates insulin gene transcription by hyperacetylation of histone h4. J Biol Chem 278:19660–19666

Motoyama H, Ogawa S, Kubo A, Miwa S, Nakayama J, Tagawa Y, Miyagawa S (2009) In vitro reprogramming of adult hepatocytes into insulin-producing cells without viral vectors. Biochem Biophys Res Commun 385:123–128

Nakajima-Nagata N, Sakurai T, Mitaka T, Katakai T, Yamato E, Miyazaki J, Tabata Y, Sugai M, Shimizu A (2004) In vitro induction of adult hepatic progenitor cells into insulin-producing cells. Biochem Biophys Res Commun 318:625–630

Naya FJ, Huang HP, Qiu Y, Mutoh H, DeMayo FJ, Leiter AB, Tasi MJ (1997) Diabetes, defective pancreatic morphogenisis, and abnormal enteroendocrine differentiation in BETA2/NeuroD-deficient mice. Genes Dev 11:2323–2334

Nerlov C, Querfurth E, Kulessa H, Graf T (2000) GATA-1 interacts with the myeloid PU.1 transcription factor and represses PU.1-dependent transcription. Blood 95:2543–2551

Nishimura W, Bonner-Weir S, Sharma A (2009) Expression of MafA in pancreatic progenitors is detrimental for pancreatic development. Dev Biol 333:108–120

Nizzardo M, Simone C, Falcone M, Riboldi G, Comi GP, Bresolin N, Corti S (2013) Direct reprogramming of adult somatic cells into other lineages: past evidence and future perspectives. Cell Transplant 22:921–944

Nussler A, Konig S, Ott M, Sokal E, Christ B, Thasler W, Brulport M, Gabelein G, Schormann W, Schulze M et al (2006) Present status and perspectives of cell-based therapies for liver diseases. J Hepatol 45:144–159

Ober EA, Field HA, Stainier DY (2003) From endoderm formation to liver and pancreas development in zebrafish. Mech Dev 120:5–18

Offield MF, Jetton TL, Labosky PA, Ray M, Stein RW, Magnuson MA, Hogan BL, Wright CV (1996) PDX-1 is required for pancreatic outgrowth and differentiation of the rostral duodenum. Development 122:983–995

Otsuka M, Hoshida Y, Kato N, Moriyama M, Taniguchi H, Arai M, Mori M, Seki N, Omata M (2003) Liver chip and gene shaving. J Gastroenterol 38:89–92

Pang ZP, Yang N, Vierbuchen T, Ostermeier A, Fuentes DR, Yang TQ, Citri A, Sebastiano V, Marro S, Sudhof TC et al (2011) Induction of human neuronal cells by defined transcription factors. Nature 476:220–223

Pennarossa G, Maffei S, Campagnol M, Tarantini L, Gandolfi F, Brevini TA (2013) Brief demethylation step allows the conversion of adult human skin fibroblasts into insulin-secreting cells. Proc Natl Acad Sci USA 110:8948–8953

Pereyra-Bonnet F, Gimeno ML, Argumedo NR, Ielpi M, Cardozo JA, Giménez CA, Hyon SH, Balzaretti M, Loreti M, Fainstein-Day P et al (2014) Skin fibroblasts from patients with type 1 diabetes (T1D) can be chemically transdifferentiated into insulin-expressing clusters: a transgene-free approach. PloS One 9(6):e100369

Pomerantz J, Blau HM (2004) Nuclear reprogramming: a key to stem cell function in regenerative medicine. Nat Cell Biol 6:810–816

Poradzka A, Wronski J, Jasik M, Karnafel W, Fiedor P (2013) Insulin replacement therapy in patients with type 1 diabetes by isolated pancreatic islet transplantation. Acta Pol Pharm 70:943–950

Rezania A, Bruin JE, Arora P, Rubin A, Batushansky I, Asadi A, O'Dwyer S, Quiskamp N, Mojibian M, Albrecht T et al (2014) Reversal of diabetes with insulin-producing cells derived in vitro from human pluripotent stem cells. Nat Biotechnol 32:1121–1133

Rezania A, Bruin JE, Xu J, Narayan K, Fox JK, O'Neil JJ, Kieffer TJ (2013) Enrichment of human embryonic stem cell-derived NKX6.1-expressing pancreatic progenitor cells accelerates the maturation of insulin-secreting cells in vivo. Stem Cells (Dayton, Ohio) 31:2432–2442

Ricordi C, Inverardi L, Dominguez-Bendala J (2012) From cellular therapies to tissue reprogramming and regenerative strategies in the treatment of diabetes. Regen Med 7:41–48

Sapir T, Shternhall K, Meivar-Levy I, Blumenfeld T, Cohen H, Skutelsky E, Eventov-Friedman S, Barshack I, Goldberg I, Pri-Chen S et al (2005) From the cover: cell-replacement therapy for diabetes: generating functional insulin-producing tissue from adult human liver cells. Proc Natl Acad Sci USA 102:7964–7969

Schulz TC, Young HY, Agulnick AD, Babin MJ, Baetge EE, Bang AG, Bhoumik A, Cepa I, Cesario RM, Haakmeester C et al (2012) A scalable system for production of functional pancreatic progenitors from human embryonic stem cells. PloS One 7:e37004

Selvaraj V, Plane JM, Williams AJ, Deng W (2010) Switching cell fate: the remarkable rise of iPS cells and lineage reprogramming technologies. Trends Biotechnol 28:214–223

Shapiro AM (2011) Strategies toward single-donor islets of Langerhans transplantation. Curr Opin Organ Transplant 16:627–631

Shen CN, Horb ME, Slack JM, Tosh D (2003) Transdifferentiation of pancreas to liver. Mech Dev 120:107–116

Shen CN, Slack JM, Tosh D (2000) Molecular basis of transdifferentiation of pancreas to liver. Nat Cell Biol 2:879–888

Shternhall-Ron K, Quintana FJ, Perl S, Meivar-Levy I, Barshack I, Cohen IR, Ferber S (2007) Ectopic PDX-1 expression in liver ameliorates type 1 diabetes. J Autoimmun 28:134–142

Simon HG, Nelson C, Goff D, Laufer E, Morgan BA, Tabin C (1995) Differential expression of myogenic regulatory genes and Msx-1 during dedifferentiation and redifferentiation of regenerating amphibian limbs. Dev Dyn 202:1–12

Slack JM, Tosh D (2001) Transdifferentiation and metaplasia—switching cell types. Curr Opin Genet Dev 11:581–586

Sokal EM (2014) Treating inborn errors of liver metabolism with stem cells: current clinical development. J Inherit Metab Dis 37:535–539

Stoffers DA, Thomas MK, Habener JF (1997) The homeodomain protein IDX-1. Trends Endocrinol Metab 8:145–151

Tamura K, Ohgo S, Yokoyama H (2010) Limb blastema cell: a stem cell for morphological regeneration. Dev Growth Differ 52:89–99

Tang DQ, Shun L, Koya V, Sun Y, Wang Q, Wang H, Li SW, Sun Y, Purich DL, Zhang C et al (2013) Genetically reprogrammed, liver-derived insulin-producing cells are glucose-responsive, but susceptible to autoimmune destruction in settings of murine model of type 1 diabetes. Am J Transl Res 5:184–199

Thorgeirsson SS (1996) Hepatic stem cells in liver regeneration. Faseb J 10:1249–1256

Thule PM, Jia D, Safley S, Gordon K, Barber G, Yi H, Nalli S, Onderci M, Sharma J, Shires J et al (2014) Engineered insulin secretion from neuroendocrine cells isolated from human thyroid. World J Surg 38:1251–1261

Vierbuchen T, Ostermeier A, Pang ZP, Kokubu Y, Sudhof TC, Wernig M (2010) Direct conversion of fibroblasts to functional neurons by defined factors. Nature 463:1035–1041

Visvader JE, Crossley M, Hill J, Orkin SH, Adams JM (1995) The C-terminal zinc finger of GATA-1 or GATA-2 is sufficient to induce megakaryocytic differentiation of an early myeloid cell line. Mole Cell Biol 15:634–641

Weintraub H, Tapscott SJ, Davis RL, Thayer MJ, Adam MA, Lassar AB, Miller AD (1989) Activation of muscle-specific genes in pigment, nerve, fat, liver, and fibroblast cell lines by forced expression of MyoD. Proc Natl Acad Sci USA 86:5434–5438

Wilson ME, Scheel D, German MS (2003) Gene expression cascades in pancreatic development. Mech Dev 120:65–80

Xie H, Ye M, Feng R, Graf T (2004) Stepwise reprogramming of B cells into macrophages. Cell 117:663–676

Xu CR, Zaret KS (2012) Chromatin "pre-pattern" and epigenetic modulation in the cell fate choice of liver over pancreas in the endoderm. Nucleus (Austin, Tex) 3:150–154

Yamanaka S (2008) Induction of pluripotent stem cells from mouse fibroblasts by four transcription factors. Cell Prolif 41:51–56

Yamanaka S (2009) Elite and stochastic models for induced pluripotent stem cell generation. Nature 460:49–52

Yang H, Morrison CM, Conlon JM, Laybolt K, Wright JR Jr (1999) Immunocytochemical characterization of the pancreatic islet cells of the Nile Tilapia (Oreochromis niloticus). Gen Comp Endocrinol 114:47–56

Yang Y, Akinci E, Dutton JR, Banga A, Slack JM (2013) Stage specific reprogramming of mouse embryo liver cells to a beta cell-like phenotype. Mech Dev 130(11–12):602–612

Yang Z, Liu Q, Mannix RJ, Xu X, Li H, Ma Z, Ingber DE, Allen PD, Wang Y (2014) Mononuclear cells from dedifferentiation of mouse myotubes display remarkable regenerative capability. Stem Cells (Dayton, Ohio) 32:2492–2501

Yechoor V, Liu V, Espiritu C, Paul A, Oka K, Kojima H, Chan L (2009) Neurogenin3 is sufficient for transdetermination of hepatic progenitor cells into neo-islets in vivo but not transdifferentiation of hepatocytes. Dev Cell 16:358–373

Yoo AS, Sun AX, Li L, Shcheglovitov A, Portmann T, Li Y, Lee-Messer C, Dolmetsch RE, Tsien RW, Crabtree GR (2011) MicroRNA-mediated conversion of human fibroblasts to neurons. Nature 476:228–231

Zalzman M, Anker-Kitai L, Efrat S (2005) Differentiation of human liver-derived, insulin-producing cells toward the {beta}-cell phenotype. Diabetes 54:2568–2575

Zalzman M, Gupta S, Giri RK, Berkovich I, Sappal BS, Karnieli O, Zern MA, Fleischer N, Efrat S (2003) Reversal of hyperglycemia in mice by using human expandable insulin-producing cells differentiated from fetal liver progenitor cells. Proc Natl Acad Sci USA 100:7253–7258

Zaret KS (2008) Genetic programming of liver and pancreas progenitors: lessons for stem-cell differentiation. Nat Rev Genet 9:329–340

Zaret KS, Carroll JS (2011) Pioneer transcription factors: establishing competence for gene expression. Genes Dev 25:2227–2241

Zhou Q, Brown J, Kanarek A, Rajagopal J, Melton DA (2008) In vivo reprogramming of adult pancreatic exocrine cells to beta-cells. Nature 455:627–632

Zhou Q, Melton DA (2008) Extreme makeover: converting one cell into another. Cell Stem Cell 3:382–388

Chapter 9
Generation of Human Islet Progenitor Cells via Epithelial-to-Mesenchymal Transition

Wilson Wong, Anandwardhan A. Hardikar and Mugdha V. Joglekar

Introduction

Epithelial to mesenchymal transition (EMT) has been shown to occur during generation of human islet-derived progenitor cells (hIPCs), which have been demonstrated to retain potential to differentiate into insulin-producing cells. EMT is a biological process where epithelial cells go through a phenotypic change to transition into mesenchymal cells. EMT is reported to form the basis of three distinct physiological and pathological processes: i) embryo formation/implantation, ii) tissue repair and iii) carcinoma/metastasis. Human islets undergo EMT when exposed to growth-promoting conditions in vitro. Here, we provide an overview of EMT, generation of hIPCs and other stem cells with this phenomenon, the debate surrounding the origin of lineage-committed progenitor cells and finally the role of microRNAs in regulating EMT in human islet cells.

Alternatives to Pancreatic Islet β-Cells

Diabetes is characterized by hyperglycaemia either due to beta cell loss (type 1 diabetes) or peripheral tissue resistance (type 2 diabetes). Pancreatic islet transplantation is the currently practiced cell-based therapy for diabetes; however, it has

W. Wong · A.A. Hardikar · M.V. Joglekar (✉)
Diabetes and Islet Biology Group, NHMRC Clinical Trials Centre, Faculty of Medicine, The University of Sydney, Level 6, Medical Foundation Building, 92-94 Parramatta Road, Camperdown, NSW 2050, Australia
e-mail: mugdha.joglekar@ctc.usyd.edu.au
URL: http://www.isletbiology.info/

A.A. Hardikar (ed.), *Pancreatic Islet Biology*, Stem Cell Biology and Regenerative Medicine, DOI 10.1007/978-3-319-45307-1_9

its own limitations including scarcity of good quality islets for transplantation and graft failure, due to fewer numbers of insulin-producing cells, immune rejection and graft survival in (new) host environment. The two major roadblocks to successful outcomes of islet transplantation therapy are (1) sufficient numbers of insulin-producing and glucose-responsive cells and (2) efficient ways to immune-isolate such cells for transplantation. This chapter discusses efforts towards addressing the first road-block towards achieving success in cell replacement therapy for diabetes. It is therefore important to look for alternate sources of insulin-producing cells. We demonstrated for the first time that human islet-derived progenitor cells (hIPCs) are generated by epithelial-to-mesenchymal transition of pancreatic beta cells; a process regulated by a specific family (miR30 family) of islet-enriched non-coding RNAs (Gershengorn et al. 2004, 2005; Joglekar and Hardikar 2012; Joglekar et al. 2009a, b). Other stem cells derived from the embryonic (Lumelsky et al. 2001; Assady et al. 2001; Jiang et al. 2007; Bernardo et al. 2009; Inada et al. 2006), induced pluripotent (iPSC) (Takahashi and Yamanaka 2006; Takahashi et al. 2007; Aoi et al. 2008; Nakagawa et al. 2008; Okita et al. 2007; Wong et al. 2014; Wang et al. 2014; Lahmy et al. 2014), bone-marrow (Ianus et al. 2003; Hess et al. 2003; Phadnis et al. 2011; Lechner et al. 2004; Jafarian et al. 2014), exocrine pancreas (Seaberg et al. 2004) or ducts (Bonner-Weir et al. 2000), epithelial cell types (Zalzman et al. 2005) and trans-differentiation from different cell types (Meivar-Levy and Ferber 2003; Ber et al. 2003; van der Meulen and Huising 2015; Ye et al. 2015; Sapir et al. 2005; Meivar-Levy and Ferber 2006), have also been used in research to generate human pancreatic beta-like cells. It has been suggested (Gershengorn et al. 2005) that embryonic stem (ES) cells are the most difficult to differentiate into mature pancreatic β-cells, since they have to pass through multiple stages of development including endoderm commitment, foregut/pancreatic epithelium development prior to maturing into insulin-producing cells. These are followed by bone-marrow-derived stem cells and trans-differentiation of other epithelial cell types (see chapter by Ferber et al. in this book). Human islet-derived progenitor cells (hIPCs) are recognized to be better precursors (Gershengorn et al. 2005), since they originate from human islets and thereby retain the chromatin conformation (epigenetic memory) that is necessary for efficient insulin production.

Recently, two independent groups demonstrated differentiation of embryonic stem cells (ESCs) into insulin-producing cells (Pagliuca et al. 2014; Rezania et al. 2014). Both of them involve multi-stage differentiation protocols, and resultant cells are not completely equivalent of mature beta cells. However, both the groups were able to demonstrate long-term reversal of diabetes in vivo. Use of ESCs still suffers from ethical constraints, teratoma formations and immune rejection. Nonetheless, current research in this area is promising.

Yamanaka's reprogramming approach involves forced expression of key transcription factors (initially-Oct3/4, Sox2, c-Myc and Klf4) allowing to convert fibroblasts/somatic cells to ES-like cells, known as iPSCs (Takahashi and Yamanaka 2006; Takahashi et al. 2007; Aoi et al. 2008; Nakagawa et al. 2008; Okita et al. 2007; Wong et al. 2014). iPSCs have functional similarities to ESCs

(Takahashi et al. 2007); however, much like ES cells, they are difficult to efficiently differentiate into islet β-cells; unless accompanied by genetic manipulation (Wong et al. 2014). Wang et al. (2014) demonstrated differentiation of iPSCs into insulin-producing β-cells through over-expression of three pancreatic transcription factors-pdx-1, NeruoD1 and MafA using adenoviral infection. Another study has suggested miR-375, which is abundantly expressed in human islets, could also be a prominent target for studies involving human iPSC differentiation into pancreatic islet β-cells (Lahmy et al. 2014). A study by Bar-Nur et al. (2011) utilized human islet cells to generate lineage-committed iPSCs that have epigenetic memory of insulin production and a better potential for differentiation into insulin-producing β-cells. This study along with findings in hIPCs presented by our group and the group of Professor Shimon Efrat emphasizes the importance of using lineage-committed cell types for further differentiation studies.

Human bone-marrow-derived adult stem cells (BMCs) have been reported by several groups, mainly because of their potential for autologous transplants (Ianus et al. 2003; Hess et al. 2003; Phadnis et al. 2011; Weissman 2000). However, the actual differentiation of these BMCs into insulin-producing beta-like cells is controversial. In a study by Hess et al. (2003), adult bone-marrow-derived cells were transplanted in mice with streptozotocin (STZ)-induced hyperglycaemia. Adult BMCs were present in STZ-damaged pancreatic tissues; however, it was shown that these BMCs induced endogenous pancreatic replication thereby restoring the insulin production. In another study, almost negligible proportion of beta cells were found to be differentiated from the donor BMCs during pancreatic endocrine regeneration in mice (Lechner et al. 2004). Overall, BMCs are considered to have more of a supportive role in pancreatic tissue repair rather than actual trans-differentiation. On the other hand, in vivo differentiation of BMCs into glucose-responsive pancreatic islet cells was demonstrated using a lineage tracing Cre-LoxP system (Ianus et al. 2003). We have observed that human BMCs-derived islet-like cell aggregates mature into functional insulin-producing cell clusters in vivo, through paracrine factors secreted during pancreas regeneration (Phadnis et al. 2011), suggesting that identification of specific factors would aid in improving the efficiency of cellular differentiation in BMCs. Furthermore, a recent study has developed a stepwise protocol to enhance the yield of β-cells through human BMCs in vitro (Jafarian et al. 2014). It is, however, important to note several stages of differentiation along with the requirement of growth/differentiation factors were needed for relatively less successful β-cells generation by BMCs.

Trans-differentiation is a process that involves significant changes mainly at the epigenetic level, leading to a mature cell type to switch into a different cell type (Slack and Tosh 2001). Studies have attempted to use liver cells, acinar cells or α-cells to trans-differentiate into β-cells types (Meivar-Levy and Ferber 2003; Ber et al. 2003; van der Meulen and Huising 2015; Ye et al. 2015; Sapir et al. 2005; Meivar-Levy and Ferber 2006). Certain pancreatic transcription factors (such as pdx-1) and adenovirus recombination are used to achieve trans-differentiation for turning a non-islet cell type to an islet-like or β-cell type. A study by Sapir et al. (2005), studied adenovirus-mediated pdx-1 over-expression with soluble factors during trans-differentiation of adult human liver to functional β-like insulin-producing cells. However, the major problem with these gene modulation approaches using adenovirus/lentivirus is that these cannot be translated to humans with their potential to persist and induce tumours in humans (Inada et al. 2006).

We have extensively researched the differentiation potential of hIPCs for past several years. When human islets are exposed to growth-promoting conditions in vitro, cells within the islets migrate out and start to proliferate. The islet epithelial cells undergo EMT and generate highly proliferative population of mesenchymal-like cells; the human islet-derived progenitor cells. We have demonstrated using seven different techniques that human beta cells also proliferate and contribute in generation of hIPCs that retain an "open" conformation at the insulin promoter region even after hundreds of cellular replication events in vitro (Joglekar et al. 2009a). As these cells from the islet retain epigenetic memory; they have better potential to differentiate back to insulin-producing islet-like cell aggregates even after 1000-fold expansion in vitro (Joglekar et al. 2009a; Wong et al. 2014; Russ et al. 2008; Davani et al. 2009; Efrat 2008; Joglekar and Hardikar 2010; Russ et al. 2009).

What is EMT?

Epithelial to mesenchymal transition (EMT) is a biological phenomenon where epithelial cells change their phenotype to become more mesenchymal-like cells. This process involves complex cellular reprogramming that includes loss of cell–cell interactions and epithelial (apico-basal) polarity, acquisition of mesenchymal proteins and enhanced migratory and invasive properties (Fig. 1) (Kalluri and Weinberg 2009; Nieto and Cano 2012). EMT is an important mechanism in embryonic development, tissue repair/wound healing and also in pathological conditions such as in cancer and organ fibrosis (Kalluri and Weinberg 2009; Nieto and Cano 2012; Gonzalez and Medici 2014; Cieslik et al. 2013). To understand the mechanism on how a cell passages through EMT, certain key molecules/proteins have been identified and are now outlined in Fig. 9.1 (Kalluri and Weinberg 2009).

Multiple distinct molecular processes are involved in order to initiate and complete EMT. These include transcription factor activation, alteration of specific gene expression, dysregulated micro(mi)RNA expression, changes in cell-surface protein expression, cytoskeletal protein reorganization as well as production of extracellular matrix (ECM)-degrading enzymes. Various signalling pathways are also activated during this complex phenomenon of EMT. Transforming growth factor-β (TGF-β) signalling is one of the first pathways, which triggers other downstream processes leading to EMT (Arumugam et al. 2009; Steinestel et al. 2014). TGF-β binds to its cell-surface receptors (type I-III), activates downstream molecular complexes (SMAD, PI3K-AKT, mTORC1, Rho-like GPTases) that collectively lead to transcription of EMT genes such as Snail, Twist (bHLH-basic helix-loop-helix factors) and Zinc finger E-box binding (ZEB) family members (Steinestel

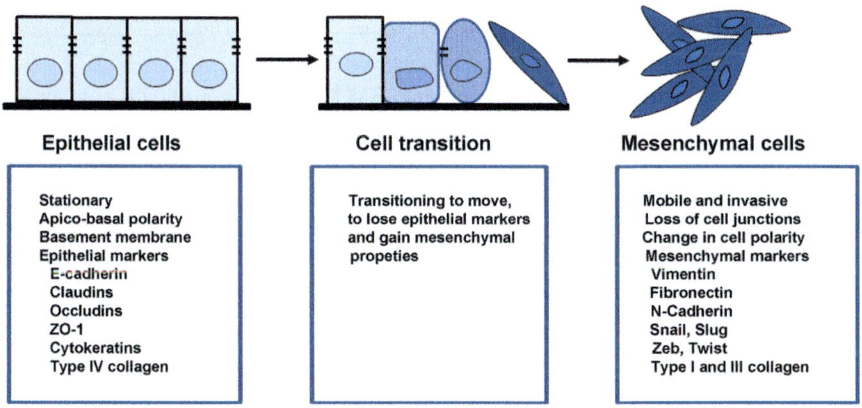

Fig. 9.1 EMT. Epithelial to mesenchymal transition is schematically presented here. Epithelial cells initially exhibit polarity, gap junctions and different epithelial proteins. During the transition phase, they lose their epithelial characteristics and gain more mesenchymal features that include mobility, invasiveness and a range of mesenchymal proteins and transcription factors. Figure adapted from Kalluri and Weinberg (2009), Tiwari et al. (2012), Gonzalez and Medici (2014), Lamouille et al. (2014)

et al. 2014; Shi and Massague 2003; Lamouille et al. 2014). These transcription factors (Snail, Twist and ZEB) repress epithelial genes such as claudins, occludins, E(epithelial)-cadherin and Zona Occludens 1(ZO-1), and simultaneously activate genes associated with the mesenchymal phenotype such as fibronectin, vitronectin, N-cadherin and matrix metalloproteinases (MMPs) (Lamouille et al. 2014). Other signalling molecules downstream of TGF-β increase cell size, cell junction dissolution, induce cytoskeletal changes as well as contribute to enhanced motility and extracellular matrix invasion (Gonzalez and Medici 2014; Steinestel et al. 2014; Lamouille et al. 2012, 2014; Lamouille and Derynck 2007).

Receptors of tyrosine kinases (RTKs) activation is another signalling pathway, which is observed in the progression of EMT. RTKs activate a major downstream complex known as RAS-RAF-MEK-ERK MAPK signalling and functions similar to TGF-β signalling (Gonzalez and Medici 2014; Lamouille et al. 2014). Wnt, Notch and Hedgehog (HH) signalling are other extracellular pathways, which are also known to be activated during EMT. Wnt signalling leads to translocation of β-catenin to the nucleus, which then promotes mesenchymal gene expression (Lamouille et al. 2014). The Notch signalling pathway controls the expression of Snail2 (slug) to promote EMT (Leong et al. 2007). Hedgehog signalling activates the family of Gli transcription factors that induce Snail1 expression (Gonzalez and Medici 2014; Lamouille et al. 2014; Li et al. 2006). Both Notch and HH signalling are known to decrease E-cadherin expression during EMT (Leong et al. 2007; Li et al. 2006, 2007).

The phenomenon of epithelial-to-mesenchymal transition is observed during embryonic development, tissue repair/wound healing and cancer progression/metastasis (Kalluri and Weinberg 2009; Tiwari et al. 2012; Zeisberg and Neilson 2009).

EMT in Embryonic Development

In early stages of embryonic development, EMT plays a significant role in the implantation of the embryo and initiation of placenta formation in the primitive endoderm to form the parietal endoderm (Kalluri and Weinberg 2009; Vicovac and Aplin 1996; Lim and Thiery 2012). The reverse process of EMT, which is mesenchymal to epithelial transition (MET), is typically seen during embryogenesis (Kalluri and Weinberg 2009). In the later stages, gastrulation, germ layer separation and formation of multiple tissues are associated with EMT and MET (Lim and Thiery 2012; Acloque et al. 2009; Hay 1990; Thiery and Sleeman 2006).

EMT in Tissue Repair and Fibrosis

EMT is the major mechanism in differentiated adult tissues to facilitate tissue reconstruction following trauma and inflammatory injuries (Zeisberg and Neilson 2009; Okada et al. 1997; Kalluri and Neilson 2003; Lopez-Novoa and Nieto 2009). There are four prominent steps during any wound healing process; (1) haemostasis/coagulation, (2) inflammation, (3) proliferation and (4) maturation (Velnar et al. 2009). Haemostasis/coagulation is an immediate physiological response to stop the bleeding by clot formation (Stadelmann et al. 1998). During inflammation, wound is populated by neutrophils and macrophages, which prevent further tissue damage, while removing necrotic tissue, bacteria and other cell debris (Eming et al. 2007). The proliferation phase includes formation of granulation tissue, angiogenesis, new ECM disposition and re-epithelialization. EMT-like process takes place during the re-epithelialization stage, where the localized epithelial cells migrate to the wound edges (Nakamura and Tokura 2011; Weber et al. 2012). TGF-β signalling in EMT acts like a chemotactic factor for fibroblasts, secreting collagen to strengthen the healing wound (Weber et al. 2012). Final phase of maturation of wound healing involves contraction and remodelling process of the wound (Velnar et al. 2009). A number of organs such as the kidney, liver, lung, heart, skin and intestine have been reported to exhibit EMT during their development of fibrosis in chronic inflammation (Nakamura and Tokura 2011; Zeisberg et al. 2007a, b; Rastaldi et al. 2002; Kim et al. 2006; Flier et al. 2010).

EMT in Cancer

EMT has been observed in epithelial cancerous cells, which can de-differentiate and gain mesenchymal phenotype to dislodge from the origin of cancer, migrating to form secondary tumours in other tissues/sites in the host (Garg 2013). EMT in cancer has been shown to have similar phenotypic properties to EMT during development and wound healing, including the expression of primitive markers of EMT (Zeisberg and Neilson 2009; Lopez-Novoa and Nieto 2009; Garg 2013). During cancer, certain epithelial cells undergo unexpected genetic/epigenetic changes usually with specific genes that favour clonal outgrowth, leading to a localized tumour development (Kalluri and Weinberg 2009). These cells later on exhibit EMT and acquire mesenchymal characteristics such as expression of vimentin, α-SMA or FSP1, which mark them to invade and metastasize thus leading to cancer progression (Kalluri and Weinberg 2009; Zeisberg and Neilson 2009; Garg 2013; Lan et al. 2013; Lih et al. 2013; Katz et al. 2011; Fischer et al. 2007).

EMT in hIPCs

While EMT plays a central role in embryo development, tissue repair and in cancer/metastasis, we demonstrated that EMT is also responsible for generating human islet-derived progenitor cells (hIPCs) (Gershengorn et al. 2004; Joglekar et al. 2009a; Joglekar and Hardikar 2010). Human cadaveric islet transplantation is the only established cell-based therapy for treating diabetes; however, scarcity of the number of donors and low yield of islets after isolation has always been an issue (Sahu et al. 2009). Research to find alternative sources for potential insulin-producing cells has led to promising approaches to generate these important β-like cells.

The concept of EMT in human islets was first introduced in 2004 (Gershengorn et al. 2004). Later on, different studies were carried out to generate, expand and characterize hIPCs (Gershengorn et al. 2004; Joglekar and Hardikar 2012; Lechner et al. 2005; Ouziel-Yahalom et al. 2006). We first demonstrated clonal human pancreatic cancer cells (PANC-1) can transition into hormone-expressing islet-like cell aggregates (ICAs) (Hardikar et al. 2003). This observation was later on extended to human islets, and we found that after exposing adult human islets to a specific set of growth factors, islet cells migrate out, proliferate and generate mesenchymal-like cells in vitro (hIPCs). These mesenchymal-like cells exhibit properties of progenitor cells in terms of high proliferative potential as well as capacity to easily differentiate back to insulin-producing cells. Further studies confirmed that de-differentiation and expansion of adult and foetal human islets into hIPCs and re-differentiation of hIPCs back to a β-like cell phenotype occurs via EMT and MET, respectively (Joglekar et al. 2009a, b; Davani et al. 2009; Russ et al. 2009). There are multiple evidences of EMT during generation of hIPCs: visible cellular phenotypic changes, increased expression of mesenchymal proteins and transcription factors (Vimentin, SMA, Snail1 and 2, Zeb, etc.), simultaneous reduction in epithelial markers (E-cadherin, Claudins, etc.) as well as translocation of β-catenin from cell membrane to the nucleus. We also observe that hIPCs are derived directly from transition of islet cells to mesenchymal phenotype in the presence of serum or certain defined growth factors (such as EGF); while removal of serum (serum-free medium-SFM) after brief exposure to trypsin leads to reverse-EMT (MET), generating islet-like cell aggregates that eventually mature into pro-insulin-expressing endocrine cells. Further studies aimed at dissecting molecular signals that induce cellular proliferation are important to identify molecular events leading to generation of lineage-committed islet progenitor cells. This marks the plasticity of differentiated adult endocrine cells, which can be a potential source for generating β-cells in vitro (Gershengorn et al. 2004; Joglekar and Hardikar 2010; Yamada and Kojima 2005; Gallo et al. 2007).

Lineage tracing studies by our group and others have demonstrated that adult and foetal human β-cells proliferate in vitro and also undergo EMT (Joglekar et al. 2009a, b; Russ et al. 2008; Davani et al. 2007, 2009; 2009). We used multiple approaches to confirm human β-cell proliferation and mesenchymal transition. Pro-insulin transcript has a longer half-life than insulin protein. Using TaqMan-based real time PCR, we observed that proliferating populations of hIPCs (up to passage 10, >40 days in vitro) contain pro-insulin transcript along with several mesenchymal transcripts and proteins. Combined florescence in situ hybridization (FISH) and immunocytochemistry (ICC) on proliferating islets indicated the presence of pro-insulin transcript tagged with fluorescent probe in the cells containing vimentin and smooth muscle actin proteins. We also observed Ki67 (a protein expressed in cells preparing for or undergoing cellular proliferation) and insulin double immunopositive cells in these cultures. Single cell PCR is a powerful technique to identify multiple transcripts co-expressed in the same cell. Using this method, optimized in our laboratory (Joglekar et al. 2010), we determined that pro-insulin transcript-containing cells are present in the expanding populations of hIPCs and they also start co-expressing different mesenchymal transcripts. We then used a DNA-analogue-based lineage tracing system developed by Teta et al. (2007) to understand beta cell proliferation in vitro. Using 2 different thymidine analogues (CldU and IdU) for two different pulse periods, we demonstrated that insulin-containing cells incorporate either one or both the analogues in a period of 7 days. This unbiased approach confirms that the human beta cells proliferate in vitro (Joglekar et al. 2009a). Finally, we assessed the insulin promoter region in human islets and hIPCs at different passages. We found that the histone marks indicating active promoter region are retained in the proliferating hIPCs, thus confirming contribution of β-cells in generating hIPCs. Another group also demonstrated this epigenetically active insulin promoter region in hIPCs (Mutskov et al. 2007). Professor Shimon Efrat's group infected human islets with two lentiviruses encoding RIP-Cre (cre recombinase driven by insulin promoter) and a reporter cassette (CMV promoter-loxP-DsRed2-loxP-eGFP), respectively. This genetic labelling yielded all infected beta cells with eGFP tag and all other infected cells with DsRed2 tag. Labelled cells were then easily followed, and it was observed that human beta cells proliferate in vitro (Russ et al. 2008). This group later on also confirmed that human beta cells, labelled with the dual lentiviral system mentioned above, undergo EMT (Russ et al. 2009).

Another study by Ouziel-Yahalom et al. (2006) developed in vitro culturing conditions suitable for the proliferation/expansion of adult cadaveric human islet β-cells following de-differentiation. In comparison with previous studies

(Gershengorn et al. 2004) using intact cadaveric human islets, their study cultured single cells suspension from the dissociated human islets. These single cell cultures, however, had a much slower rate of replication and proliferation/expansion than the intact islets. The differences between replication rates, re-differentiation potential and insulin content in re-differentiated cells were considered to be mainly due to different ways of initial culturing of islets (intact versus single cells). Nonetheless, this study (Ouziel-Yahalom et al. 2006) also reported the presence of mesenchymal markers in beta cells in vitro, thus supporting the phenomenon of EMT, as a preferred default mechanism for generation of islet progenitor cells.

Furthermore, hIPCs have been suggested to be better precursors generated by EMT than other stem/precursor cells (Gershengorn et al. 2005). This case has been suggested since hIPCs are initially derived from β-cells, they would inherit epigenetic marks which define an active insulin promoter region, and thus retain memory to know how to transcribe and produce insulin. Later studies have confirmed that insulin promoter region in hIPCs retains open chromatin conformation after expansion/proliferation in vitro (Joglekar et al. 2009a; Mutskov et al. 2007). hIPCs transition to epithelial islet-like cell aggregates in serum-free conditions following MET, and these aggregates have potential to mature into insulin-producing cells (Gershengorn et al. 2004; Davani et al. 2009; Ouziel-Yahalom et al. 2006).

EMT in Islet β-Cells of Mouse and Human Origin

Despite supporting evidence of EMT in pancreatic islet β-cells in humans, there were a number of studies which showed that this concept is not true in mouse islets (Chase et al. 2007; Atouf et al. 2007; Morton et al. 2007; Weinberg et al. 2007). These studies used transgenic mice having genetically labelled β-cells (with cre/loxP) for lineage tracing analysis and to evaluate if adult mouse pancreatic β-cells undergo EMT. In vitro culture of β-cells from these transgenic mice demonstrated that labelled cells are not present after expansion/proliferation in vitro. This suggests that mouse β-cells do not proliferate (Chase et al. 2007; Atouf et al. 2007; Morton et al. 2007; Weinberg et al. 2007), even though they could de-differentiate as shown in some of them (Morton et al. 2007; Weinberg et al. 2007). It was proposed that the mesenchymal or fibroblast-like cells present within their cultures are not originated or derived from β-cells, but instead are generated from the expansion of original, contaminating mesenchymal cells in the culture.

The origin of hIPCs from human islet β-cells has remained controversial. There are studies that have opposed the concept of EMT in human pancreatic islet β-cells (Parnaud et al. 2008; Kayali et al. 2007). One of them examined purified human and adult rat islet β-cells isolated using a Newport Green dye, and grown under similar culture conditions and found that the human β-cells did not proliferate, however, in contrast the rat β-cells did (Parnaud et al. 2008). However, in this study, a zinc-binding Newport Green dye was used that would label only the primary insulin-positive cells but not the proliferated, insulin protein-negative cells which were originally derived from β-cells. Study by Kayali et al. (2007), which attempted to reproduce cell re-differentiation from de-differentiated human islet β-cells in vitro using SFM, was unable to consistently recover β-cell phenotypes. Thus, they refuted the concept of proliferation and EMT of human β-cells and their contribution to generation of hIPCs. A later work by the group of Dr. Marvin Gershengorn expressed questions in regard to their initial suggestion of generation of hIPCs by EMT of β-cells (Davani et al. 2007). This later work argued that hIPCs can be a specific type of pancreas-derived mesenchymal stromal cells (MSCs) capable of differentiating into mesodermal cell types such as adipocytes, chondrocytes and osteocytes in vitro. Other studies have also showed the presence of MSCs in human islet preparations (Gallo et al. 2007; Carlotti et al. 2010). The criteria of defining a MSC have been described (Dominici et al. 2006), where a MSCs must have the following (1) plastic adherence; (2) express CD105, CD73 and CD90 and lack expression of CD45, CD34, CD14 or CD11b, CD79α or CD19 and HLA-DR surface molecules; (3) differentiate to osteoblasts, adipocytes and chondroblasts in vitro. A study by Sordi et al. (2010) has suggested that the hIPCs in the human pancreatic islet cultures originate from non-β-cells including the contaminating pancreatic or extra-pancreatic MSCs from bone-marrow. However, the contribution of bone-marrow MSCs was conflicted by another study that analysed 31 human pancreata obtained from bone-marrow transplant recipients who had received transplant from a donor of opposite sex (Butler et al. 2007). This study found no conversion of any bone-marrow mesenchymal cells into insulin-producing cells.

Our DNA-analogue-based lineage tracing (Joglekar et al. 2009a) and Cre/loxP-based cell-lineage tracing from the group of Shimon Efrat (Russ et al. 2008, 2009) have clearly and undoubtedly demonstrated that the hIPCs are derived from human islet β-cells through EMT. hIPCs express MSC-associated markers, but did not differentiate into all mesodermal cell types. This suggests EMT does not induce multi-potency in cells derived from primary human β-cells. In addition, a more recent work by Russ et al. (2011) demonstrated that the expanded de-differentiated β-cells can be induced to re-differentiate in culture by a combination of soluble factors in SFM in vitro.

In our hands, we have observed that the re-differentiation potential of hIPCs is highly dependent on the original islet cells, their purity, the presence of non-islet fractions, number of passages in vitro as well as the protocol and SFM culture conditions employed for re-differentiation. Even though hIPCs generated by culturing intact islets contain cells derived from β-cells, the rate at which the derivatives of β-cells and non-β-cells replicate is different that eventually dilutes out the cells of β-cell origin in the entire populations of hIPCs. We think this is the reason for observing fewer active epigenetic marks at the insulin promoter region at later passages of hIPCs as compared to those at earlier passages (Joglekar et al. 2009a). It may be interesting to assess the insulin promoter region in the populations generated following the method of Ouziel-Yahalom et al. (2006).

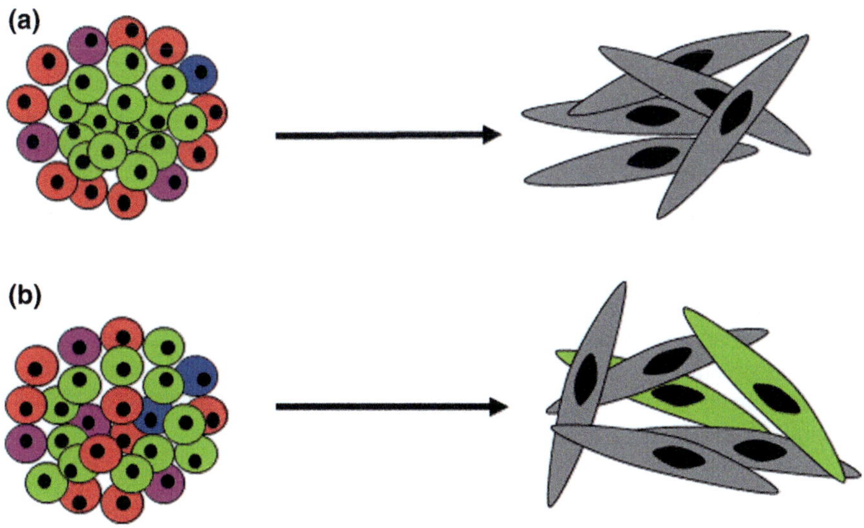

Fig. 9.2 Difference in mouse and human islet culture. Apart from the architectural differences seen in mouse (*top panel*) and human (*bottom panel*) islets, they also behave differently when cultured in vitro. The mesenchymal-like cells originating in the mouse islet culture have non-β-cell origin. Whereas the mesenchymal cells in the human islet cultures are derived from β-cells by the process of EMT, however, contribution of other non-β-cells in these cultures cannot be refuted

Comprehensively, these findings show that β-cells from different species have a different capacity to proliferate. While mouse β-cells do not appear to proliferate/have considerable rare proliferation potential (Russ et al. 2008), human β-cells proliferate and undergo EMT in vitro (Joglekar et al. 2009a, b; Russ et al. 2008; Joglekar and Hardikar 2010) (Fig. 9.2). This highlights another crucial difference between islets of mice and men and underscores the importance of using human islets for research rather than extrapolating findings from mouse islets to human studies.

EMT and Other Stem Cells

Properties of the cells generated through EMT have been found to be similar to the stem cells (Mani et al. 2008; Radisky and LaBarge 2008). A study by Mani et al. (2008), which examined immortalized human mammary epithelial cell (HMLEs), reported that through EMT HMLEs acquire mesenchymal traits and also expressed stem cell properties. They further induced EMT in HMLEs by exposing them to TGF-β1 or activating the Snail or Twist and observed that it resulted in generation of CD44^{high}/CD24^{low} cells with mammosphere-forming ability. Studies have shown that CD44^{high}/CD24^{low} is a phenotype associated with breast cancer progenitors/stem cells (Al-Hajj et al. 2003; Sleeman et al. 2006; Yu et al. 2009) that exhibit enhanced stem cell features such as mammosphere-forming ability and self-renewal in vitro (Mani et al. 2008; Morel et al. 2008; Ponti et al. 2005). The significance of the work by Mani et al. (2008) is that they were for the first time able to link EMT with generation of cancer stem cells (CSCs). Radisky and LaBarge (2008) raise the possibilities that EMT either generate CSCs from pre-existing epithelial cells (non-CSCs) or induce proliferation of already existing minor fraction of CSCs. These possibilities can be similarly correlated back to the origin of hIPCs that they are generated by EMT of human pancreatic islet β-cells or actually MSCs in pancreas have proliferated. Our studies indicate that hIPCs have properties of mesenchymal-like stem cells and are derived through EMT of human pancreatic islet β-cells. Nonetheless, it is important to note that EMT sufficiently induces characteristics of stem cells in epithelial population (Radisky and LaBarge 2008).

EMT and MicroRNAs

miRNAs are small, noncoding, single stranded RNA molecules that are widely recognized as one of the gene regulators acting at post-transcriptional level (Joglekar et al. 2011). In recent years, miRNAs are emerging as potent biomarkers to assess a broad range of physiological and pathological processes (Hardikar et al. 2012). With technology advancement, various approaches are developed to assess miRNAs-real-time qPCR, gene arrays and sequencing (Hardikar et al. 2014), quantification of multiple miRNA and gene transcripts in a single cell (Joglekar et al. 2010) as well as simultaneous imaging of miRNA and proteins at a single cell resolution (Ranjan et al. 2012).

A single microRNA can often target multiple mRNAs in a pathway to increase their effectiveness on cellular process such as EMT (Subramanyam and Blelloch 2011). There have been several miRNAs identified to target transcription factors which are associated with EMT (Steinestel et al. 2014; Lamouille et al. 2014; Zaravinos 2015; Lamouille et al. 2013). MiRNAs such as the miR-200 family (which includes miR-200a, b, c, as well as miR-141, miR-429), -9, -24, -34a, -155, -192/215, -204, -491-5p, -302/372 have been described to play a central role in EMT and can be used as markers of carcinoma metastasis (refer to Table 9.1). Different miRNAs can either act to promote, suppress or have both effects on EMT and carcinoma/metastasis (refer to Table 9.1). miRNAs such as miR-1, miR-29b, miR-153 and miR-194 have been shown to be related to suppressing EMT, while miRNAs such as miR-9, miR-24, miR-27 and miR-181a have been shown to promote EMT.

The role of miR-200 family in EMT and cancer/metastasis has been widely studied (Gregory et al. 2008; Park et al. 2008; Korpal et al. 2008; Burk et al. 2008; Bracken et al. 2008). Expression of the miRNAs from the miR-200 family along with miR-205 in conjunction regulates E-cadherin expression, while

Table 9.1 miRNAs targeting EMT

miRNA	Targeted transcription factor/pathway associated with EMT	References
miR-1	Targets Slug	Liu et al. (2013)
miR-9	Down-regulates E-cadherin, activating β-catenin signalling-contributing to VEGF expression	Ma et al. (2010)
miR-24	Down-regulates Net1A, for late TGF-β induction	Papadimitriou et al. (2012)
miR-27	Up-regulates ZEB1, ZEB2, Slug and Vimentin. Down-regulates E-cadherin and APC	Zhang et al. (2011)
miR-29b	Down-regulates N-cadherin, Twist and Snail expressions. Also targets AKT2 to regulate TGF-β1	Ru et al. (2012), Li et al. (2014)
miR-30a	Targets Snail1	Kumarswamy et al. (2012)
miR-30a-5p, -30b, -30c, -30d and -30e-5p	Induce EMT and target in insulin gene expression in pancreatic islet β-cells	Joglekar et al. (2009), Ozcan (2009), Tang et al. (2009)
miR-31	Target RhoA	Valastyan et al. (2009, 2010)
miR-34a	Targets Snail and Snail regulating expressing transcription factors (e.g. ZEB1). Represses Slug and ZEB	Siemens et al. (2011)
miR-103/107	Inhibits Dicer	Martello et al. (2010)
miR-124	Targeting rho-kniase2 (ROCK2) and enhance zeste homologue 2 (EZH2) genes	Zheng et al. (2012)
miR-153	Targets Snail1 and ZEB2	Xu et al. (2013)
miR-155	Targets RhoA	Kong et al. (2008)
miR-181a	Regulates TGF-β signalling	Taylor et al. (2013)
miR-192/215	Associated with ZEB2 in TGF-β/connective tissue growth factor (CTGF). Increase E-cadherin expression	Wang et al. (2010)
miR-194	Decreases N-cadherin	Song et al. (2012), Meng et al. (2010)
miR-197	Down-regulates p120 catenin (an E-cadherin interacting protein)	Hamada et al. (2013)
miR-200a, 200b, 200c, miR-141 and miR-429	Represses ZEB1/2	Gregory et al. (2008), Bracken et al. (2008), Diaz-Lopez et al. (2015), Huang et al. (2015)
miR-203	Targets Slug, down-regulating Wnt signalling	Liao et al. (2015), Taube et al. (2013)
miR-204	Targets TGF-βR2 and Snail2. Increases E-cadherin and decreases vimentin	Wang et al. (2010), Zhang et al. (2013a)
miR-205	Represses ZEB1/2. Regulates E-cadherin	Gregory et al. (2008)

(continued)

Table 9.1 (continued)

miRNA	Targeted transcription factor/pathway associated with EMT	References
miR-216a/217	Targets Phosphatase and tensin homologue (PTEN) and SMAD7. Down-regulates epithelial hepatocellular carcinoma (HCC)	Xia et al. (2013)
miR-221/222	Regulates basal-like factor FOSL1. Represses trichorhitional repressors (TRPS1)-repressing ZEB2	Stinson et al. (2011)
miR-302/miR-372	Represses multiple target genes of EMT. Inhibits TGF-β	Subramanyam et al. (2011)
miR-375	Targets short stature homeobox 2 (SHOX2). Targets E-cadherin	Ward et al. (2013), Hong et al. (2014, Shen et al. (2014)
miR-490-3p	Regulates endoplasmic reticulum–Golgi intermediate compartment protein 3 (ERGIC3)	Zhang et al. (2013b),
miR-491-5p	Down-regulates of Par3. Regulates TGF-β	Zhou et al. (2010)
miR-612	Down-regulates AKT2 and suppresses Wnt/β-catenin signalling	Tao et al. (2013), Tang et al. (2014)
miR-661	Targets cell–cell adhesion protein Nectin-2 and lipid transferase StarD10. Up-regulates Slug	Vetter et al. (2010)

repressing ZEB1/2 in EMT to maintain epithelial cell phenotype (Gregory et al. 2008). Bracken et al. (2008) identified a double-negative feedback loop controlling ZEB1/2 and the miR-200 family, which acts as internal control of EMT in carcinoma progression. The double-negative feedback loop specifies the epithelial and mesenchymal phenotype, and it has also been described to represent a form of epigenetic memory (Inui et al. 2010). miR-200 family targets and supresses ZEB1/2 levels and EMT progression; also evidently by up-regulating E-cadherin to reduce motility of the cancer cell line (Park et al. 2008). Other miRNAs related to suppressing EMT have been shown to target a different regulator in EMT; for example miR-1, miR-29b, miR-30a and miR-31. miR-1 along with miR-200 has been identified to target Slug expression, a conserved regulator for mesenchymal expression in EMT (Liu et al. 2013). As shown in prostate adenocarcinoma progression, Slug acts as a repressor of miR-1 and miR-200 to reduce their expression. Forced expression of miR-1 and miR-200 expression is able to suppress EMT and tumorigenesis, by depleting Slug. miR-29b and miR-30a have shown to target Snail1 expression in EMT (Ru et al. 2012; Li et al. 2014; Kumarswamy et al. 2012). In prostate cancer when miR-29b expression is increased, it leads to the down-regulation of Snail, as well as N-cadherin and Twist (Ru et al. 2012). While in retinal pigment epithelial cells, miR-29b has also been shown to target and down-regulate Akt2 for TGF-β1-mediated EMT suppression (Li et al. 2014). Similarly, miR-30a expression in non-small cell lung cancer has been shown to target and down-regulate Snail1; leading to the up-regulation of E-cadherin and EMT suppression (Kumarswamy et al. 2012).

Multiple miRNAs have shown to promote specific regulators triggering EMT during carcinoma progression. For example, miR-9 has been shown to directly target CDH1 (which is the E-cadherin-encoding messenger RNA) (Ma et al. 2010). Targeting CDH1 leads to the down-regulation of E-cadherin, activation of β-catenin signalling as well as up-regulation of VEGF. In breast cancer, these processes together contribute to increase in motility and invasiveness for metastasis. miR-24 has been shown to promote EMT through down-regulating APC gene expression, thus activating the Wnt/β-catenin pathway (Zhang et al. 2011). In gastric cancer, miR-27 expression increases the level of EMT-associated regulators ZEB1/2, Slug and Vimentin, thereby promoting metastasis. Other miRNAs known to promote EMT include miR-155 that disrupts RhoA expression (associated with tight junctions) (Kong et al. 2008); miR-24 that down-regulates Net1 isoform2 (Net1A-required for TGF-β-mediated RhoA activation) (Papadimitriou et al. 2012) and many more (refer to Table 9.1).

EMT, hIPCs and miR-30 Family

Studies have also shown that miRNAs belonging to the miR-30 family are involved in the process of EMT during generation of hIPCs from human pancreatic islet β-cells and insulin gene transcription within the pancreatic islets (Joglekar et al. 2009b; Ozcan 2009; Tang et al. 2009). We observed that as islets transition to hIPCs in vitro, the abundance of miR-30 family miRNAs decreases significantly. Interestingly, we found that these miRNAs target several mesenchymal transcripts including vimentin. All these mesenchymal transcripts are abundantly expressed in human islets even though their proteins are not seen. This leads us to investigate the role of miR-30 family during EMT of human islets and also in maintaining the epithelial phenotype of islets. NF-YC, a transcription factor known to regulate E-cadherin expression in epithelial cells, has been shown to decrease as EMT occurs in human islets. This is accompanied with a simultaneous decrease in miR-30c and -30e expression. The NF-YC gene harbours these 2 miRNAs; miR-30c and -30e within its introns (Joglekar et al. 2009b). The inhibition of the two miRNAs (miR-30c and -30e) induces EMT of human pancreatic islet β-cells. Together, we find that miR-30 family members along with NF-YC are transcribed in human islets and they maintain epithelial phenotype by targeting mesenchymal transcripts along with expressing E-cadherin. However, when islets are exposed to serum or other growth factors, the NF-YC along with miR-30 family miRNAs are down-regulated, thereby allowing the EMT to occur in islet cells.

Additionally, another study showed miR-30d expression is up-regulated by glucose and increases insulin transcription, yet there is no effect on insulin secretion in the β-cell line MIN6. Despite insulin secretion is not effected by miR-30d expression, their study suggests that miR-30d can be a target for enhancing insulin gene expression (Tang et al. 2009).

Conclusion

EMT plays an essential role in embryonic development, tissue repair, fibrosis, cancer/metastasis and is regulated by multiple signalling pathways. We observed that EMT is the underlying mechanism for generation and derivation of hIPCs from human islets in vitro. Controversies in past regarding the origin of hIPCs have been resolved since lineage tracing studies offer conclusive evidence that adult as well as foetal human β-cells proliferate in vitro and undergo EMT to contribute in the generation of hIPCs. Since hIPCs are the direct (de-differentiated) progeny of β-cells, they retain open chromatin conformation at the insulin promoter region and thereby the memory to transcribe insulin gene more efficiently than in non-beta/embryonic stem cells. hIPCs are therefore perceived to be better precursors with significant potential for cell replacement therapy in diabetes.

miRNAs are novel gene regulators that mostly fine-tune several physiological and pathological process. They have been shown to regulate EMT in cancer cells as well as in primary human islets. The miR-30 family microRNAs in particular have been linked to EMT in human islets leading to generation of hIPCs.

In conclusion, hIPCs have a potential to differentiate into insulin-producing cells. More strategies employing regulators of EMT (signalling pathways/miRNAs) need to be investigated for efficient de-differentiation and proliferation of β-cells as well as for the re-differentiation of hIPCs into insulin-producing cell aggregates.

References

Acloque H, Adams MS, Fishwick K, Bronner-Fraser M, Nieto MA (2009) Epithelial-mesenchymal transitions: the importance of changing cell state in development and disease. J Clin Invest 119(6):1438–1449

Al-Hajj M, Wicha MS, Benito-Hernandez A, Morrison SJ, Clarke MF (2003) Prospective identification of tumorigenic breast cancer cells. Proc Natl Acad Sci USA 100(7):3983–3988

Aoi T, Yae K, Nakagawa M, Ichisaka T, Okita K, Takahashi K, Chiba T, Yamanaka S (2008) Generation of pluripotent stem cells from adult mouse liver and stomach cells. Science 321(5889):699–702

Arumugam T, Ramachandran V, Fournier KF, Wang H, Marquis L, Abbruzzese JL, Gallick GE, Logsdon CD, McConkey DJ, Choi W (2009) Epithelial to mesenchymal transition contributes to drug resistance in pancreatic cancer. Cancer Res 69(14):5820–5828

Assady S, Maor G, Amit M, Itskovitz-Eldor J, Skorecki KL, Tzukerman M (2001) Insulin production by human embryonic stem cells. Diabetes 50(8):1691–1697

Atouf F, Park CH, Pechhold K, Ta M, Choi Y, Lumelsky NL (2007) No evidence for mouse pancreatic beta-cell epithelial-mesenchymal transition in vitro. Diabetes 56(3):699–702

Bar-Nur O, Russ HA, Efrat S, Benvenisty N (2011) Epigenetic memory and preferential lineage-specific differentiation in induced pluripotent stem cells derived from human pancreatic islet beta cells. Cell Stem Cell 9(1):17–23

Ber I, Shternhall K, Perl S, Ohanuna Z, Goldberg I, Barshack I, Benvenisti-Zarum L, Meivar-Levy I, Ferber S (2003) Functional, persistent, and extended liver to pancreas transdifferentiation. J Biol Chem 278(34):31950–31957

Bernardo AS, Cho CH, Mason S, Docherty HM, Pedersen RA, Vallier L, Docherty K (2009) Biphasic induction of Pdx1 in mouse and human embryonic stem cells can mimic development of pancreatic beta-cells. Stem Cells 27(2):341–351

Bonner-Weir S, Taneja M, Weir GC, Tatarkiewicz K, Song KH, Sharma A, O'Neil JJ (2000) In vitro cultivation of human islets from expanded ductal tissue. Proc Natl Acad Sci USA 97(14):7999–8004

Bracken CP, Gregory PA, Kolesnikoff N, Bert AG, Wang J, Shannon MF, Goodall GJ (2008) A double-negative feedback loop between ZEB1-SIP1 and the microRNA-200 family regulates epithelial-mesenchymal transition. Cancer Res 68(19):7846–7854

Burk U, Schubert J, Wellner U, Schmalhofer O, Vincan E, Spaderna S, Brabletz T (2008) A reciprocal repression between ZEB1 and members of the miR-200 family promotes EMT and invasion in cancer cells. EMBO Rep 9(6):582–589

Butler AE, Huang A, Rao PN, Bhushan A, Hogan WJ, Rizza RA, Butler PC (2007) Hematopoietic stem cells derived from adult donors are not a source of pancreatic beta-cells in adult nondiabetic humans. Diabetes 56(7):1810–1816

Carlotti F, Zaldumbide A, Loomans CJ, van Rossenberg E, Engelse M, de Koning EJ, Hoeben RC (2010) Isolated human islets contain a distinct population of mesenchymal stem cells. Islets 2(3):164–173

Chase LG, Ulloa-Montoya F, Kidder BL, Verfaillie CM (2007) Islet-derived fibroblast-like cells are not derived via epithelial-mesenchymal transition from Pdx-1 or insulin-positive cells. Diabetes 56(1):3–7

Cieslik M, Hoang SA, Baranova N, Chodaparambil S, Kumar M, Allison DF, Xu X, Wamsley JJ, Gray L, Jones DR et al (2013) Epigenetic coordination of signaling pathways during the epithelial-mesenchymal transition. Epigenet Chromatin 6(1):28

Davani B, Ikonomou L, Raaka BM, Geras-Raaka E, Morton RA, Marcus-Samuels B, Gershengorn MC (2007) Human islet-derived precursor cells are mesenchymal stromal cells that differentiate and mature to hormone-expressing cells in vivo. Stem Cells 25(12):3215–3222

Davani B, Ariely S, Ikonomou L, Oron Y, Gershengorn MC (2009) Human islet-derived precursor cells can cycle between epithelial clusters and mesenchymal phenotypes. J Cell Mol Med 13(8B):2570–2581

Diaz-Lopez A, Diaz-Martin J, Moreno-Bueno G, Cuevas EP, Santos V, Olmeda D, Portillo F, Palacios J, Cano A (2015) Zeb1 and Snail1 engage miR-200f transcriptional and epigenetic regulation during EMT. Int J Cancer 136(4):E62–E73

Dominici M, Le Blanc K, Mueller I, Slaper-Cortenbach I, Marini F, Krause D, Deans R, Keating A, Prockop D, Horwitz E (2006) Minimal criteria for defining multipotent mesenchymal stromal cells. The international society for cellular therapy position statement. Cytotherapy 8(4):315–317

Efrat S (2008) In vitro expansion of human beta cells. Diabetologia 51(7):1338–1339

Eming SA, Krieg T, Davidson JM (2007) Inflammation in wound repair: molecular and cellular mechanisms. J Invest Dermatol 127(3):514–525

Fischer AN, Fuchs E, Mikula M, Huber H, Beug H, Mikulits W (2007) PDGF essentially links TGF-beta signaling to nuclear beta-catenin accumulation in hepatocellular carcinoma progression. Oncogene 26(23):3395–3405

Flier SN, Tanjore H, Kokkotou EG, Sugimoto H, Zeisberg M, Kalluri R (2010) Identification of epithelial to mesenchymal transition as a novel source of fibroblasts in intestinal fibrosis. J Biol Chem 285(26):20202–20212

Gallo R, Gambelli F, Gava B, Sasdelli F, Tellone V, Masini M, Marchetti P, Dotta F, Sorrentino V (2007) Generation and expansion of multipotent mesenchymal progenitor cells from cultured human pancreatic islets. Cell Death Differ 14(11):1860–1871

Garg M (2013) Epithelial-mesenchymal transition—activating transcription factors—multifunctional regulators in cancer. World J Stem Cells 5(4):188–195

Gershengorn MC, Hardikar AA, Wei C, Geras-Raaka E, Marcus-Samuels B, Raaka BM (2004) Epithelial-to-mesenchymal transition generates proliferative human islet precursor cells. Science 306(5705):2261–2264

Gershengorn MC, Geras-Raaka E, Hardikar AA, Raaka BM (2005) Are better islet cell precursors generated by epithelial-to-mesenchymal transition? Cell Cycle 4(3):380–382
Gonzalez DM, Medici D (2014) Signaling mechanisms of the epithelial-mesenchymal transition. Sci Signal 7(344):re8
Gregory PA, Bert AG, Paterson EL, Barry SC, Tsykin A, Farshid G, Vadas MA, Khew-Goodall Y, Goodall GJ (2008) The miR-200 family and miR-205 regulate epithelial to mesenchymal transition by targeting ZEB1 and SIP1. Nat Cell Biol 10(5):593–601
Hamada S, Satoh K, Miura S, Hirota M, Kanno A, Masamune A, Kikuta K, Kume K, Unno J, Egawa S et al (2013) miR-197 induces epithelial-mesenchymal transition in pancreatic cancer cells by targeting p120 catenin. J Cell Physiol 228(6):1255–1263
Hardikar AA, Marcus-Samuels B, Geras-Raaka E, Raaka BM, Gershengorn MC (2003) Human pancreatic precursor cells secrete FGF2 to stimulate clustering into hormone-expressing islet-like cell aggregates. Proc Natl Acad Sci USA 100(12):7117–7122
Hardikar AA, Walker MD, Lynn F (2012) Noncoding RNAs. Exp Diabetes Res 2012:629249
Hardikar AA, Farr RJ, Joglekar MV (2014) Circulating microRNAs: understanding the limits for quantitative measurement by real-time PCR. J Am Heart Assoc 3(1):e000792
Hay ED (1990) Role of cell-matrix contacts in cell migration and epithelial-mesenchymal transformation. Cell Differ Dev 32(3):367–375
Hess D, Li L, Martin M, Sakano S, Hill D, Strutt B, Thyssen S, Gray DA, Bhatia M (2003) Bone marrow-derived stem cells initiate pancreatic regeneration. Nat Biotechnol 21(7):763–770
Hong S, Noh H, Teng Y, Shao J, Rehmani H, Ding HF, Dong Z, Su SB, Shi H, Kim J et al (2014) SHOX2 is a direct miR-375 target and a novel epithelial-to-mesenchymal transition inducer in breast cancer cells. Neoplasia 16(4):279–290 (e271–275)
Huang Y, Tong J, He F, Yu X, Fan L, Hu J, Tan J, Chen Z (2015) miR-141 regulates TGF-beta1-induced epithelial-mesenchymal transition through repression of HIPK2 expression in renal tubular epithelial cells. Int J Mol Med 35(2):311–318
Ianus A, Holz GG, Theise ND, Hussain MA (2003) In vivo derivation of glucose-competent pancreatic endocrine cells from bone marrow without evidence of cell fusion. J Clin Invest 111(6):843–850
Inada A, Bonner-Weir S, Toschi E (2006) How can we get more beta cells? Curr Diab Rep 6(2):96–101
Inui M, Martello G, Piccolo S (2010) MicroRNA control of signal transduction. Nat Rev Mol Cell Biol 11(4):252–263
Jafarian A, Taghikhani M, Abroun S, Pourpak Z, Allahverdi A, Soleimani M (2014) Generation of high-yield insulin producing cells from human bone marrow mesenchymal stem cells. Mol Biol Rep 41(7):4783–4794
Jiang J, Au M, Lu K, Eshpeter A, Korbutt G, Fisk G, Majumdar AS (2007) Generation of insulin-producing islet-like clusters from human embryonic stem cells. Stem Cells 25(8):1940–1953
Joglekar MV, Hardikar AA (2010) Epithelial-to-mesenchymal transition in pancreatic islet beta cells. Cell Cycle 9(20):4077–4079
Joglekar MV, Hardikar AA (2012) Isolation, expansion, and characterization of human islet-derived progenitor cells. Methods Mol Biol 879:351–366
Joglekar MV, Joglekar VM, Joglekar SV, Hardikar AA (2009a) Human fetal pancreatic insulin-producing cells proliferate in vitro. J Endocrinol 201(1):27–36
Joglekar MV, Patil D, Joglekar VM, Rao GV, Reddy DN, Sasikala M, Shouche Y, Hardikar AA (2009b) The miR-30 family microRNAs confer epithelial phenotype to human pancreatic cells. Islets 1(2):137–147
Joglekar MV, Wei C, Hardikar AA (2010) Quantitative estimation of multiple miRNAs and mRNAs from a single cell. Cold Spring Harb Protoc 2010(8):pdb prot5478
Joglekar MV, Parekh VS, Hardikar AA (2011) Islet-specific microRNAs in pancreas development, regeneration and diabetes. Indian J Exp Biol 49(6):401–408
Kalluri R, Neilson EG (2003) Epithelial-mesenchymal transition and its implications for fibrosis. J Clin Invest 112(12):1776–1784

Kalluri R, Weinberg RA (2009) The basics of epithelial-mesenchymal transition. J Clin Invest 119(6):1420–1428

Katz E, Dubois-Marshall S, Sims AH, Gautier P, Caldwell H, Meehan RR, Harrison DJ (2011) An in vitro model that recapitulates the epithelial to mesenchymal transition (EMT) in human breast cancer. PLoS One 6(2):e17083

Kayali AG, Flores LE, Lopez AD, Kutlu B, Baetge E, Kitamura R, Hao E, Beattie GM, Hayek A (2007) Limited capacity of human adult islets expanded in vitro to redifferentiate into insulin-producing beta-cells. Diabetes 56(3):703–708

Kim KK, Kugler MC, Wolters PJ, Robillard L, Galvez MG, Brumwell AN, Sheppard D, Chapman HA (2006) Alveolar epithelial cell mesenchymal transition develops in vivo during pulmonary fibrosis and is regulated by the extracellular matrix. Proc Natl Acad Sci USA 103(35):13180–13185

Kong W, Yang H, He L, Zhao JJ, Coppola D, Dalton WS, Cheng JQ (2008) MicroRNA-155 is regulated by the transforming growth factor beta/Smad pathway and contributes to epithelial cell plasticity by targeting RhoA. Mol Cell Biol 28(22):6773–6784

Korpal M, Lee ES, Hu G, Kang Y (2008) The miR-200 family inhibits epithelial-mesenchymal transition and cancer cell migration by direct targeting of E-cadherin transcriptional repressors ZEB1 and ZEB2. J Biol Chem 283(22):14910–14914

Kumarswamy R, Mudduluru G, Ceppi P, Muppala S, Kozlowski M, Niklinski J, Papotti M, Allgayer H (2012) MicroRNA-30a inhibits epithelial-to-mesenchymal transition by targeting Snail and is downregulated in non-small cell lung cancer. Int J Cancer 130(9):2044–2053

Lahmy R, Soleimani M, Sanati MH, Behmanesh M, Kouhkan F, Mobarra N (2014) MiRNA-375 promotes beta pancreatic differentiation in human induced pluripotent stem (hiPS) cells. Mol Biol Rep 41(4):2055–2066

Lamouille S, Derynck R (2007) Cell size and invasion in TGF-beta-induced epithelial to mesenchymal transition is regulated by activation of the mTOR pathway. J Cell Biol 178(3):437–451

Lamouille S, Connolly E, Smyth JW, Akhurst RJ, Derynck R (2012) TGF-beta-induced activation of mTOR complex 2 drives epithelial-mesenchymal transition and cell invasion. J Cell Sci 125(Pt 5):1259–1273

Lamouille S, Subramanyam D, Blelloch R, Derynck R (2013) Regulation of epithelial-mesenchymal and mesenchymal-epithelial transitions by microRNAs. Curr Opin Cell Biol 25(2):200–207

Lamouille S, Xu J, Derynck R (2014) Molecular mechanisms of epithelial-mesenchymal transition. Nat Rev Mol Cell Biol 15(3):178–196

Lan L, Luo Y, Cui D, Shi BY, Deng W, Huo LL, Chen HL, Zhang GY, Deng LL (2013) Epithelial-mesenchymal transition triggers cancer stem cell generation in human thyroid cancer cells. Int J Oncol 43(1):113–120

Lechner A, Yang YG, Blacken RA, Wang L, Nolan AL, Habener JF (2004) No evidence for significant transdifferentiation of bone marrow into pancreatic beta-cells in vivo. Diabetes 53(3):616–623

Lechner A, Nolan AL, Blacken RA, Habener JF (2005) Redifferentiation of insulin-secreting cells after in vitro expansion of adult human pancreatic islet tissue. Biochem Biophys Res Commun 327(2):581–588

Leong KG, Niessen K, Kulic I, Raouf A, Eaves C, Pollet I, Karsan A (2007) Jagged1-mediated Notch activation induces epithelial-to-mesenchymal transition through Slug-induced repression of E-cadherin. J Exp Med 204(12):2935–2948

Li X, Deng W, Nail CD, Bailey SK, Kraus MH, Ruppert JM, Lobo-Ruppert SM (2006) Snail induction is an early response to Gli1 that determines the efficiency of epithelial transformation. Oncogene 25(4):609–621

Li X, Deng W, Lobo-Ruppert SM, Ruppert JM (2007) Gli1 acts through Snail and E-cadherin to promote nuclear signaling by beta-catenin. Oncogene 26(31):4489–4498

Li M, Li H, Liu X, Xu D, Wang F (2014) MicroRNA-29b regulates TGF-beta1-mediated epithelial-mesenchymal transition of retinal pigment epithelial cells by targeting AKT2. Exp Cell Res 345(2):115–124

Liao H, Bai Y, Qiu S, Zheng L, Huang L, Liu T, Wang X, Liu Y, Xu N, Yan X et al (2015) MiR-203 downregulation is responsible for chemoresistance in human glioblastoma by promoting epithelial-mesenchymal transition via SNAI2. Oncotarget 6(11):8914–8928

Lili LN, Matyunina LV, Walker LD, Wells SL, Benigno BB, McDonald JF (2013) Molecular profiling supports the role of epithelial-to-mesenchymal transition (EMT) in ovarian cancer metastasis. J Ovarian Res 6(1):49

Lim J, Thiery JP (2012) Epithelial-mesenchymal transitions: insights from development. Development 139(19):3471–3486

Liu YN, Yin JJ, Abou-Kheir W, Hynes PG, Casey OM, Fang L, Yi M, Stephens RM, Seng V, Sheppard-Tillman H et al (2013) MiR-1 and miR-200 inhibit EMT via Slug-dependent and tumorigenesis via Slug-independent mechanisms. Oncogene 32(3):296–306

Lopez-Novoa JM, Nieto MA (2009) Inflammation and EMT: an alliance towards organ fibrosis and cancer progression. EMBO Mol Med 1(6–7):303–314

Lumelsky N, Blondel O, Laeng P, Velasco I, Ravin R, McKay R (2001) Differentiation of embryonic stem cells to insulin-secreting structures similar to pancreatic islets. Science 292(5520):1389–1394

Ma L, Young J, Prabhala H, Pan E, Mestdagh P, Muth D, Teruya-Feldstein J, Reinhardt F, Onder TT, Valastyan S et al (2010) miR-9, a MYC/MYCN-activated microRNA, regulates E-cadherin and cancer metastasis. Nat Cell Biol 12(3):247–256

Mani SA, Guo W, Liao MJ, Eaton EN, Ayyanan A, Zhou AY, Brooks M, Reinhard F, Zhang CC, Shipitsin M et al (2008) The epithelial-mesenchymal transition generates cells with properties of stem cells. Cell 133(4):704–715

Martello G, Rosato A, Ferrari F, Manfrin A, Cordenonsi M, Dupont S, Enzo E, Guzzardo V, Rondina M, Spruce T et al (2010) A MicroRNA targeting dicer for metastasis control. Cell 141(7):1195–1207

Meivar-Levy I, Ferber S (2003) New organs from our own tissues: liver-to-pancreas transdifferentiation. Trends Endocrinol Metab 14(10):460–466

Meivar-Levy I, Ferber S (2006) Regenerative medicine: using liver to generate pancreas for treating diabetes. Isr Med Assoc J 8(6):430–434

Meng Z, Fu X, Chen X, Zeng S, Tian Y, Jove R, Xu R, Huang W (2010) miR-194 is a marker of hepatic epithelial cells and suppresses metastasis of liver cancer cells in mice. Hepatology 52(6):2148–2157

Morel AP, Lievre M, Thomas C, Hinkal G, Ansieau S, Puisieux A (2008) Generation of breast cancer stem cells through epithelial-mesenchymal transition. PLoS One 3(8):e2888

Morton RA, Geras-Raaka E, Wilson LM, Raaka BM, Gershengorn MC (2007) Endocrine precursor cells from mouse islets are not generated by epithelial-to-mesenchymal transition of mature beta cells. Mol Cell Endocrinol 270(1–2):87–93

Mutskov V, Raaka BM, Felsenfeld G, Gershengorn MC (2007) The human insulin gene displays transcriptionally active epigenetic marks in islet-derived mesenchymal precursor cells in the absence of insulin expression. Stem Cells 25(12):3223–3233

Nakagawa M, Koyanagi M, Tanabe K, Takahashi K, Ichisaka T, Aoi T, Okita K, Mochiduki Y, Takizawa N, Yamanaka S (2008) Generation of induced pluripotent stem cells without Myc from mouse and human fibroblasts. Nat Biotechnol 26(1):101–106

Nakamura M, Tokura Y (2011) Epithelial-mesenchymal transition in the skin. J Dermatol Sci 61(1):7–13

Nieto MA, Cano A (2012) The epithelial-mesenchymal transition under control: global programs to regulate epithelial plasticity. Semin Cancer Biol 22(5–6):361–368

Okada H, Danoff TM, Kalluri R, Neilson EG (1997) Early role of Fsp1 in epithelial-mesenchymal transformation. Am J Physiol 273(4 Pt 2):F563–F574

Okita K, Ichisaka T, Yamanaka S (2007) Generation of germline-competent induced pluripotent stem cells. Nature 448(7151):313–317

Ouziel-Yahalom L, Zalzman M, Anker-Kitai L, Knoller S, Bar Y, Glandt M, Herold K, Efrat S (2006) Expansion and redifferentiation of adult human pancreatic islet cells. Biochem Biophys Res Commun 341(2):291–298

Ozcan S (2009) MiR-30 family and EMT in human fetal pancreatic islets. Islets 1(3):283–285

Pagliuca FW, Millman JR, Gurtler M, Segel M, Van Dervort A, Ryu JH, Peterson QP, Greiner D, Melton DA (2014) Generation of functional human pancreatic beta cells in vitro. Cell 159(2):428–439

Papadimitriou E, Vasilaki E, Vorvis C, Iliopoulos D, Moustakas A, Kardassis D, Stournaras C (2012) Differential regulation of the two RhoA-specific GEF isoforms Net1/Net1A by TGF-beta and miR-24: role in epithelial-to-mesenchymal transition. Oncogene 31(23):2862–2875

Park SM, Gaur AB, Lengyel E, Peter ME (2008) The miR-200 family determines the epithelial phenotype of cancer cells by targeting the E-cadherin repressors ZEB1 and ZEB2. Genes Dev 22(7):894–907

Parnaud G, Bosco D, Berney T, Pattou F, Kerr-Conte J, Donath MY, Bruun C, Mandrup-Poulsen T, Billestrup N, Halban PA (2008) Proliferation of sorted human and rat beta cells. Diabetologia 51(1):91–100

Phadnis SM, Joglekar MV, Dalvi MP, Muthyala S, Nair PD, Ghaskadbi SM, Bhonde RR, Hardikar AA (2011) Human bone marrow-derived mesenchymal cells differentiate and mature into endocrine pancreatic lineage in vivo. Cytotherapy 13(3):279–293

Ponti D, Costa A, Zaffaroni N, Pratesi G, Petrangolini G, Coradini D, Pilotti S, Pierotti MA, Daidone MG (2005) Isolation and in vitro propagation of tumorigenic breast cancer cells with stem/progenitor cell properties. Cancer Res 65(13):5506–5511

Radisky DC, LaBarge MA (2008) Epithelial-mesenchymal transition and the stem cell phenotype. Cell Stem Cell 2(6):511–512

Ranjan AK, Joglekar MV, Atre AN, Patole M, Bhonde RR, Hardikar AA (2012) Simultaneous imaging of microRNA or mRNA territories with protein territory in mammalian cells at single cell resolution. RNA Biol 9(7):949–953

Rastaldi MP, Ferrario F, Giardino L, Dell'Antonio G, Grillo C, Grillo P, Strutz F, Muller GA, Colasanti G, D'Amico G (2002) Epithelial-mesenchymal transition of tubular epithelial cells in human renal biopsies. Kidney Int 62(1):137–146

Rezania A, Bruin JE, Arora P, Rubin A, Batushansky I, Asadi A, O'Dwyer S, Quiskamp N, Mojibian M, Albrecht T et al (2014) Reversal of diabetes with insulin-producing cells derived in vitro from human pluripotent stem cells. Nat Biotechnol 32(11):1121–1133

Ru P, Steele R, Newhall P, Phillips NJ, Toth K, Ray RB (2012) miRNA-29b suppresses prostate cancer metastasis by regulating epithelial-mesenchymal transition signaling. Mol Cancer Ther 11(5):1166–1173

Russ HA, Bar Y, Ravassard P, Efrat S (2008) In vitro proliferation of cells derived from adult human beta-cells revealed by cell-lineage tracing. Diabetes 57(6):1575–1583

Russ HA, Ravassard P, Kerr-Conte J, Pattou F, Efrat S (2009) Epithelial-mesenchymal transition in cells expanded in vitro from lineage-traced adult human pancreatic beta cells. PLoS One 4(7):e6417

Russ HA, Sintov E, Anker-Kitai L, Friedman O, Lenz A, Toren G, Farhy C, Pasmanik-Chor M, Oron-Karni V, Ravassard P et al (2011) Insulin-producing cells generated from dedifferentiated human pancreatic beta cells expanded in vitro. PLoS One 6(9):e25566

Sahu S, Tosh D, Hardikar AA (2009) New sources of beta-cells for treating diabetes. J Endocrinol 202(1):13–16

Sapir T, Shternhall K, Meivar-Levy I, Blumenfeld T, Cohen H, Skutelsky E, Eventov-Friedman S, Barshack I, Goldberg I, Pri-Chen S et al (2005) Cell-replacement therapy for diabetes: generating functional insulin-producing tissue from adult human liver cells. Proc Natl Acad Sci USA 102(22):7964–7969

Seaberg RM, Smukler SR, Kieffer TJ, Enikolopov G, Asghar Z, Wheeler MB, Korbutt G, van der Kooy D (2004) Clonal identification of multipotent precursors from adult mouse pancreas that generate neural and pancreatic lineages. Nat Biotechnol 22(9):1115–1124

Shen Y, Zhou J, Li Y, Ye F, Wan X, Lu W, Xie X, Cheng X (2014) miR-375 mediated acquired chemo-resistance in cervical cancer by facilitating EMT. PLoS One 9(10):e109299

Shi Y, Massague J (2003) Mechanisms of TGF-beta signaling from cell membrane to the nucleus. Cell 113(6):685–700

Siemens H, Jackstadt R, Hunten S, Kaller M, Menssen A, Gotz U, Hermeking H (2011) miR-34 and SNAIL form a double-negative feedback loop to regulate epithelial-mesenchymal transitions. Cell Cycle 10(24):4256–4271

Slack JM, Tosh D (2001) Transdifferentiation and metaplasia—switching cell types. Curr Opin Genet Dev 11(5):581–586

Sleeman KE, Kendrick H, Ashworth A, Isacke CM, Smalley MJ (2006) CD24 staining of mouse mammary gland cells defines luminal epithelial, myoepithelial/basal and non-epithelial cells. Breast Cancer Res 8(1):R7

Song Y, Zhao F, Wang Z, Liu Z, Chiang Y, Xu Y, Gao P, Xu H (2012) Inverse association between miR-194 expression and tumor invasion in gastric cancer. Ann Surg Oncol 19(Suppl 3):S509–S517

Sordi V, Melzi R, Mercalli A, Formicola R, Doglioni C, Tiboni F, Ferrari G, Nano R, Chwalek K, Lammert E et al (2010) Mesenchymal cells appearing in pancreatic tissue culture are bone marrow-derived stem cells with the capacity to improve transplanted islet function. Stem Cells 28(1):140–151

Stadelmann WK, Digenis AG, Tobin GR (1998) Physiology and healing dynamics of chronic cutaneous wounds. Am J Surg 176(2A Suppl):26S–38S

Steinestel K, Eder S, Schrader AJ, Steinestel J (2014) Clinical significance of epithelial-mesenchymal transition. Clin Transl Med 3:17

Stinson S, Lackner MR, Adai AT, Yu N, Kim HJ, O'Brien C, Spoerke J, Jhunjhunwala S, Boyd Z, Januario T et al (2011) miR-221/222 targeting of trichorhinophalangeal 1 (TRPS1) promotes epithelial-to-mesenchymal transition in breast cancer. Sci Signal 4(186):pt5

Subramanyam D, Blelloch R (2011) From microRNAs to targets: pathway discovery in cell fate transitions. Curr Opin Genet Dev 21(4):498–503

Subramanyam D, Lamouille S, Judson RL, Liu JY, Bucay N, Derynck R, Blelloch R (2011) Multiple targets of miR-302 and miR-372 promote reprogramming of human fibroblasts to induced pluripotent stem cells. Nat Biotechnol 29(5):443–448

Takahashi K, Yamanaka S (2006) Induction of pluripotent stem cells from mouse embryonic and adult fibroblast cultures by defined factors. Cell 126(4):663–676

Takahashi K, Tanabe K, Ohnuki M, Narita M, Ichisaka T, Tomoda K, Yamanaka S (2007) Induction of pluripotent stem cells from adult human fibroblasts by defined factors. Cell 131(5):861–872

Tang X, Muniappan L, Tang G, Ozcan S (2009) Identification of glucose-regulated miRNAs from pancreatic beta cells reveals a role for miR-30d in insulin transcription. RNA 15(2):287–293

Tang J, Tao ZH, Wen D, Wan JL, Liu DL, Zhang S, Cui JF, Sun HC, Wang L, Zhou J et al (2014) MiR-612 suppresses the stemness of liver cancer via Wnt/beta-catenin signaling. Biochem Biophys Res Commun 447(1):210–215

Tao ZH, Wan JL, Zeng LY, Xie L, Sun HC, Qin LX, Wang L, Zhou J, Ren ZG, Li YX et al (2013) miR-612 suppresses the invasive-metastatic cascade in hepatocellular carcinoma. J Exp Med 210(4):789–803

Taube JH, Malouf GG, Lu E, Sphyris N, Vijay V, Ramachandran PP, Ueno KR, Gaur S, Nicoloso MS, Rossi S et al (2013) Epigenetic silencing of microRNA-203 is required for EMT and cancer stem cell properties. Sci Rep 3:2687

Taylor MA, Sossey-Alaoui K, Thompson CL, Danielpour D, Schiemann WP (2013) TGF-beta upregulates miR-181a expression to promote breast cancer metastasis. J Clin Invest 123(1):150–163

Teta M, Rankin MM, Long SY, Stein GM, Kushner JA (2007) Growth and regeneration of adult beta cells does not involve specialized progenitors. Dev Cell 12(5):817–826

Thiery JP, Sleeman JP (2006) Complex networks orchestrate epithelial-mesenchymal transitions. Nat Rev Mol Cell Biol 7(2):131–142

Tiwari N, Gheldof A, Tatari M, Christofori G (2012) EMT as the ultimate survival mechanism of cancer cells. Semin Cancer Biol 22(3):194–207

Valastyan S, Reinhardt F, Benaich N, Calogrias D, Szasz AM, Wang ZC, Brock JE, Richardson AL, Weinberg RA (2009) A pleiotropically acting microRNA, miR-31, inhibits breast cancer metastasis. Cell 137(6):1032–1046

Valastyan S, Chang A, Benaich N, Reinhardt F, Weinberg RA (2010) Concurrent suppression of integrin alpha5, radixin, and RhoA phenocopies the effects of miR-31 on metastasis. Cancer Res 70(12):5147–5154

van der Meulen T, Huising MO (2015) Role of transcription factors in the transdifferentiation of pancreatic islet cells. J Mol Endocrinol 54(2):R103–R117

Velnar T, Bailey T, Smrkolj V (2009) The wound healing process: an overview of the cellular and molecular mechanisms. J Int Med Res 37(5):1528–1542

Vetter G, Saumet A, Moes M, Vallar L, Le Bechec A, Laurini C, Sabbah M, Arar K, Theillet C, Lecellier CH et al (2010) miR-661 expression in SNAI1-induced epithelial to mesenchymal transition contributes to breast cancer cell invasion by targeting Nectin-1 and StarD10 messengers. Oncogene 29(31):4436–4448

Vicovac L, Aplin JD (1996) Epithelial-mesenchymal transition during trophoblast differentiation. Acta Anat (Basel) 156(3):202–216

Wang B, Herman-Edelstein M, Koh P, Burns W, Jandeleit-Dahm K, Watson A, Saleem M, Goodall GJ, Twigg SM, Cooper ME et al (2010a) E-cadherin expression is regulated by miR-192/215 by a mechanism that is independent of the profibrotic effects of transforming growth factor-beta. Diabetes 59(7):1794–1802

Wang FE, Zhang C, Maminishkis A, Dong L, Zhi C, Li R, Zhao J, Majerciak V, Gaur AB, Chen S et al (2010b) MicroRNA-204/211 alters epithelial physiology. FASEB J 24(5):1552–1571

Wang L, Huang Y, Guo Q, Fan X, Lu Y, Zhu S, Wang Y, Bo X, Chang X, Zhu M et al (2014) Differentiation of iPSCs into insulin-producing cells via adenoviral transfection of PDX-1, NeuroD1 and MafA. Diabetes Res Clin Pract 104(3):383–392

Ward A, Balwierz A, Zhang JD, Kublbeck M, Pawitan Y, Hielscher T, Wiemann S, Sahin O (2013) Re-expression of microRNA-375 reverses both tamoxifen resistance and accompanying EMT-like properties in breast cancer. Oncogene 32(9):1173–1182

Weber CE, Li NY, Wai PY, Kuo PC (2012) Epithelial-mesenchymal transition, TGF-beta, and osteopontin in wound healing and tissue remodeling after injury. J Burn Care Res 33(3):311–318

Weinberg N, Ouziel-Yahalom L, Knoller S, Efrat S, Dor Y (2007) Lineage tracing evidence for in vitro dedifferentiation but rare proliferation of mouse pancreatic beta-cells. Diabetes 56(5):1299–1304

Weissman IL (2000) Stem cells: units of development, units of regeneration, and units in evolution. Cell 100(1):157–168

Wong W, Joglekar MV, Satoor SN, Sahu S, Parekh VS, Hardikar AA (2014) Lineage-committed pancreatic progenitors and stem cells. In: Turksen K (ed) Adult stem cells. Springer, New York, pp 339–357

Xia H, Ooi LL, Hui KM (2013) MicroRNA-216a/217-induced epithelial-mesenchymal transition targets PTEN and SMAD7 to promote drug resistance and recurrence of liver cancer. Hepatology 58(2):629–641

Xu Q, Sun Q, Zhang J, Yu J, Chen W, Zhang Z (2013) Downregulation of miR-153 contributes to epithelial-mesenchymal transition and tumor metastasis in human epithelial cancer. Carcinogenesis 34(3):539–549

Yamada S, Kojima I (2005) Regenerative medicine of the pancreatic beta cells. J Hepatobiliary Pancreat Surg 12(3):218–226

Ye L, Robertson MA, Hesselson D, Stainier DY, Anderson RM (2015) Glucagon is essential for alpha cell transdifferentiation and beta cell neogenesis. Development 142(8):1407–1417

Yu M, Smolen GA, Zhang J, Wittner B, Schott BJ, Brachtel E, Ramaswamy S, Maheswaran S, Haber DA (2009) A developmentally regulated inducer of EMT, LBX1, contributes to breast cancer progression. Genes Dev 23(15):1737–1742

Zalzman M, Anker-Kitai L, Efrat S (2005) Differentiation of human liver-derived, insulin-producing cells toward the beta-cell phenotype. Diabetes 54(9):2568–2575

Zaravinos A (2015) The regulatory role of MicroRNAs in EMT and cancer. J Oncol 2015:865816

Zeisberg M, Neilson EG (2009) Biomarkers for epithelial-mesenchymal transitions. J Clin Invest 119(6):1429–1437

Zeisberg EM, Tarnavski O, Zeisberg M, Dorfman AL, McMullen JR, Gustafsson E, Chandraker A, Yuan X, Pu WT, Roberts AB et al (2007a) Endothelial-to-mesenchymal transition contributes to cardiac fibrosis. Nat Med 13(8):952–961

Zeisberg M, Yang C, Martino M, Duncan MB, Rieder F, Tanjore H, Kalluri R (2007b) Fibroblasts derive from hepatocytes in liver fibrosis via epithelial to mesenchymal transition. J Biol Chem 282(32):23337–23347

Zhang Z, Liu S, Shi R, Zhao G (2011) miR-27 promotes human gastric cancer cell metastasis by inducing epithelial-to-mesenchymal transition. Cancer Genet 204(9):486–491

Zhang L, Wang X, Chen P (2013a) MiR-204 down regulates SIRT1 and reverts SIRT1-induced epithelial-mesenchymal transition, anoikis resistance and invasion in gastric cancer cells. BMC Cancer 13:290

Zhang LY, Liu M, Li X, Tang H (2013b) miR-490-3p modulates cell growth and epithelial to mesenchymal transition of hepatocellular carcinoma cells by targeting endoplasmic reticulum-Golgi intermediate compartment protein 3 (ERGIC3). J Biol Chem 288(6):4035–4047

Zheng F, Liao YJ, Cai MY, Liu YH, Liu TH, Chen SP, Bian XW, Guan XY, Lin MC, Zeng YX et al (2012) The putative tumour suppressor microRNA-124 modulates hepatocellular carcinoma cell aggressiveness by repressing ROCK2 and EZH2. Gut 61(2):278–289

Zhou Q, Fan J, Ding X, Peng W, Yu X, Chen Y, Nie J (2010) TGF-{beta}-induced MiR-491-5p expression promotes Par-3 degradation in rat proximal tubular epithelial cells. J Biol Chem 285(51):40019–40027

Chapter 10
Viral-Mediated Gene Therapy for the Generation of Artificial Insulin-Producing Cells as a Therapeutic Treatment for Type 1 Diabetes Mellitus

Dario Gerace, Rosetta Martiniello-Wilks and Ann M. Simpson

Introduction

Several approaches have been employed to develop cell and gene therapy strategies that generate artificial insulin-producing cells (IPCs) for potential therapeutic applications in the treatment of type 1 diabetes mellitus (T1D). The genetic engineering of functional IPCs necessitates a broad understanding of the pancreatic developmental process and the β-cell transcription factors that govern mature β-cell differentiation and function. To successfully obtain functional IPCs, the type of vectors utilized for gene transfer and the selection of a suitable target cell for subsequent differentiation into IPCs is of fundamental importance. Techniques for manufacturing IPCs include the de-differentiation and directed differentiation of autologous or allogeneic cells ex vivo followed by transplantation and the in vivo differentiation of target tissue via viral gene transfer. Ultimately, the goal is to construct IPCs that have the capacity to process, store and secrete insulin in response to and relative to the circulating blood glucose levels, whilst avoiding the administration of immunosuppressants and recurrent autoimmune destruction, thereby restoring normoglycaemia.

Dario Gerace and Rosetta Martiniello-Wilks have contributed equally to this work.

D. Gerace · R. Martiniello-Wilks · A.M. Simpson (✉)
School of Life Sciences and the Centre for Health Technologies,
University of Technology Sydney, PO Box 123, Broadway, NSW 2007, Australia
e-mail: Ann.Simpson@uts.edu.au

R. Martiniello-Wilks
Translational Cancer Research Group, University of Technology Sydney, Sydney, Australia

A.A. Hardikar (ed.), *Pancreatic Islet Biology*, Stem Cell Biology and Regenerative Medicine, DOI 10.1007/978-3-319-45307-1_10

Table 10.1 Criteria for comparing the suitability of potential target cells for T1D gene therapy

Characteristics	Target cell				
	Liver	Pituitary	Muscle	K cells	BMSC
Derived from endodermal origin	Yes	No	No	No	Yes
Endogenous β cell transcription factors	No	No	No	No	Yes
Glucose-sensing system	Yes	No	No	Yes	No
Proinsulin processing enzymes	No	Yes	No	Yes	No
Glucose-regulatable promoter	Yes	No	No	Yes	No
Secretory system	No	Yes	No	Yes	No
Autologous use	Yes	Yes	Yes	Yes	Yes
Allogeneic use	No	No	No	No	Yes

Selecting an Ideal Viral Vector for T1D Gene Therapy

The use of viruses as tools for the correction of genetic disorders due to their ability to infect and deliver genetic material to cells has paved the way for a number of promising cell and gene therapies. By engineering viral vectors that are replication deficient and efficiently transduce genes into target cells, key challenges encountered in the generation of successful therapies can be circumvented. Selecting a suitable viral vector is contingent on the nature of gene expression required (sustained or short-term expression), whether the vector is integrative or non-integrative, and the type of cells targeted for gene transfer (dividing or non-dividing) (Table 10.1). Ideally, the generation of artificial IPCs would utilise integrating viral vectors that offer long-term gene expression over the life of the patient, resulting in a persisting therapeutic benefit.

Retroviral Vectors

Retroviral vectors are derived from disabled murine viruses and are the most commonly used gene delivery vector due to their ability to integrate chromosomally and sustain expression of the transgene (Morgan and Anderson 1993). However, the risk of site-specific insertional mutagenesis near oncogenic locations could increase predisposition to tumour development, limiting their utility (Bushman 2007; Laufs et al. 2004). This feature of retroviral vectors was demonstrated in 1999 following the treatment of nine severe combined immunodeficiency (*Scid*) patients which led to the development of leukaemia in four of the patients (Cavazzana-Calvo et al. 2000). The primary disadvantage with retroviral gene transfer is the requirement for target cells to be in an active state of division for transduction; therefore, target tissues composed predominantly of non-dividing cells represent a challenge. A successful use of retroviral vectors to generate

IPCs was performed by Xu et al. (2007), who transduced bone marrow-derived mesenchymal stem cells (BMSCs) with an insulin gene under the control of the cytomegalovirus (CMV) promoter. Following transplantation into streptozotocin (STZ)-diabetic mice, they discovered that the transduced BMSCs expressed insulin and were able to maintain normoglycaemia for at least 42 days whilst evading autoimmune destruction.

Adenoviral and Adeno-Associated Vectors

Due to their ability to transduce dividing and non-dividing cells with high efficiency, adenoviral vectors have been studied intensively, particularly for the targeted therapy of cystic fibrosis (Zabner et al. 1993; Knowles et al. 1995). However, challenges associated with the administration of adenoviral vectors such as the development of immune responses against the viral capsid proteins, and in some cases the transgene itself, have limited their use in the clinic (Wold et al. 1999; McCaffrey et al. 2008; Nayak and Herzog 2010). In addition, adenoviral vectors transfer their genetic cargo episomally and subsequently offer only transient gene expression (Morgan and Anderson 1993; Volpers and Kochanek 2004). To address the immunogenicity of the viral capsid proteins, a "gutless" adenovirus was constructed where the genes encoding the common viral capsid proteins were deleted (Alba et al. 2005). Although the new generation vectors reduce immunogenicity, the administration of immunosuppressants is still required following treatment (Zhou et al. 2004).

Adeno-associated (AAV) viral vectors are becoming an attractive alternative for gene therapy as they are able to transduce both dividing and non-dividing cells, whilst also preferentially integrating their genetic cargo at a specific site on chromosome 19 (Muzyczka 1992). Despite possessing a limited gene cargo capacity of less than 5 kb, AAV vectors have been utilised to deliver the preproinsulin gene to livers of STZ-diabetic mice (Sugiyama et al. 1997), resulting in the transient stabilisation of blood glucose levels.

Lentiviral Vectors

Lentiviral vectors (LV) are derived from the human immunodeficiency virus (HIV) and are an attractive candidate for gene therapy as they are able to transduce both dividing and non-dividing cells (Yoon and Jun 2002). Due to their derivation from HIV, biosafety was a concern for their application as therapeutics; however, engineered deletions in the long terminal repeat (LTR) promoter reduce the possibility of producing replication competent virus and thereby improve safety (Zufferey et al. 1998). LV is at present the vector of choice for gene therapy within our laboratory, and we have successfully used a lentiviral vector (HMD) to deliver furin-cleavable insulin (INS-FUR) to the livers of STZ-diabetic rats (Ren et al. 2007), non-obese diabetic (NOD) mice (Ren et al. 2013) and pancreatectomised Westran pigs (Gerace et al. 2013). In the rodent animal models, we observed liver to pancreas transdifferentiation characterised by spontaneous expression of β cell transcription factors, formation of insulin storage granules, normal glucose tolerance and permanent reversal of diabetes.

Selecting a Suitable Target Cell for T1D Gene Therapy

Somatic cell gene therapy for T1D was first performed in monkey kidney cells and fibroblasts (Laub and Rutter 1983; Iwata et al. 1993). However, due to their intrinsic lack of β cell characteristics, these cells were not able to produce biologically active insulin despite successfully transcribing and translating the insulin gene. Likewise, the targeting of muscle cells has been studied sparingly due to their lack of β cell characteristics. Nevertheless, a study utilising vascular smooth muscle cells transduced with INS-FUR was able to reduce blood glucose levels in spontaneously diabetic congenic BioBreeding (BB) rats for a period of 6 weeks (Barry et al. 2001). Of more success was the reversal of diabetes following the dual expression of insulin and glucokinase (GK) in the muscle of STZ-diabetic mice for >4 months (Mas et al. 2006). However, a suitable target cell for the production of functional artificial IPCs would possess characteristics similar to those of normal β cells (Table 2) such as a glucose-sensing system, proinsulin-processing enzymes and an exocytosis system (Zaret and Grompe 2008).

Endocrine Cells

Pituitary cells contain proinsulin-processing enzymes and secretory granules. In a study where a murine pituitary cell line (AtT20) was transfected with a recombinant plasmid containing a human preproinsulin cDNA (AtT20Ins-1.4), expression of biologically active insulin was demonstrated; however, glucose responsiveness was absent (Stewart et al. 1994). Upon the insertion of GLUT2 and glucokinase, the AtT20Ins-1.4 cells became glucose responsive; however, the secretion of adenocorticotropic hormone stimulated glucocorticoid synthesis which inhibited insulin function, and therefore limited their effectiveness (Hughes et al. 1993).

K cells, a type of endocrine cell found in the gut, possess many β cell characteristics and have drawn a significant attention as a potential target for gene therapy (Yoon and Jun 2002). An in vivo study targeting K cells with a construct of human insulin under the control of the glucose-dependent insulinotropic polypeptide (GIP) regulatory promoter resulted in the production of biologically active insulin and the restoration of normoglycaemia (Cheung et al. 2000). Despite this encouraging result, an efficient method of delivery to these cells is yet to be established.

Liver Cells

Considering liver cells are derived from the same endodermal origin as the pancreas (Zaret 2008), increasing effort has been focused on engineering hepatocytes to function as artificial IPCs. The possession of GLUT2 and glucokinase in hepatocytes permits a response to fluctuating glucose concentrations similar to that in normal β cells and hence their preferred use in gene therapy protocols for the treatment of T1D. Although hepatocytes do not contain proinsulin-processing enzymes and lack secretory granules, these functions can be induced in certain circumstances (e.g. via the expression of β cell transcription factors or INS-FUR) (Ren et al. 2007, 2013; Gerace et al. 2013; Tuch et al. 2003).

Mesenchymal Stem Cells (MSCs)

Within the last decade, the targeting of MSCs for genetic manipulation has become increasingly popular owing to their wide-ranging immunomodulatory properties and reported ability to elude immune detection (Gebler et al. 2012; Vija et al. 2009; Abdi et al. 2008; da Silva Meirelles et al. 2009). Due to the immunomodulatory capacities of MSC, their predominant use in the correction of diabetes has been in the form of transplantation of native MSCs intended to control immune responses (Lee et al. 2006; Ezquer et al. 2008). However, at the present time, MSCs are more commonly becoming utilised as target cells for the generation of artificial IPCs via chemically induced differentiation protocols or direct transfer of genetic material (Tang et al. 2004). With regard to the chemical induction of MSCs to generate IPCs, it has been shown that BMSCs cultured in high-glucose medium (Oh et al. 2004) or nicotinamide-enriched medium are inclined to differentiate into IPCs (Wu et al. 2007). Similarly, BMSCs subjected to a three-step chemically induced differentiation protocol produced glucose-responsive IPCs with high expression levels of *Pdx-1*, insulin and glucagon (Sun et al. 2007). On the other hand, BMSCs targeted for gene therapy with a retroviral vector containing the β cell transcription factor *Pdx-1* generated IPCs that reduced blood glucose concentrations after transplantation in STZ-diabetic/*Scid* mice and exhibited a normal glucose tolerance until 6–8 weeks post-transplantation (Karnieli et al. 2007).

Viral-Mediated Transfer of β Cell Transcription Factors

Currently, pancreas and islet transplantation are the only "cures" for diabetes mellitus. However, the limitations associated with the current therapeutic options necessitate the requirement for an alternative IPC that is also capable of evading recurrent immune destruction. During normal pancreatic development, islet cell differentiation is regulated by the expression of β cell transcription factors (Fig. 10.1), and during

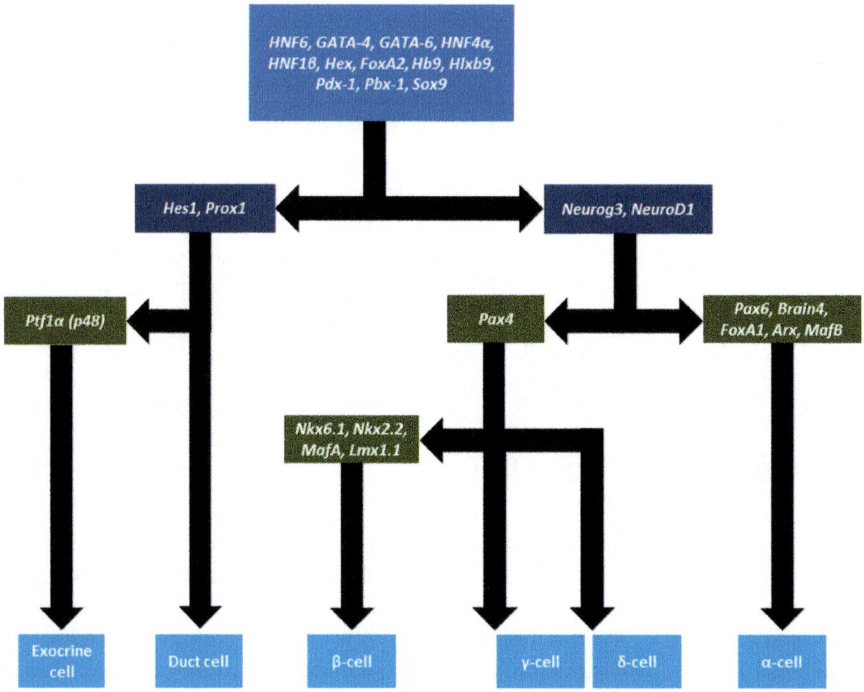

Fig. 10.1 Hierarchical representation of the transcription factors involved in pancreatic islet cell differentiation. Signals from the mesenchyme and notochord induce the early gut endoderm to form the pancreatic buds, where expression of *Pdx-1* then drives differentiation of the pancreatic precursor cells. Exocrine and endocrine differentiation are then specified by the expression of *Neurog3* and *Hes1*, respectively, with subsequent expression of *NeuroD1* maintaining the endocrine cell lineage. α-Cell and β cell differentiation is mediated by *Pax6* and *Pax4,* respectively. Final differentiation of β cells is governed by the expression of *Nkx2.2*, *Nkx6.1* and *MafA*

adult life, the transcription factors regulate the expression of pancreatic hormones. As a result, the utilisation of β cell transcription factors as mediators of IPC generation became of considerable interest as an alternative therapy for the treatment of diabetes mellitus. The production of IPCs for diabetes reversal via viral-mediated transfer of β cell transcription factors has been comprehensively examined in liver tissue as it is derived from the same endodermal origin as the pancreas (Zaret and Grompe 2008), making it more amenable to the transdifferentiation process.

Pdx-1

The direct in vivo delivery of the transcription factor *Pdx-1* to the livers of diabetic mice was studied by Ferber et al. (2000) as a potential method of correcting hyperglycaemia via the induced expression of insulin. The study reported the restoration of normoglycaemia in the mice for a period of 8 days as a consequence of the expression of *Pdx-1* which induced insulin expression and secretion. However, uncontrolled transdifferentiation led to the undesirable development of exocrine tissue which resulted in hepatitis in the liver (Ferber et al. 2000; Kojima et al. 2003).

Exocrine differentiation following delivery of *Pdx-1* to the livers of STZ-diabetic mice was also reported by Kojima et al. (2003) and was presumably a consequence of the continuous expression of high levels of *Pdx-1*. Thus far, *Pdx-1*-mediated transdifferentiation from hepatocytes to pancreatic tissue has been accomplished on numerous occasions with varying degrees of success (Fodor et al. 2007; Ber et al. 2003; Wang et al. 2007; Sapir et al. 2005; Nagaya et al. 2009). Other target cells for direct delivery of *Pdx-1* include mouse pancreas via the bile duct (Taniguchi et al. 2003), rat intestinal epithelium-derived cells (IEC-6) (Yoshida et al. 2002) and primary duct cells (Noguchi et al. 2006). A combinatorial approach utilising the transcription factors *Pdx-1*, *Neurog3* and *MafA* successfully converted pancreatic exocrine cells into IPCs in vivo, providing support for the delivery of combinations of pancreatic transcription factors (Zhou et al. 2008). However, glucose tolerance in the animals differed significantly from normal.

As a result of the success of utilising *Pdx-1* as a mediator of pancreatic transdifferentiation, stem cells emerged as potential targets for gene transfer due to their regenerative capabilities and plasticity. As mentioned in "Mesenchymal Stem Cells (MSCs)" section, MSCs became of particular interest due to their unique immune-evading capabilities. MSCs from a variety of sources have been targeted for *Pdx-1* delivery, including BMSC (Karnieli et al. 2007; Sun et al. 2006; Limbert et al. 2011; Moriscot et al. 2005; Li et al. 2007, 2008), umbilical cord MSC (UC-MSC) (He et al. 2011) and adipose-derived MSC (AMSC) (Baer 2011; Lin et al. 2009; Kajiyama et al. 2010). A study by Miyazaki et al. (2004) targeted a murine embryonic stem cell (ESC) line (EB3) for *Pdx-1* gene transfer and found that it could be induced to differentiate into IPCs; however, the cells lacked expression of pancreatic genes in vivo; therefore, the cells would ultimately be of no therapeutic potential. Consequently, an effort was made to improve the generation of IPCs in a number of other ESC lines that could be of more benefit for the treatment of diabetes mellitus (Lavon et al. 2006; Vincent et al. 2006; Raikwar and Zavazava 2012).

Neurog3

The implications of *Neurog3* in specifying endocrine cell lineage suggest that it could potentially overcome the undesirable development of exocrine differentiation associated with the use of *Pdx-1* as a mediator of pancreatic transdifferentiation. Unfortunately, most studies employing the transfer of *Neurog3* have reported low levels of insulin production and a lack of glucose responsiveness (Noguchi et al. 2006; Kaneto et al. 2005; Wang et al. 2007; Song et al. 2007; Heremans et al. 2002). Adenoviral transfer of *Neurog3* and betacellulin to oval cells did result in the production of insulin and pancreatic transdifferentiation (Yechoor et al. 2009); however, the most effective use of *Neurog3*-induced transdifferentiation was in combination with other transcription factors (Zhou et al. 2008).

NeuroD1

Kojima et al. (2003) expressed *NeuroD1* and betacellulin in the livers of STZ-treated diabetic mice and showed a return to normoglycaemia for greater than 120 days, in addition to the expression of a number of pancreatic transcription factors. Most importantly, there was no evidence of any exocrine differentiation or hepatotoxicity. Further studies have also shown the ability of *NeuroD1* to strongly induce insulin expression, supporting its utility as a means for the production of IPCs (Noguchi et al. 2006; Yatoh et al. 2007).

Success utilising *NeuroD1* has also been reported within our laboratory, where the endocrine-specific transcription factor has been delivered to a genetically modified rat liver cell line (H4IIE) which does not endogenously express β cell transcription factors. Following the delivery of both insulin and *NeuroD1* to the H4IIE cell line, the cells were able to transcribe and translate insulin, process the translated protein into its mature form and package the mature insulin with storage granules (Swan MA 2009). After transplantation in NOD/*Scid* mice, the cells secreted insulin in response to increasing concentrations of glucose and restored normoglycaemia. Most importantly, they also displayed the hallmark characteristics of pancreatic transdifferentiation with expression of *Pdx-1*, *NeuroD1*, *Pax6*, *Nkx2.2* and *Nkx6.1*, in addition to rat insulin 1 and 2, glucagon, somatostatin, proconvertase 1 and 2 (PC1/2) and pancreatic polypeptide. Consequently, this study supports the already available evidence for the prospective use of *NeuroD1* to produce IPCs that are safe for use as a therapeutic for diabetes mellitus.

Pax4

Pax4 is a lower-hierarchy transcription factor that is essential for directing β cell differentiation and therefore is a suitable candidate for the generation of IPCs. Liew et al. (2008) demonstrated that human ESCs engineered to overexpress *Pax4* have an enhanced propensity to form putative β cells. A study supporting this finding showed that IPCs created from mouse ESCs via the overexpression of *Pax4* and selected for nestin expression were proficient in restoring normoglycaemia for 14 days (Blyszczuk et al. 2003). However, the propensity for ESCs to form teratomas limits their potential for clinical application (Stachelscheid et al. 2013; Hentze et al. 2009).

Nkx6.1

The disruption in the differentiation of β cells in mice has been demonstrated by knockouts of the *Nkx6.1* transcription factor, implicating it as an essential part of the β cell developmental pathway. As a result, it has the potential to be a successful mediator of IPC generation. However, ectopic expression of *Nkx6.1* alone has been shown to poorly induce the expression of vital upper-hierarchy β cell transcription factors. Only upon co-expression with *Pdx-1* are the upper-hierarchy transcription factors expressed, which subsequently leads to insulin expression and glucose-responsive insulin release (Gefen-Halevi et al. 2010). As a result of the poor capacity of *Nkx6.1* to induce the expression of the full repertoire of β cell transcription factors, it is not a promising choice for the generation of IPCs.

Viral-Mediated Transfer of Insulin

As mentioned in section "Lentiviral Vectors", our laboratory has successfully employed the use of lentiviral vectors to transfer insulin to hepatocytes as an alternative therapeutic strategy for the treatment of T1D. The development of exocrine differentiation associated with liver-directed gene therapy using *Pdx-1* (Ferber et al. 2000; Kojima et al. 2003) has never been observed in our studies, presumably due to the diverse choice of genes utilised for gene delivery. In the human liver cell line (Huh7), which endogenously expresses β cell transcription factors, we were able to induce the development of insulin storage granules and glucose-responsive insulin secretion following the delivery of the insulin gene (Huh7ins). Diabetes in NOD/*Scid* mice was corrected following transplantation of the Huh7ins cells (Tuch et al. 2003).

Insulin Transfer in Rodent Models

Permanent reversal of diabetes in STZ-diabetic rats (Ren et al. 2007) and spontaneously diabetic NOD mice (Ren et al. 2013) following liver-directed lentiviral delivery of INS-FUR has been demonstrated within our laboratory. In these rodent models, we showed the spontaneous expression of the essential upper-hierarchy β cell transcription factors (*Pdx-1*, *Neurog3* and *NeuroD1*), some lower-hierarchy β cell transcription factors (*Pax4* and *Nkx2.2*) and the development of glucose-responsive insulin secretion (Nathwani et al. 2011; Lisowski et al. 2014; Apelqvist et al. 1999; Sommer et al. 1996). Permanent reversal of diabetes was characterised by normal intravenous glucose tolerance tests in the STZ-diabetic rat and NOD mouse study (Fig. 10.2). The insulin-secreting liver cells and NOD mouse livers engineered to express insulin generated within our laboratory are also resistant to the detrimental effects of β cell cytotoxins and proinflammatory cytokines that play a principle role in the pathogenesis of T1D (Ren et al. 2013; Tabiin et al. 2001, 2004; Tuch et al. 1997).

Due to the success shown by the viral delivery of INS-FUR within our laboratory, a number of successive studies utilising viral delivery of INS-FUR showing amelioration of hyperglycaemia in rodent models were reported (Han et al. 2011; Tatake

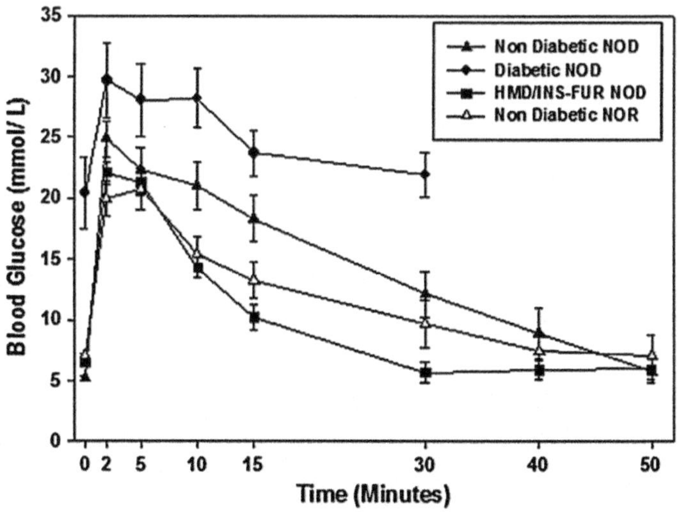

Fig. 10.2 Plasma glucose levels following an intravenous glucose tolerance test (IVGTT) in diabetic non-obese diabetic (NOD) mice treated with furin-cleavable human insulin (INS-FUR) within a lentiviral vector (HMD). An IVGTT was performed on non-diabetic NOD (12–16 weeks), non-obese-resistant (NOR) mice and HMD/INS-FUR-treated NOD mice, 5 months after reversal of diabetes. ($n = 5$, data were examined by one-way analysis of variance after log transformation of data and expressed as the mean ± SEM). [Reproduced from Ren et al. (2013)]

et al. 2007; Tudurí et al. 2012; Hsu et al. 2008). In contrast to our research, Elsner et al. (2012) applied lentiviral delivery of INS-FUR to the liver of diabetic rats and observed reversal of hyperglycaemia; however, pancreatic transdifferentiation and development of secretory granules were absent. Their gene therapy approach also lacked closely regulated control of blood glucose observed within normally functioning β cells, therefore requiring islet supplementation for clinical application.

Insulin Transfer in a Porcine Model

In order to more appropriately translate the rodent studies performed within our laboratory to the clinical setting, we also reversed diabetes in a large animal model that more closely resembles human physiology. Reversal of diabetes following liver-directed lentiviral delivery of INS-FUR was characterised by the expression of β cell transcription factors and normal glucose tolerance (Gerace et al. 2013). Unexpectedly, the complexity of applying the surgical procedure (which isolates the liver from the circulation) in a large animal model led to difficulties in reproducing the successful reversal of diabetes. This suggested that translation of the surgical procedure in the clinical setting may also raise a number of challenges; therefore, it would likely be of more benefit to transplant cells modified ex vivo so as to overcome the surgical obstacles.

Conclusion

Viral-mediated gene transfer of β cell transcription factors and insulin represents an alternative approach to generating artificial IPCs for the treatment of T1D. Currently, one of the challenges facing the translation of these potential alternative cell therapies to the clinic is the generation of sufficient quantities of IPCs on a large scale. The success of liver-directed gene therapy in the past decade is nowadays becoming overshadowed by the greater understanding of the regenerative and therapeutic potential of stem cells. It is understandable that the emerging cell and gene therapy approaches are targeting stem cells for ex vivo modification as they overcome the surgical difficulties associated with in vivo gene therapy. The source of stem cells is also to be considered, as autologous cell therapies would require considerable effort to generate a single therapy. In addition, the increased likelihood of developing a full repertoire of β cell autoantigens from autologous cell modification would increase susceptibility of the grafts to recurrent autoimmunity. As discussed, the success of generating IPCs via gene therapy that are functionally equivalent to normal β cells is closely related to the choice of β cell transcription factor, viral vector and gene promoter, as unwanted exocrine differentiation can lead to tissue destruction. Ideally, targeting of an allogeneic source of cells that are capable of circumventing the autoimmune response for ex vivo gene therapy and subsequent differentiation into IPCs would overcome these limitations.

References

Abdi R, Fiorina P, Adra CN, Atkinson M, Sayegh MH (2008) Immunomodulation by mesenchymal stem cells: a potential therapeutic strategy for type 1 diabetes. Diabetes 57(7):1759–1767

Alba R, Bosch A, Chillon M (2005) Gutless adenovirus: last-generation adenovirus for gene therapy. Gene Ther 12:18–27

Apelqvist A, Li H, Sommer L, Beatus P, Anderson DJ, Honjo T et al (1999) Notch signalling controls pancreatic cell differentiation. Nature 400(6747):877–881

Baer PC (2011) Adipose-derived stem cells and their potential to differentiate into the epithelial lineage. Stem Cells Dev 20(10):1805–1816

Barry SC, Ramesh N, Lejnieks D, Simonson WT, Kemper L, Lernmark A et al (2001) Glucose-regulated insulin expression in diabetic rats. Hum Gene Therapy 12(2):131–139

Ber I, Shternhall K, Perl S, Ohanuna Z, Goldberg I, Barshack I et al (2003) Functional, persistent, and extended liver to pancreas transdifferentiation. J Biol Chem 278(34):31950–31957

Blyszczuk P, Czyz J, Kania G, Wagner M, Roll U, St Onge L et al (2003) Expression of Pax4 in embryonic stem cells promotes differentiation of nestin-positive progenitor and insulin-producing cells. Proc Natl Acad Sci USA 100(3):998–1003

Bushman FD (2007) Retroviral integration and human gene therapy. J Clin Investig 117(8):2083–2086

Cavazzana-Calvo M, Hacein-Bey S, de Saint Basile G, Gross F, Yvon E, Nusbaum P et al (2000) Gene therapy of human severe combined immunodeficiency (SCID) X1 disease. Science 288(5466):669–672

Cheung AT, Dayanandan B, Lewis JT, Korbutt GS, Rajotte RV, Bryer-Ash M et al (2000) Glucose-dependent insulin release from genetically engineered K cells. Science 290(5498):1959–1962 da Silva Meirelles L, Fontes AM, Covas DT, Caplan AI (2009) Mechanisms involved in the therapeutic properties of mesenchymal stem cells. Cytokine Growth Factor Rev 20(5):419–427

Elsner M, Terbish T, Jorns A, Naujok O, Wedekind D, Hedrich H-J et al (2012) Reversal of diabetes through gene therapy of diabetic rats by hepatic insulin expression via lentiviral transduction. Mol Ther 20(5):918–926

Ezquer FE, Ezquer ME, Parrau DB, Carpio D, Yanez AJ, Conget PA (2008) Systemic administration of multipotent mesenchymal stromal cells reverts hyperglycemia and prevents nephropathy in type 1 diabetic mice. Biol Blood Marrow Transpl 14(6):631–640

Ferber S, Halkin A, Cohen H, Ber I, Einav Y, Goldberg I et al (2000) Pancreatic and duodenal homeobox gene 1 induces expression of insulin genes in liver and ameliorates streptozotocin-induced hyperglycemia. Nat Med 6(5):568–572

Fodor A, Harel C, Fodor L, Armoni M, Salmon P, Trono D et al (2007) Adult rat liver cells transdifferentiated with lentiviral IPF1 vectors reverse diabetes in mice: an ex vivo gene therapy approach. Diabetologia 50(1):121–130

Gebler A, Zabel O, Seliger B (2012) The immunomodulatory capacity of mesenchymal stem cells. Trends Mol Med 18(2):128–134

Gefen-Halevi S, Rachmut IH, Molakandov K, Berneman D, Mor E, Meivar-Levy I et al (2010) NKX6.1 promotes PDX-1-induced liver to pancreatic beta-cells reprogramming. Cell Reprogram 12(6):655–664

Gerace D, Ren B, Hawthorne WJ, Byrne MR, Phillips PM, O'Brien BA et al (2013) Pancreatic transdifferentiation in porcine liver following lentiviral delivery of human furin–cleavable insulin. Transpl Proc 45(5):1869–1874

Han J, McLane B, Kim E-H, Yoon J-W, Jun H-S (2011) Remission of diabetes by insulin gene therapy using a hepatocyte-specific and glucose-responsive synthetic promoter. Mol Ther 19(3):470–478

He D, Wang J, Gao Y, Zhang Y (2011) Differentiation of PDX1 gene-modified human umbilical cord mesenchymal stem cells into insulin-producing cells in vitro. Int J Mol Med 28(6):1019–1024

Hentze H, Soong PL, Wang ST, Phillips BW, Putti TC, Dunn NR (2009) Teratoma formation by human embryonic stem cells: evaluation of essential parameters for future safety studies. Stem Cell Res 2(3):198–210

Heremans Y, Van De Casteele M, Gradwohl G, Serup P, Madsen O et al (2002) Recapitulation of embryonic neuroendocrine differentiation in adult human pancreatic duct cells expressing neurogenin 3. J Cell Biol 159(2):303–312

Hsu P-J, Kotin R, Yang Y-W (2008) Glucose- and metabolically regulated hepatic insulin gene therapy for diabetes. Pharm Res 25(6):1460–1468

Hughes SD, Quaade C, Johnson JH, Ferber S, Newgard CB (1993) Transfection of AtT-20ins cells with GLUT-2 but not GLUT-1 confers glucose-stimulated insulin secretion. Relationship to glucose metabolism. J Biol Chem 268(20):15205–15212

Iwata H, Ogawa N, Takagi T, Mizoguchi J (1993) Preparation of insulin-releasing Chinese hamster ovary cell by transfection of human insulin gene. In: Polymers of biological and biomedical significance, vol 540. American Chemical Society, pp 306–313

Kajiyama H, Hamazaki TS, Tokuhara M, Masui S, Okabayashi K, Ohnuma K et al (2010) Pdx1-transfected adipose tissue-derived stem cells differentiate into insulin-producing cells in vivo and reduce hyperglycemia in diabetic mice. Int J Dev Biol 54(4):699–705

Kaneto H, Nakatani Y, Miyatsuka T, Matsuoka TA, Matsuhisa M, Hori M et al (2005) PDX-1/VP16 fusion protein, together with NeuroD or Ngn3, markedly induces insulin gene transcription and ameliorates glucose tolerance. Diabetes 54(4):1009–1022

Karnieli O, Izhar-Prato Y, Bulvik S, Efrat S (2007) Generation of insulin-producing cells from human bone marrow mesenchymal stem cells by genetic manipulation. Stem Cells 25(11):2837–2844

Knowles MR, Hohneker KW, Zhou Z, Olsen JC, Noah TL, Hu P-C et al (1995) A controlled study of adenoviral-vector–mediated gene transfer in the nasal epithelium of patients with cystic fibrosis. N Engl J Med 333(13):823–831

Kojima H, Fujimiya M, Matsumura K, Younan P, Imaeda H, Maeda M et al (2003) NeuroD-betacellulin gene therapy induces islet neogenesis in the liver and reverses diabetes in mice. Nat Med 9(5):596–603

Laub O, Rutter WJ (1983) Expression of the human insulin gene and cDNA in a heterologous mammalian system. J Biol Chem 258(10):6043–6050

Laufs S, Nagy KZ, Giordano FA, Hotz-Wagenblatt A, Zeller WJ, Fruehauf S (2004) Insertion of retroviral vectors in NOD/SCID repopulating human peripheral blood progenitor cells occurs preferentially in the vicinity of transcription start regions and in introns. Mol Ther 10(5):874–881

Lavon N, Yanuka O, Benvenisty N (2006) The effect of overexpression of Pdx1 and Foxa2 on the differentiation of human embryonic stem cells into pancreatic cells. Stem Cells 24(8):1923–1930

Lee RH, Seo MJ, Reger RL, Spees JL, Pulin AA, Olson SD et al (2006) Multipotent stromal cells from human marrow home to and promote repair of pancreatic islets and renal glomeruli in diabetic NOD/scid mice. Proc Natl Acad Sci USA 103(46):17438–17443

Li L, Li F, Qi H, Feng G, Yuan K, Deng H et al (2008) Coexpression of Pdx1 and betacellulin in mesenchymal stem cells could promote the differentiation of nestin-positive epithelium-like progenitors and pancreatic islet-like spheroids. Stem Cells Dev 17(4):815–823

Li Y, Zhang R, Qiao H, Zhang H, Wang Y, Yuan H et al (2007) Generation of insulin-producing cells from PDX-1 gene-modified human mesenchymal stem cells. J Cell Physiol 211(1):36–44

Liew CG, Shah NN, Briston SJ, Shepherd RM, Khoo CP, Dunne MJ et al (2008) PAX4 enhances beta-cell differentiation of human embryonic stem cells. PLoS One 3(3):e1783

Limbert C, Path G, Ebert R, Rothhammer V, Kassem M, Jakob F et al (2011) PDX1- and NGN3-mediated in vitro reprogramming of human bone marrow-derived mesenchymal stromal cells into pancreatic endocrine lineages. Cytotherapy 13(7):802–813

Lin G, Wang G, Liu G, Yang LJ, Chang LJ, Lue TF et al (2009) Treatment of type 1 diabetes with adipose tissue-derived stem cells expressing pancreatic duodenal homeobox 1. Stem Cells Dev 18(10):1399–1406

Lisowski L, Dane AP, Chu K, Zhang Y, Cunningham SC, Wilson EM et al (2014) Selection and evaluation of clinically relevant AAV variants in a xenograft liver model. Nature 506(7488):382–386

Mas A, Montané J, Anguela XM, Muñoz S, Douar AM, Riu E et al (2006) Reversal of type 1 diabetes by engineering a glucose sensor in skeletal muscle. Diabetes 55(6):1546–1553

McCaffrey AP, Fawcett P, Nakai H, McCaffrey RL, Ehrhardt A, Pham TT et al (2008) The host response to adenovirus, helper-dependent adenovirus, and adeno-associated virus in mouse liver. Mol Ther 16(5):931–941

Miyazaki S, Yamato E, Miyazaki JI (2004) Regulated expression of pdx-1 promotes in vitro differentiation of insulin-producing cells from embryonic stem cells. Diabetes 53(4):1030–1037

Morgan RA, Anderson WF (1993) Human gene therapy. Annu Rev Biochem 62:191–217

Moriscot C, de Fraipont F, Richard M-J, Marchand M, Savatier P, Bosco D et al (2005) Human bone marrow mesenchymal stem cells can express insulin and key transcription factors of the endocrine pancreas developmental pathway upon genetic and/or microenvironmental manipulation in vitro. Stem Cells 23(4):594–603

Muzyczka N (1992) Use of adeno-associated virus as a general transduction vector for mammalian cells. In: Muzyczka N (ed) Viral expression vectors, vol 158. Springer, Berlin, pp 97–129

Nagaya M, Katsuta H, Kaneto H, Bonner-Weir S, Weir GC (2009) Adult mouse intrahepatic biliary epithelial cells induced in vitro to become insulin-producing cells. J Endocrinol 201(1):37–47

Nathwani AC, Tuddenham EG, Rangarajan S, Rosales C, McIntosh J, Linch DC et al (2011) Adenovirus-associated virus vector-mediated gene transfer in hemophilia B. N Engl J Med 365(25):2357–2365

Nayak S, Herzog RW (2010) Progress and prospects: immune responses to viral vectors. Gene Ther 17(3):295–304

Noguchi H, Xu G, Matsumoto S, Kaneto H, Kobayashi N, Bonner-Weir S et al (2006) Induction of pancreatic stem/progenitor cells into insulin-producing cells by adenoviral-mediated gene transfer technology. Cell Transpl 15(10):929–938

Oh S-H, Muzzonigro TM, Bae S-H, LaPlante JM, Hatch HM, Petersen BE (2004) Adult bone marrow-derived cells trans-differentiating into insulin-producing cells for the treatment of type I diabetes. Lab Invest 84(5):607–617

Raikwar SP, Zavazava N (2012) PDX1-engineered embryonic stem cell-derived insulin producing cells regulate hyperglycemia in diabetic mice. Transpl Res 1(1):1440–2047

Ren B, O'Brien BA, Byrne MR, Ch'ng E, Gatt PN, Swan MA et al (2013) Long-term reversal of diabetes in non-obese diabetic mice by liver-directed gene therapy. J Gene Med 15(1):28–41

Ren B, O'Brien BA, Swan MA, Koina ME, Nassif N, Wei MQ et al (2007) Long-term correction of diabetes in rats after lentiviral hepatic insulin gene therapy. Diabetologia 50(9):1910–1920

Sapir T, Shternhall K, Meivar-Levy I, Blumenfeld T, Cohen H, Skutelsky E et al (2005) Cell-replacement therapy for diabetes: generating functional insulin-producing tissue from adult human liver cells. Proc Natl Acad Sci USA 102(22):7964–7969

Simpson AM, Tao C, Swan MA, B R, O'Brien BA (2009) An engineered rat liver cell line H4IIEins/ND reverses diabetes in mice. In: International diabetes federation world diabetes congress, Montreal, Abstract no. MT-0996

Sommer L, Ma Q, Anderson DJ (1996) neurogenins, a novel family of atonal-related bHLH transcription factors, are putative mammalian neuronal determination genes that reveal progenitor cell heterogeneity in the developing CNS and PNS. Mol Cell Neurosci 8(4):221–241

Song YD, Lee EJ, Yashar P, Pfaff LE, Kim SY, Jameson JL (2007) Islet cell differentiation in liver by combinatorial expression of transcription factors neurogenin-3, BETA2, and RIPE3b1. Biochem Biophys Res Commun 354(2):334–339

Stachelscheid H, Wulf-Goldenberg A, Eckert K, Jensen J, Edsbagge J, Bjorquist P et al (2013) Teratoma formation of human embryonic stem cells in three-dimensional perfusion culture bioreactors. J Tissue Eng Regen Med 7(9):729–741

Stewart C, Taylor NA, Green IC, Docherty K, Bailey CJ (1994) Insulin-releasing pituitary cells as a model for somatic cell gene therapy in diabetes mellitus. J Endocrinol 142(2):339–343

Sugiyama A, Hattori S, Tanaka S, Isoda F, Kleopoulos S, Rosenfeld M et al (1997) Defective adenoassociated viral-mediated transfection of insulin gene by direct injection into liver parenchyma decreases blood glucose of diabetic mice. Horm Metab Res 29(12):599–603

Sun Y, Chen L, Hou XG, Hou WK, Dong JJ, Sun L et al (2007) Differentiation of bone marrow-derived mesenchymal stem cells from diabetic patients into insulin-producing cells in vitro. Chin Med J 120(9):771–776

Sun J, Yang Y, Wang X, Song J, Jia Y (2006) Expression of Pdx-1 in bone marrow mesenchymal stem cells promotes differentiation of islet-like cells in vitro. Sci China C Life Sci 49(5):480–489

Tabiin MT, Tuch BE, Bai L, Han XG, Simpson AM (2001) Susceptibility of insulin-secreting hepatocytes to the toxicity of pro-inflammatory cytokines. J Autoimmun 17(3):229–242

Tabiin MT, White CP, Morahan G, Tuch BE (2004) Insulin expressing hepatocytes not destroyed in transgenic NOD mice. J Autoimmune Dis 1(1):3

Tang DQ, Cao LZ, Burkhardt BR, Xia CQ, Litherland SA, Atkinson MA et al (2004) In vivo and in vitro characterization of insulin-producing cells obtained from murine bone marrow. Diabetes 53(7):1721–1732

Taniguchi H, Yamato E, Tashiro F, Ikegami H, Ogihara T, Miyazaki J (2003) beta-cell neogenesis induced by adenovirus-mediated gene delivery of transcription factor pdx-1 into mouse pancreas. Gene Ther 10(1):15–23

Tatake RJ, O'Neill MM, Kennedy CA, Reale VD, Runyan JD, Monaco K-AD et al (2007) Glucose-regulated insulin production from genetically engineered human non-beta cells. Life Sci 81(17–18):1346–1354

Tuch BE, Beynon S, Tabiin MT, Sassoon R, Goodman RJ, Simpson AM (1997) Effect of beta-cell toxins on genetically engineered insulin-secreting cells. J Autoimmun 10(3):239–244

Tuch BE, Szymanska B, Yao M, Tabiin MT, Gross DJ, Holman S et al (2003) Function of a genetically modified human liver cell line that stores, processes and secretes insulin. Gene Ther 10(6):490–503

Tudurí E, Bruin JE, Kieffer TJ (2012) Restoring insulin production for type 1 diabetes. J Diabetes 4(4):319–331

Vija L, Farge D, Gautier JF, Vexiau P, Dumitrache C, Bourgarit A et al (2009) Mesenchymal stem cells: stem cell therapy perspectives for type 1 diabetes. Diabetes Metab 35(2):85–93

Vincent R, Treff N, Budde M, Kastenberg Z, Odorico J (2006) Generation and characterization of novel tetracycline-inducible pancreatic transcription factor-expressing murine embryonic stem cell lines. Stem Cells Dev 15(6):953–962

Volpers C, Kochanek S (2004) Adenoviral vectors for gene transfer and therapy. J Gene Med 6(S1):S164–S171

Wang AY, Ehrhardt A, Xu H, Kay MA (2007a) Adenovirus transduction is required for the correction of diabetes using Pdx-1 or Neurogenin-3 in the liver. Mol Ther J Am Soc Gene Ther 15(2):255–263

Wang AY, Ehrhardt A, Xu H, Kay MA (2007b) Adenovirus transduction is required for the correction of diabetes using Pdx-1 or Neurogenin-3 in the liver. Mol Ther 15(2):255–263

Wold WS, Doronin K, Toth K, Kuppuswamy M, Lichtenstein DL, Tollefson AE (1999) Immune responses to adenoviruses: viral evasion mechanisms and their implications for the clinic. Curr Opin Immunol 11(4):380–386

Wu XH, Liu CP, Xu KF, Mao XD, Zhu J, Jiang JJ et al (2007) Reversal of hyperglycemia in diabetic rats by portal vein transplantation of islet-like cells generated from bone marrow mesenchymal stem cells. World J Gastroenterol 13(24):3342–3349

Xu J, Lu Y, Ding F, Zhan X, Zhu M, Wang Z (2007) Reversal of diabetes in mice by intrahepatic injection of bone-derived GFP-murine mesenchymal stem cells infected with the recombinant retrovirus-carrying human insulin gene. World J Surg 31(9):1872–1882

Yatoh S, Akashi T, Chan PP, Kaneto H, Sharma A, Bonner-Weir S et al (2007) NeuroD and reaggregation induce β-cell specific gene expression in cultured hepatocytes. Diabetes/Metab Res Rev 23(3):239–249

Yechoor V, Liu V, Espiritu C, Paul A, Oka K, Kojima H et al (2009) Neurogenin3 is sufficient for transdetermination of hepatic progenitor cells into neo-islets in vivo but not transdifferentiation of hepatocytes. Dev Cell 16(3):358–373

Yoon JW, Jun HS (2002) Recent advances in insulin gene therapy for type 1 diabetes. Trends Mol Med 8(2):62–68

Yoshida S, Kajimoto Y, Yasuda T, Watada H, Fujitani Y, Kosaka H et al (2002) PDX-1 induces differentiation of intestinal epithelioid IEC-6 into insulin-producing cells. Diabetes 51(8):2505–2513

Zabner J, Couture LA, Gregory RJ, Graham SM, Smith AE, Welsh MJ (1993) Adenovirus-mediated gene transfer transiently corrects the chloride transport defect in nasal epithelia of patients with cystic fibrosis. Cell 75(2):207–216

Zaret KS (2008) Genetic programming of liver and pancreas progenitors: lessons for stem-cell differentiation. Nat Rev Genet 9(5):329–340

Zaret KS, Grompe M (2008) Generation and regeneration of cells of the liver and pancreas. Science 322(5907):1490–1494

Zhou Q, Brown J, Kanarek A, Rajagopal J, Melton DA (2008) In vivo reprogramming of adult pancreatic exocrine cells to b-cells. Nature 455(7213):627–632

Zhou HS, Liu DP, Liang CC (2004) Challenges and strategies: the immune responses in gene therapy. Med Res Rev 24(6):748–761

Zufferey R, Dull T, Mandel RJ, Bukovsky A, Quiroz D, Naldini L et al (1998) Self-inactivating lentivirus vector for safe and efficient in vivo gene delivery. J Virol 72(12):9873–9880

Chapter 11
Regenerative Medicine: Clinical Islet Transplantation

Helen E. Thomas, Kate L. Graham, Thomas Loudovaris and Thomas W.H. Kay

Type 1 Diabetes

Type 1 diabetes is the result of autoimmune destruction of pancreatic beta cells, resulting in insulin deficiency and the need for lifelong insulin therapy. Even though insulin replacement is lifesaving for patients, it cannot match the glycemic control provided by beta cells. There is significant morbidity associated with type 1 diabetes, with a high risk of developing complications such as vascular disease, blindness, and kidney failure. Attempts to provide tight glycemic control reduce the risk of complications; however, the risk of hypoglycemia is high with intensive insulin therapy. Because of these risks, the replacement of beta cells offers an alternative to insulin therapy.

Islet Transplantation

Transplantation of either whole pancreas or islets is the current method of beta cell replacement (Niclauss et al. 2014). Whole-pancreas transplantation was first achieved in 1966, and although there were many technical failures at first, advances in surgical procedures and immunosuppression have increased the success rate

H.E. Thomas (✉) · K.L. Graham · T. Loudovaris · T.W.H. Kay
St. Vincent's Institute of Medical Research, 41 Victoria Parade, Fitzroy, VIC 3065, Australia
e-mail: hthomas@svi.edu.au

H.E. Thomas · K.L. Graham · T.W.H. Kay
Department of Medicine, St. Vincent's Hospital, The University of Melbourne, Fitzroy, VIC, Australia

A.A. Hardikar (ed.), *Pancreatic Islet Biology*, Stem Cell Biology and Regenerative Medicine, DOI 10.1007/978-3-319-45307-1_11

to around 80 % survival at 1 year and 55 % survival at 5 years posttransplant (International Pancreas Transplant Registry data for 2006–2010) (Gruessner 2011; Gruessner et al. 2012). Islets were first transplanted in 1974; however, the success rate of clinically transplanted islets was very low until the Edmonton protocol, published in 2000, reported a success rate of 80 % after 1 year (Shapiro et al. 2000). Since then, over 750 islet transplants have been done in over 30 international centers (Bruni et al. 2014; Vantyghem et al. 2014). The success rate in the 2007–2010 period, based on insulin independence, is 66 % at 1 year, dropping to 44 % at 3 years (Collaborative Islet Transplant Registry data) (Barton et al. 2012). These rates are approaching the success of pancreas transplantation, without the complication and morbidity associated with whole-organ transplantation.

Recipients of islet grafts are patients with severe hypoglycemia or hypoglycemia unawareness. The best recipients are older patients with lower insulin requirements. In this group of patients, excellent metabolic control can be restored without the morbidity associated with whole-pancreas transplantation, which is most often performed together with a kidney transplant in patients with end-stage nephropathy that is secondary to type 1 diabetes.

Islets for transplantation are isolated from donated pancreases. The best donors are younger and heavier individuals, who usually have a greater number of islets that are easily purified from the exocrine pancreas. Donors with a short stay in intensive care and pancreases retrieved with short warm/cold ischemia times are more suitable for transplantation. Islets are isolated using mechanical and enzymatic digestion, followed by density gradient purification (Fig. 11.1). There are many excellent recent reviews about the history and methods of islet

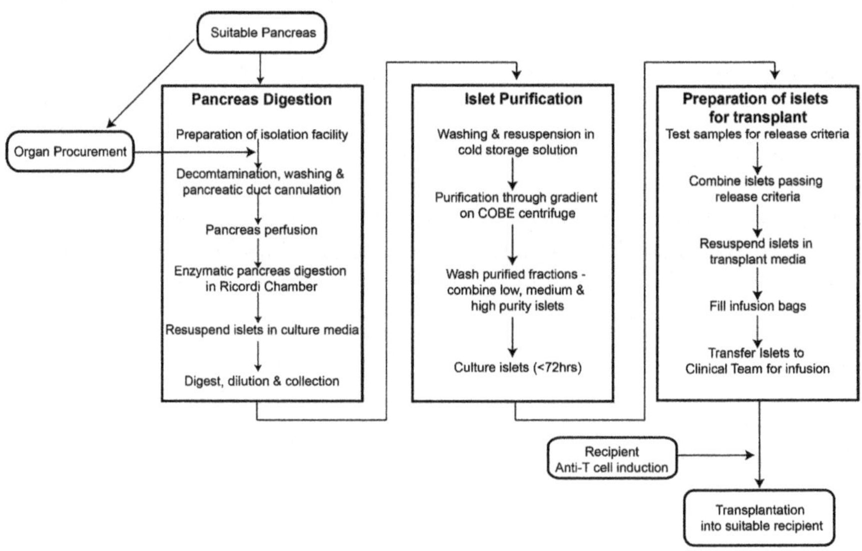

Fig. 11.1 Schematic overview of the islet isolation process

transplantation, so these will not be covered in detail here (Bruni et al. 2014; Niclauss et al. 2014; Vantyghem et al. 2014).

Mechanisms of Islet Graft Rejection

Despite the increasing long-term survival rates of islet allografts, one of the main factors restricting transplantation of allogeneic islets as a common therapy for type 1 diabetes is rejection. Rejection of islets is multifactorial (Fig. 11.2). There is primary loss of graft function in the early time period after transplantation, due to a process known as the instant blood-mediated inflammatory reaction (Bennet et al. 1999), as well as hypoxia due to loss of vascularization during the islet isolation process. Once engrafted, islets are subject to immune-mediated rejection due to allogeneic or autoimmune processes.

Instant Blood-Mediated Inflammatory Reaction

The instant blood-mediated inflammatory reaction (IBMIR) occurs in the first 3 h of islet infusion, and it is estimated that up to 50 % islets are lost in this time

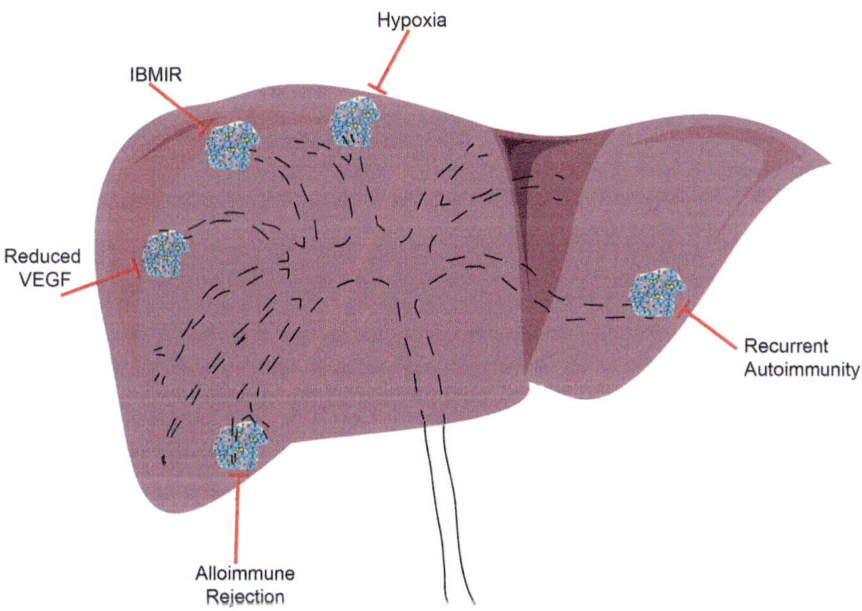

Fig. 11.2 Rejection mechanisms involved in islet graft rejection and transplant failure

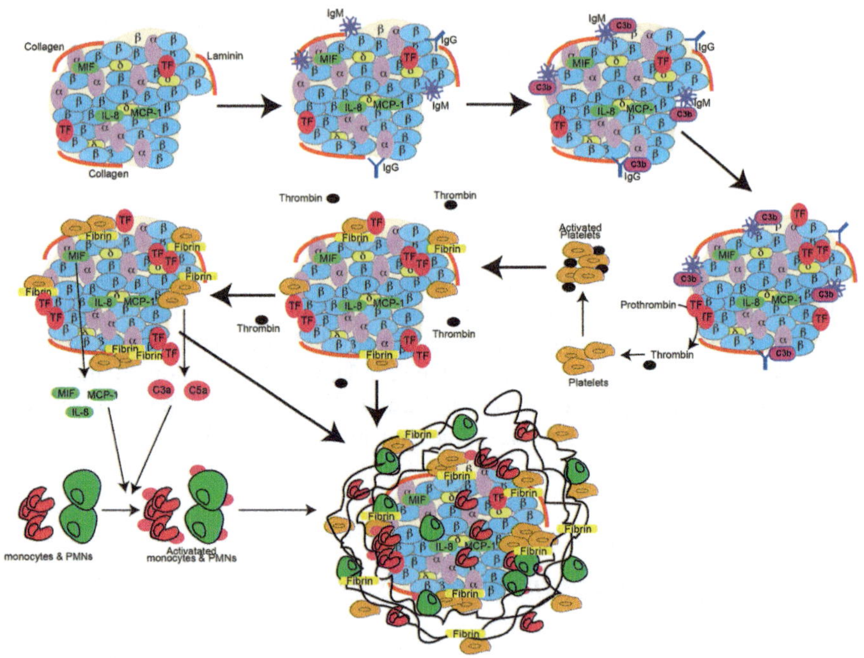

Fig. 11.3 Model of instant blood-mediated inflammatory reaction (IBMIR). When isolated islets are exposed to blood, IgG and IgM antibodies bind to proteins on the islet surface including collagen and laminins. This activates complement pathways resulting in deposition of C3b on the surface. Tissue factor (TF) then activates the coagulation system generating thrombin, which activates platelets and leads to fibrin deposition on islet surface and the binding of platelets. This is further amplified by C3b deposition, which increases the level of tissue factor. Platelet activation also promotes the release of cytokines/chemokines (MCP-1, MIF and IL-8) and complement proteins C3a and C5a. These act to recruit monocytes and polymorphonuclear leukocytes (PMNs), which also deposit on the islet. Platelet activation continues to amplify the amount of thrombin released and generates more fibrin. This eventually forms a capsule around the islet consisting of platelets, fibrin, monocytes and PMNs, which disrupts the islet integrity and ultimately leads to islet loss (Nilsson et al. 2011)

period (Fig. 11.3) (Moberg et al. 2002; Ozmen et al. 2002). This islet loss has been demonstrated in allogeneic, xenogeneic, and autologous islet transplantation (Goto et al. 2008; Naziruddin et al. 2014). IBMIR is an innate thrombotic and inflammatory response characterized by coagulation, complement activation, infiltration of immune cells, and platelet adhesion (Moberg et al. 2005). Thrombosis results in the release of inflammatory mediators including IL-8, MCP-1, and macrophage migration inhibitory factor. Thrombin promotes the activation of monocytes, neutrophils, and platelets. A major trigger of IBMIR is tissue factor, which is expressed by islets (Ji et al. 2011; Johansson et al. 2005; Moberg et al. 2002). Blocking tissue factor with inhibitory monoclonal antibodies or inhibiting its expression prevents the thrombotic response in vitro (Moberg et al. 2005; Ozmen et al. 2002).

Strategies for the prevention of IBMIR include treatment of the recipient with heparin to prevent thrombosis, and this method is used in the current islet allotransplantation protocols. Other strategies include thrombin inhibitors and complement inhibitors (Contreras et al. 2004; Ozmen et al. 2002). However, these all involve systemic treatment of the recipient, and this places them at risk of bleeding or infection from complement inhibition and anti-coagulation. Another approach is modulation of islets, for example reduction of tissue factor expression using siRNA. Islet pretreatment or coating with substances such as nicotinamide or heparin may prevent thrombosis, as well as providing an anchor for VEGF-A, which promotes revascularization (Zhang et al. 2004). However, all of these require manipulation of islets prior to transplantation, some of which are not yet feasible for clinical transplantation.

Hypoxia

During islet isolation and transplantation, the islet vasculature is lost, so the islets must rely on diffusion of oxygen and nutrients from cell culture medium and blood vessels in the transplant environment. This prolonged hypoxia is likely to be one of the major factors in early graft loss. The negative effect of hypoxia on islet survival is thought to occur through hypoxia-inducible factor (HIF)-1α and activation of its target genes that can result in impaired islet function and apoptosis (Cantley et al. 2010; Cheng et al. 2010; Moritz et al. 2002).

Approaches to increase the oxygen supply to islets after transplantation include in situ oxygen generation and improved revascularization. For in situ oxygen generation, a subcutaneous implantable device with controlled and adequate oxygen supply has been reported; however, it requires daily replenishment of the oxygen compartment (Ludwig et al. 2010, 2012). Alternatively hydrolytically reactive solid peroxides can prevent hypoxia-induced islet cell death by providing a continuous chemical release of oxygen (Pedraza et al. 2012). However, oxygen is only provided for around 40 days, and once exhausted, the islets may be left only with oxygen obtained from the vasculature or surrounding tissue. Implantable oxygen generators are currently being investigated (Klearchos Papas, personal communication); however, they may not be able to deliver sufficient oxygen for islets to function, even though they are able to prevent hypoxia (Dionne et al. 1989). It is also possible to improve oxygenation of the donor pancreas before islet isolation with oxygenation buffers such as perfluorocarbon (Brandhorst et al. 2008), or gaseous oxygen perfusion using the persufflation device that has been developed for heart transplantation (Suszynski et al. 2013).

Vascular endothelial growth factor (VEGF) stimulates proliferation and survival of endothelial cells resulting in the formation of new blood vessels, which takes from 2 to 14 days after islet infusion. There is reduced VEGF expression in islets in the first few days after infusion that may limit the extent of vascularization during this period (Jansson and Carlsson 2002; Mattsson et al. 2002). In addition, hyperglycemia can inhibit angiogenesis (Dubois et al. 2010). Overexpression of VEGF can improve vascularization and function of islet grafts (Shimoda et al. 2010). A more clinically applicable approach may be to deliver oxygen or pro-angiogenic factors to the host liver before islet infusion, for example using ultrasound-targeted microbubbles (Shimoda et al. 2010).

Cellular Rejection

The recurrence of autoimmunity occurs due to reactivation of memory islet-reactive T cells after exposure to islets (Abreu and Roep 2013). This is known to be a problem in pancreas and islet transplantation from studies in HLA-identical twins, where a rapid return to hyperglycemia occurred after pancreas transplantation (Sutherland et al. 1984). Alloimmune and autoimmune rejection (Fig. 11.4) are

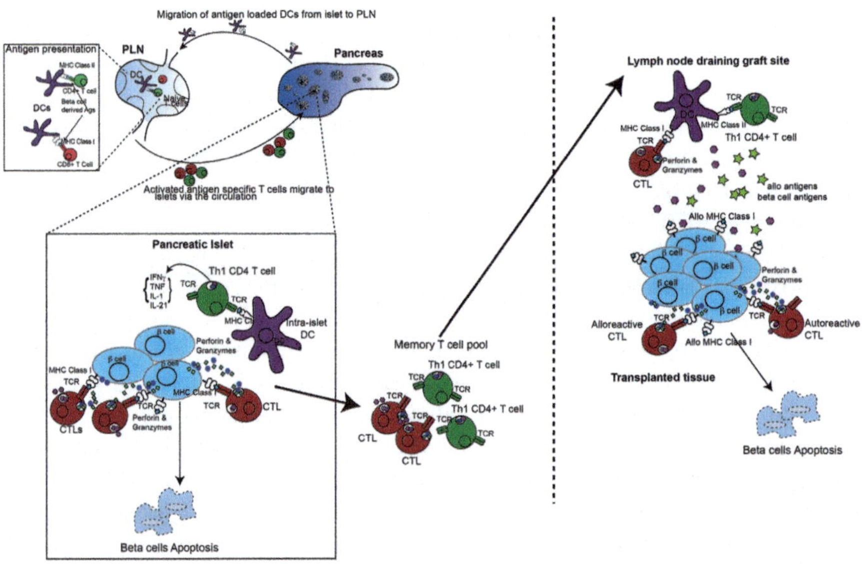

Fig. 11.4 Cellular mechanisms of islet destruction in type 1 diabetes and allogeneic and autoimmune rejection of transplanted islet tissue. In type 1 diabetes dendritic cells loaded with beta cell antigens migrate from islets to pancreatic lymph nodes and present antigen to naïve autoreactive $CD4^+$ and $CD8^+$ T cells. These T cells undergo initial activation and migrate toward the pancreas via the circulation. Once in the islet, cytotoxic $CD8^+$ T cells (CTLs) are the main mediators of beta cell death through the production of perforin and granzymes. Th1 $CD4^+$ T cells do not directly kill beta cells, but secrete proinflammatory cytokines that promote CTL function and activate macrophages. Within the islet a pool of memory CTLs and Th1 $CD4^+$ T cells forms, which then emigrate from the islets and reside in the periphery. In islet transplantation memory T cells that recognize both allogeneic antigens and also beta cell autoantigens are reactivated in the lymph node draining the graft site. These cells then infiltrate the graft to destroy the transplanted islet cells

dealt with to some extent with immunosuppression; however, monitoring immune-mediated rejection still remains very difficult, and there is a need for new biomarkers to be developed (Abreu and Roep 2013).

Several studies have described an association between the presence of autoantibodies or islet-reactive T cells and graft failure or failure to achieve insulin independence after islet transplantation (Monti et al. 2008; Piemonti et al. 2013; Vendrame et al. 2010). Humoral immunity can be measured in the form of donor-specific antibodies and an increase in autoantibodies, and both of these correlate with poor graft function (Piemonti et al. 2013). It is much more difficult to measure cellular rejection, partly because patients are immunosuppressed after transplantation, and partly due to limitations of the assays developed for measuring cellular immunity. We have recently shown that the presence of autoreactive effector-memory T cells in peripheral blood report pathology in the islet, suggesting that assessing peripheral T cell responses may be a useful marker of islet graft rejection (Chee et al. 2014). Huurman et al. showed that pretransplant cellular reactivity was strongly associated with a delay in insulin independence after transplantation (Huurman et al. 2008).

There is much to be done to improve the survival and function of islets after transplantation. Current areas of research include the use of alternative sources of beta cells (e.g., stem cells or xenotransplantation), encapsulation of islets in devices to provide immune isolation, the use of scaffolds to provide support and blood supply to islets, development of targeted immunosuppressive agents, and co-transplantation with cells to prolong islet survival. In this review, we will focus on some of these methods that are being developed to improve the survival and function of islets.

Imaging

Currently we know very little about the process of islet graft rejection in humans. Assessment of direct damage to islets in real time in the early peri-transplant period, as well as immune destruction of islet grafts, would greatly enhance our knowledge of islet transplantation. Current monitoring of transplant recipients includes measurements of blood glucose levels, serum C peptide, glucose tolerance tests, and HbA1c. These are all tests of metabolic function, and changes in these parameters are likely to reflect damage to islets that cannot be rescued. If we had early biomarkers of islet cell damage or immune infiltration, intervention to prevent further loss of graft function may be possible.

Several methods for imaging islet grafts have been tested, and although none are yet in routine clinical use, they show promise for future assessment of islet graft survival. Needle biopsies, although an invasive procedure, are perhaps the most direct way to assess islet graft destruction and immune infiltration (Toso et al. 2009). However, the chance of locating an islet by biopsy is low, and often only a small number of islets can be assessed. Therefore, noninvasive methods that can assess all islets need to be developed. Possible methods include positron emission tomography (PET), magnetic resonance imaging (MRI), and ultrasonography (Sakata et al. 2013).

PET is noninvasive imaging of cellular consumption of positron-emitting isotopes. It is safe and has high resolution and good sensitivity. A trial of PET imaging during clinical islet transplantation after labeling islets with 2-[^{18}F]-fluoro-2deoxy-D-glucose demonstrated radioactivity in the liver (Eriksson et al. 2009). There was a wide variation between patients in location and concentration of islets. The peak radioactivity was 19 min after the start of infusion, but was only 75 % of what was expected, indicating that there was some loss of islets during infusion (Eriksson et al. 2009). Other trials of PET in preclinical models suggest that there may be more radioactive probe uptake when there is better islet function (Kim et al. 2006). However, this technique requires the labeling of islets prior to transplantation, or genetic modification of islets to enable them to retain probe, and therefore currently cannot be applied to posttransplant islets. The development of novel probes to specifically label islets after intravenous injection, with a long half-life for long-term visualization of islet grafts will make PET a useful technique in the future.

So far MRI has not been useful for visualization of loss of islet graft function, even though it has high resolution and good penetration depth. This may be because there is a similar intensity between transplanted islets and liver tissue. Therefore, similar to PET, islets need to be pretreated with a labeling agent before imaging. Several preclinical studies suggest that with more development, MRI may be useful for visualization of islet engraftment (Kim et al. 2012; Kriz et al. 2005; Wang et al. 2011). MRI using supramagnetic iron oxide nanoparticles for high-resolution detection of microvascular changes in the endogenous pancreas was useful for detecting vascular leakage with accompanying inflammation that precedes experimental diabetes (Denis et al. 2004). While MRI may also be useful for imaging vascular changes around the transplanted islets, it is difficult to see how this technique could easily be applied to islets transplanted into the liver.

Other more experimental ways to visualize graft function include delivery of reporter genes to islets, Ramen spectroscopy based on the scattering of monochromatic light by different tissues (Hilderink et al. 2013), or hetero-bivalent probes that improve specificity of labeling based on targeting two or more receptors on the beta cell surface (Hart et al. 2014). Preclinical studies using intravital imaging have visualized the infiltration of effector T cells and vascularization of islet grafts in real time (Abdulreda et al. 2011). While these techniques are not yet possible for clinical transplantation, their future development holds promise for understanding more about islet graft rejection.

Immunomodulation

Providing islets with an environment that suppresses immune-mediated rejection and promotes beta cell survival will prolong insulin independence in recipients. There are many new developments in the area of immunomodulation including highly specific immunosuppression regimens that directly target the mechanism of graft rejection, the co-transplantation of islets with cells that have immunomodulatory properties, and targeting islet survival.

Immunosuppression

Anti-coagulation and intensive insulin therapy in the immediate posttransplant period are used to control early graft loss (Koh et al. 2010). Current immunosuppression includes anti-T cell induction (e.g., with anti-IL-2R or anti-thymocyte globulin), high-dose mTOR inhibitor, low-dose calcineurin inhibitors for maintenance as well as avoiding steroids, which are toxic to islets (Gala-Lopez et al. 2013).

Combining anti-inflammatory biologics such as TNFα blockade with maintenance immunosuppression has led to improved single-donor success rates (Hering et al. 2005; Shapiro and Ricordi 2004). The addition of peri-transplant insulin and heparin increases success rate of single-donor islet transplants from 10 to 40 % insulin independence (Koh et al. 2010). Additional therapies may provide more targeted depletion or blockade of specific immune populations, protection against beta cell apoptosis, and stimulation of islet proliferation, and these have shown clinical success in single-donor islet engraftment (Hering et al. 2005; Shapiro and Ricordi 2004).

Inflammation

TNFα has been shown to be detrimental to islet engraftment, so trials of TNFα blockade along with a standard immunosuppression regimen of daclizumab, sirolimus, and tacrolimus are ongoing (Qi et al. 2014; Rickels et al. 2013). The IL-8/CXCR1–CXCR2 signaling cascade is involved in posttransplant inflammation. Reparixin, a CXCR1/2 allosteric inhibitor, resulted in improved glycemic control and higher C peptide than controls (Citro et al. 2012). Other pathways that may be the target of future anti-inflammatory immunosuppression include chemokine blockade in the donor islets to reduce recognition of graft tissue and prevent trafficking of immune cells to the graft site, NFκB inhibition to reduce inflammation and free radical production, or neutralizing the inflammatory molecule HMGB1.

Targeting T Cells

Current immunosuppression targets T cells in a very general way, by depleting T cells or preventing their activation. Anti-CD3 has been shown to induce tolerance in non-autoimmune models of allograft transplantation and can reverse autoimmunity in NOD mice and humans (Chatenoud 2003; Keymeulen et al. 2005). It efficiently depletes effector T cells, drives remaining T cells to become Th2-like, and promotes expansion of Treg cells (Chatenoud 2003). Strong adverse effects of anti-CD3 may limit its use in islet transplantation.

Thymoglobulin or anti-thymocyte globulin (ATG) is the purified IgG fraction of sera from rabbits, horses, or goats that have been immunized with human T cells. ATG depletes peripheral lymphocytes through complement dependent lysis and modulation of surface adhesion molecule or chemokine receptor expression. It can also promote expansion of antigen-specific T reg cells (Gabardi et al. 2011). ATG has been used as induction therapy for islet transplantation (Bellin et al. 2012; O'Connell et al. 2013b).

Anti-CD25 antibody is directed against the low affinity IL-2 receptor α chain. This targets activated T cells and has similar properties to ATG in terms of preventing alloreactivity, but is considered less potent than T cell depletion antibodies. Anti-CD25 (Daclizumab) was one of the critical components of the Edmonton protocol (Shapiro et al. 2000).

Anti-CD52 (Campath-1, Alemtuzumab) is used in solid organ transplantation. CD52 is present on virtually all B and T cells as well as macrophages, NK cells, and some granulocytes. When anti-CD52 binds, it triggers antibody-dependent lysis of these cells, resulting in profound and long-lasting T cell lymphopenia (Morris and Russell 2006). Because of its action on multiple inflammatory cell types, including macrophages, anti-CD52 may reduce early islet loss by preventing the production of inflammatory mediators by intrahepatic macrophages.

Anti-CD2 (Alefacept), which binds all T cells, has been effective in clinical trials especially for targeting memory T cells, an effect likely to be beneficial to suppress existing islet autoimmunity. It was also in clinical development for psoriasis but has been withdrawn from further clinical development at this time.

Co-stimulation Blockade

Presentation of MHC-peptide to T cells in the absence of co-stimulation results in apoptosis or anergy of effector T cells or the generation of Treg cells (Li et al. 1999; Rothstein and Sayegh 2003; Wekerle et al. 2002). Co-stimulation blockade is a non-depleting therapy that has advantages because lymphocyte populations are not compromised. Belatacept (CTLA4-Ig) has achieved regulatory approval after promising trials in kidney transplantation (Durrbach et al. 2010). This is a non-depleting, well-tolerated agent and has shown encouraging results in a few clinical trials (Lowe et al. 2013; Posselt et al. 2010). Humanized anti-CD154 has also been implemented in autoimmune clinical trials, but is potentially too risky, so anti-CD40 monoclonal antibodies might be more attractive (Halloran 2004). The PD1, ICOS, and OX40 pathways may also be useful pathways to test.

Migration

Anti-LFA-1 (Efalizumab) is a CD11a-specific monoclonal antibody. LFA-1 binds ICAM-1 to prevent lymphocyte diapedesis and disrupt adhesion that is required for optimal T cell function. Anti-LFA-1 prolongs the survival of islet allografts and kidney grafts (Turgeon et al. 2010). However, it has recently been withdrawn because of concern about progressive multifocal myeloencephalopathy in a handful of patients who received the drug for psoriasis. FTY-720 (Fingolimod) is a sphingosine 1 phosphate receptor modulator that sequesters lymphocytes in lymph nodes and therefore could be used for prevention of lymphocyte trafficking (Halloran 2004; Yamasaki et al. 1998).

Targeting B Cells

Anti-CD20 (Rituximab) is currently approved for refractory non-Hodgkin's B cell lymphomas, but has also been used in transplantation in immunosuppression combinations. It may be a candidate for reducing the presentation of antigen to T cells and prevention of B cell expansion and autoantibody production, and therefore it could reduce the autoimmune response to an islet allograft (Faye et al. 1998). B cell-activating factor (BAFF) is a co-stimulatory factor for B cells, and its blockade has been successful in preventing diabetes in NOD mice (Marino et al. 2014).

Complement Blockade

It would be useful to inhibit complement in islet transplantation because islets are transplanted into the circulation, and many are thought to be lost due to IBMIR. Anti-C5 monoclonal antibody and the C1 esterase inhibitor Berinert both inhibit activation of complement and may be useful in islet transplantation (Gala-Lopez et al. 2013).

Co-transplantation with Immunomodulatory Cells

Mesenchymal stem cells (MSC) reside mainly in the bone marrow, and are non-hemopoietic multipotent stromal cells that can differentiate into a variety of tissues. They may be useful for islet transplantation for two reasons (Figliuzzi et al. 2014). Firstly, MSC secrete VEGF and other growth factors that promote proliferation of endothelial cells. They also secrete cytokines that result in increased blood flow to ischemic tissue and provide regenerative signals to promote islet graft revascularization (Berman et al. 2010; Figliuzzi et al. 2009; Ito et al. 2010). Secondly, MSCs have immunomodulatory properties. They can reduce inflammatory damage in the early transplant period (Ding et al. 2009) and may be able to reduce immune rejection of islets by inhibiting proliferation of T cells and suppressing maturation of DCs (Di Nicola et al. 2002; Nauta et al. 2006).

Regulatory T (Treg) cells prevent the activation of autoreactive cells and have dominant immunomodulatory capabilities. They have been shown to prevent autoimmunity and rejection of islet transplants (Wu et al. 2013). Interest has mainly focused on $CD4^+CD25^+FoxP3^+$ and $CD4^+CD25^-IL10^+$ Treg cells (Walsh et al. 2004). Rapamycin and IL-10 induce direct expansion of Treg cells in vitro and in animal models of type 1 diabetes and clinical trials with such expanded Treg cells are underway for transplantation (Putnam et al. 2013). However, there is the potential that FoxP3 expression is unstable in vivo and that Treg cells may lose their suppressive capacity over time (Bailey-Bucktrout et al. 2013; Komatsu et al. 2014).

For cells to exert their immunomodulatory properties for improved islet engraftment and survival, it is necessary for them to be co-localized with the islets. This would be difficult using current intraportal delivery of islets, so the preparation of co-aggregates of cells before transplantation, or physical containment of islets with other cells in a device or on a scaffold would be required.

Scaffolds and Immunoprotection

In their native environment in the pancreas, islets are embedded in extracellular matrix (ECM) that is a dynamic 3D structure (Coronel and Stabler 2013; Veriter et al. 2013). The ECM has an essential role in islet survival, function, and proliferation. It also facilitates the growth of vasculature and promotes the migration and attachment of endothelial cells. This matrix is disrupted during islet isolation (Stendahl et al. 2009). A 3D scaffold can provide an environment for the islets that mimics the native islet environment as well as providing a structure for the combination of islets and other cell types, even islet distribution, and mechanical protection.

The pore size of a scaffold is important. A microporous scaffold is ideal because it does not let cells in or out; however, it is difficult to engineer. Macroporous scaffolds have pores >50 μm and provide an open environment that supports efficient delivery of nutrients and allows infiltration of host ECM cells into the scaffold. However, such a scaffold would not exclude the infiltration of immune cells. Scaffolds are made from synthetic polymers that are stable, reproducible, and easy to use. Biodegradable polymers such as PLGA promote host ECM formation (Borg and Bonifacio 2011).

The introduction of proteins or ligands responsible for cell–ECM or cell–cell interactions onto the 3D scaffold may help promote formation of an environment that is closer to that of the native pancreas, and reduce the mean time to euglycemia in animal models (e.g., islets in PLGA macroporous scaffolds or in PEG hydrogels) (Lin and Anseth 2011; Salvay et al. 2008). Scaffolds also allow for the addition of supplemental cells, such as mesenchymal stem cells or Treg cells, or the tethering of bioactive agents. One such example is the combination of ECDI-fixed splenocytes with islets in PLGA scaffolds that resulted in reversal of diabetes without immunosuppression in a rodent model (Kheradmand et al. 2011). Such scaffolds are yet to be tested in human clinical trials.

Encapsulation

Encapsulation within cell-impermeable barriers has been developed as an alternative to chronic immunosuppression. Encapsulation is a way of immune isolation of grafted islets in a device that lets in the nutrients and oxygen without allowing the infiltration of immune cells in, or transplanted cells out (Dolgin 2014; Scharp and Marchetti 2014). There are two categories of cell-containing implantable devices: (A) microencapsulation, in which cells are encapsulated in gel spheres ranging from a few millimeters in diameter down to a thin (conformal) coating around the cells, or (B) macroencapsulation, in which cells are encapsulated in devices that are either attached to the vascular system, anastomosed to the avascular system as arteriovenous shunts (Maki et al. 1991), or implanted into soft tissues. The use of encapsulation devices in the clinic faces hurdles including biocompatibility, membrane permeability, and immune protection of grafted tissue.

Biocompatibility

When any material is implanted, the host will try to isolate it and immunologically destroy it. Most implanted materials are non-living and are impervious to most cytotoxic factors. However, the host can isolate the implanted material with an avascular fibrotic capsule. This response is referred to as the foreign body response and is an indicator of biocompatibility (Anderson et al. 2008). The foreign body response to biomaterials begins with non-specific protein adsorption, which does not occur in the normal physiological process of wound healing. This is followed by the adherence of monocytes, leukocytes, and platelets and then the upregulation of cytokines and proinflammatory processes. In the chronic inflammation that follows the adherent macrophages cannot phagocytose the large foreign body and fuse together to form multinucleated foreign body giant cells that may persist for the lifetime of the implant (Salthouse 1984). The final result is a device that is walled off by an avascular fibrous layer of connective tissue that can be greater than 200 μm thick.

Altering the surface chemical properties of the implant material to reduce the non-specific protein adsorption and monocyte adhesion may positively influence biocompatibility (Collier and Anderson 2002). However, this is difficult to prevent in vivo and an absolutely inert biomaterial may not be possible (Basmadjian et al. 1997). Polyethylene glycol (PEG) is resistant to both protein and cell adhesion and has been used as a component in both micro- and macrocapsules (Lee et al. 2004; Xie et al. 2005).

Some membranes promote vascularization and disrupt the formation of the multinucleated giant cells and the fibrous avascular layer of the foreign body response (Brauker et al. 1996; Padera and Colton 1996). These have demonstrated efficacy in humans and rats, and have been incorporated in devices for transplantation of cells at high packing densities (Brauker et al. 1995, 1996; Loudovaris et al. 1996). However, because of their vascularizing property, these encapsulating membranes allow cells to infiltrate the membrane and therefore cannot be used as a cell-impermeable barrier. The solution to this problem is a dual or bilayer membrane, with the outermost layer providing vascularization and an inner layer being cell impermeable. This is the basis for the multilayer TheraCyte and Viacyte devices. Both support high densities of insulin-producing cells in vivo (Scharp and Marchetti 2014).

Many encapsulation devices are designed to allow maximal oxygen delivery from the vasculature of the host tissue. It remains unclear whether this provides sufficient oxygen for beta cell function, as beta cells consume large amounts of oxygen to produce insulin (Sato et al. 2011). Beta-O2 recently began a pilot clinical study of an implantable macroencapsulation system composed of an immune protection unit connected to ports through which oxygen can be periodically injected to support the survival of the enclosed beta cells or islets. It remains to be determined whether this will improve the survival and/or function of the encapsulated cells.

Site of Implantation

The site of implantation is important to the outcome of the device (Coronel and Stabler 2013; Veriter et al. 2013). It must be large enough for the volume of islets to be transplanted, close to abundant vascularization with good oxygen supply, and preferably allow easy access for future removal of the device. The current site for infusion of free islets is the liver portal vein, although other sites have been suggested including the pancreas, spleen, peritoneum/omentum, gastric submucosal space, kidney capsule, and striated muscle. The site of implantation of encapsulated islets is more limited. Current encapsulation devices are too big for the liver, but the intraperitoneal/omental pouch is a suitable site that allows portal insulin delivery, enough space, is highly vascularized and has high oxygen levels. However, this site is pro-inflammatory and implants are difficult to retrieve. Subcutaneous transplantation is another possible site that is minimally invasive. However, this site has poor blood supply and islets may be subject to mechanical stress because of the superficial location.

Individual islets or small groups of islets can be encapsulated in an envelope made of a semipermeable substance such as alginate hydrogel. These are spherical, which provides a better surface/volume ratio and for diffusion. Microencapsulated islets are also mechanically stable and easy to implant. However, they can form clusters after transplantation, which limits oxygen diffusion to the islets in the middle of the cluster. In addition, the size of microencapsulated islets is a concern. Islets are around 150 μm in diameter, and with the capsules they are 300–500 μm, so their total volume is much greater than that of islets alone (Scharp and Marchetti 2014).

Intraperitoneal transplantation of microencapsulated islets can restore normoglycemia in diabetic animals (Fritschy et al. 1991; O'Shea and Sun 1986; Soon-Shiong et al. 1994). The transport of insulin from the peritoneal cavity to the blood stream is mainly by passive diffusion (De Vos et al. 1996), and therefore animals with intraperitoneal capsules have abnormal insulin responses to physiological stimulus. Additionally, the nutrition provided by the peritoneal fluid is not optimal, the microcapsules are difficult to retrieve, and because they are fragile, they can cause irritation and blockages of lymphatic or other ducts. Biodegradable microcapsules represent a possible alternative.

Unlike microcapsules, macroencapsulation devices are less fragile and are easily retrievable. Intravascular devices have been shown to function for at least 267 days (Sullivan et al. 1991). However, placement of these devices requires vascular surgery and exposure of the patient to potential thrombosis. Extravascular devices are usually implanted into soft tissues and suffer from the foreign body response. Islets placed in hollow fibers have functioned for extended periods with transient correction of diabetes in animal models (Lanza et al. 1991).

Immune Protection

The permeability of the encapsulating membrane provides a balance between immune protection and nutrition of the encapsulated tissues. The membrane needs to restrict direct contact with cells of the immune system, as well as exclude cytotoxic molecules, including antibodies, components of the complement system, cytokines and other cytotoxic substances. Encapsulation devices that have a high molecular permeability also will have the greatest capacity for diffusion of nutrients and therapeutic proteins. To restrict antibodies and components of the complement system, a molecular mass restriction of 50,000 Daltons or higher is required, and to restrict cytokines for protection of xenografts, a molecular mass restriction of 10,000–14,000 Daltons is required (Brissova et al. 1998; Broadhead et al. 2002; Risbud et al. 2003). Inhibition of cellular contact with the host provides long-term protection of allografts in humans (Tibell et al. 2001).

Alternative Sources of Beta Cells

Currently, insulin independence is achieved with the infusion of islets from more than one donor. CITR data indicate that >60 % of islet recipients have two or more infusions. In addition, only half of all donated pancreases are processed for islets, and only about half of these are used for transplantation. These numbers mean that there is a severe shortfall in the number of pancreases for islet transplantation. An alternative to allogeneic islet transplantation is to replace the beta cell mass through expansion of the native pancreatic beta cell pool, through the use of exogenously grafted stem cells, or transplantation of islets from an alternative source, such as the pig. As the differentiation of beta cells and the use of stem cell-derived beta cells in diabetes research are covered in other chapters of this book, they will not be discussed in detail here.

Expansion/Transdifferentiation

It is known that expansion of beta cell mass can occur in situations of increased metabolic demand such as pregnancy, hyperinsulinaemia that results from insulin resistance, or in the postnatal period (Brelje et al. 1993; Karnik et al. 2007; Porat et al. 2011; Sorenson et al. 1993). Residual beta cells have been observed in the pancreas of type 1 diabetic subjects even 20 years after diagnosis (Keenan et al. 2010), and it is possible that these could be expanded. There is debate about the origin of these new beta cells—ductal, acinar, or endocrine. Reprogramming all of these cell types into insulin-producing cells has been described (Bonner-Weir et al. 2000; Dor et al. 2004; Xu et al. 2008). However, these are rare populations that are difficult to expand in culture. Therefore, it is unlikely that expansion of the endogenous beta cell pool with current methods will replace islet transplantation.

Embryonic Stem Cells (ESCs)

ESCs are pluripotent stem cells that theoretically can give rise to any tissue. Despite the possible ethical issues with use of human ESCs, generation of insulin positive cells from ESCs has been achieved (Assady et al. 2001; Lumelsky et al. 2001). The process that has been developed by ViaCyte has just begun in a clinical trial (Kroon et al. 2008). The main bottleneck for the use of ESCs is in the ability to generate functional beta cells in vitro. ViaCyte plan to transplant cells at the progenitor stage, before they are able to produce insulin, so that the recipient's body provides a microenvironment suitable for functional differentiation. The maturation in vivo may take several months. Recently methods for purification of insulin-producing cells in vitro have been described, paving the way for further research in this area (Pagliuca et al. 2014; Rezania et al. 2014).

Immune rejection of insulin-producing cells made from ESCs is still an issue, and ViaCyte plan to use macroencapsulation devices without immunosuppression. The use of such devices will also enable easy removal of cells should there be issues with teratoma formation.

Induced Pluripotent Stem Cells (iPSCs)

These cells were developed to overcome the ethical problems with ESCs. They were first developed in 2006 by expression of pluripotency genes in somatic cells. The use of iPSCs circumvents the problem of alloimmune rejection because the recipient's own cells can be used. Pancreas endocrine cells can be generated in vitro using iPSCs (Alipio et al. 2010). However, there are concerns about their stability, as well as their ability to maintain epigenetic memory of the parent cell.

There is also the possibility of oncogenic mutations forming. So the use of iPSC for transplantation, although promising, requires more work (Teo et al. 2013).

Adult Stem Cells

Adult-derived stem cells, such as MSCs, may also be differentiated into insulin positive cells (Dalvi et al. 2009), and these are being tested in type 1 and type 2 diabetes. MSCs have shown partial success in preclinical models, but it is likely that the best use of these cells is for promoting revascularization and immunomodulation (see above).

Xenotransplantation

The use of pigs as donors for islet xenotransplantation has many advantages. Pigs have many litters with many offspring, they have organs relatively similar in size and physiologic capacity as humans and they produce insulin that is biologically active in humans. The most significant benefit of using pig islets is the potential for genetic modification. Pig-to-primate models have been established for preclinical safety and efficacy testing before human trials (Cooper et al. 2013; O'Connell et al. 2013a).

The major barrier to clinical use of pig islets for transplantation is the issue of rejection. Hyperacute rejection occurs because of the presence of preformed antibodies to the oligosaccharide galactose α1–3 galactose (αGal), expressed at high levels on neonatal pig islet cell clusters (Basnet et al. 2010). Antibody binding to islet clusters immediately after transplantation results in complement activation. In addition, IBMIR is likely to be a problem in xenotransplantation because of incompatibilities between pig regulatory molecules and human thrombotic factors. Cellular xenograft rejection is also a major hurdle that needs to be overcome before clinical islet xenotransplantation can be successful.

$CD4^+$ T cells are the predominant cell type involved in xenograft rejection in rodent models and humanized mice (Gill et al. 1994). Large numbers of activated $CD4^+$ T cells infiltrate the rejecting pig xenograft, which results in IFNγ-mediated activation and infiltration of macrophages and NK cells (Yi et al. 2002). Because of molecular incompatibility between species, the human T cell response to pig tissue is greater than an allo-immune response (O'Connell et al. 2013a). Human T cells respond to pig MHC (swine leukocyte antigen, SLA) in a similar way to allogeneic HLA, with similar molecular interactions required for APC stimulation, but a greater precursor frequency of anti pig T cells (Koulmanda et al. 2004, Olack et al. 2002). Effector mechanisms include direct T cell-mediated killing, cytokine production by T cells, and recruitment of other cytotoxic cells such as macrophages and NK cells, and providing B cell help for production of xenoreactive antibodies (O'Connell et al. 2013a). To prevent this response, a greater amount of immunosuppression is required than that used for allogeneic transplantation, which is currently not clinically feasible. Other possibilities include encapsulation of pig islets to provide a physical barrier, or the use of genetically modified pig islets to prevent immune rejection.

Techniques for genetic modification of the pig genome have become increasingly sophisticated, such as rapid targeted knockouts using TALENs, and efficient co-expression of multiple transgenes (Carlson et al. 2012; Fisicaro et al. 2011). To prevent hyperacute rejection and IBMIR, pigs with deletion of the αGal xenoantigen (GalT gene knockout) (Lutz et al. 2013; Thompson et al. 2011) and pigs with transgenic expression of human complement regulatory proteins (including CD46, CD55 and CD59) and anti-thrombotic molecules (including CD39, thrombomodulin, tissue factor pathway inhibitor), and combinations of these have been generated and tested in preclinical models (Dwyer et al. 2006; van der Windt et al. 2009; Yazaki et al. 2012). The future generation of pigs expressing anti-inflammatory genes such as hemeoxygenase-1 or A20 may also be useful for the IBMIR response.

Conventional immunosuppression with co-stimulation blockade has yielded progressively improved survival and outcomes in islet xenotransplantation. The most successful is the use of anti-CD154 antibodies; however, these are unlikely to be used in humans due to toxicity (Cardona et al. 2006; Hering et al. 2006). As yet, no regimen in non-human primates has been reported in which all the immunosuppressive agents required can be used in a human clinical trial. Because of the requirement for three to four immunosuppressive drugs, the risks currently outweigh the benefits (Samy et al. 2014). In islet allotransplantation, the current regimens we use are likely to be at the edge of tolerability in terms of infection risk and off-target side effects. So the use of multiple immunosuppressive agents for xenotransplantation raises the question of what is tolerable long-term in patients with type 1 diabetes versus all other available therapeutic options.

Conclusion

In conclusion, islet transplantation is approaching the success rates of whole-pancreas transplantation, without the associated morbidity and complications. There is much to improve before beta cell replacement can become routine therapy for sufferers of type 1 diabetes, including prevention of graft rejection and improving supply of insulin-producing cells. A huge worldwide effort is aimed at improving the outcomes of beta cell replacement, including better monitoring of graft survival, more targeted approaches to immunosuppression and alternative sources of beta cells, and many new approaches are being tested or about to enter clinical trials. This effort should make beta cell replacement available to more patients with type 1 diabetes in years to come.

Acknowledgments The authors thank Mr. William Stanley for assistance with figures. This work was funded by a Program Grant (APP1037321), fellowship (HET), and Clinical Centre of Research Excellence: Clinical Science in Diabetes (APP454461) from the National Health and Medical Research Council of Australia; the Department of Health and Ageing, Commonwealth of Australia administered by the Juvenile Diabetes Research Foundation; and the Susan Alberti Medical Research Foundation. The St. Vincent's Institute receives support from the Operational Infrastructure Support Scheme of the Government of Victoria.

References

Abdulreda MH, Faleo G, Molano RD, Lopez-Cabezas M, Molina J, Tan Y, Echeverria OA, Zahr-Akrawi E, Rodriguez-Diaz R, Edlund PK et al (2011) High-resolution, noninvasive longitudinal live imaging of immune responses. Proc Natl Acad Sci USA 108:12863–12868

Abreu JR, Roep BO (2013) Immune monitoring of islet and pancreas transplant recipients. Curr Diab Rep 13:704–712

Alipio Z, Liao W, Roemer EJ, Waner M, Fink LM, Ward DC, Ma Y (2010) Reversal of hyperglycemia in diabetic mouse models using induced-pluripotent stem (iPS)-derived pancreatic beta-like cells. Proc Natl Acad Sci USA 107:13426–13431

Anderson JM, Rodriguez A, Chang DT (2008) Foreign body reaction to biomaterials. Semin Immunol 20:86–100

Assady S, Maor G, Amit M, Itskovitz-Eldor J, Skorecki KL, Tzukerman M (2001) Insulin production by human embryonic stem cells. Diabetes 50:1691–1697

Bailey-Bucktrout SL, Martinez-Llordella M, Zhou X, Anthony B, Rosenthal W, Luche H, Fehling HJ, Bluestone JA (2013) Self-antigen-driven activation induces instability of regulatory T cells during an inflammatory autoimmune response. Immunity 39:949–962

Barton FB, Rickels MR, Alejandro R, Hering BJ, Wease S, Naziruddin B, Oberholzer J, Odorico JS, Garfinkel MR, Levy M et al (2012) Improvement in outcomes of clinical islet transplantation: 1999–2010. Diabetes Care 35:1436–1445

Basmadjian D, Sefton MV, Baldwin SA (1997) Coagulation on biomaterials in flowing blood: some theoretical considerations. Biomaterials 18:1511–1522

Basnet NB, Ide K, Tahara H, Tanaka Y, Ohdan H (2010) Deficiency of *N*-glycolylneuraminic acid and Galalpha1-3Galbeta1-4GlcNAc epitopes in xenogeneic cells attenuates cytotoxicity of human natural antibodies. Xenotransplantation 17:440–448

Bellin MD, Barton FB, Heitman A, Harmon JV, Kandaswamy R, Balamurugan AN, Sutherland DE, Alejandro R, Hering BJ (2012) Potent induction immunotherapy promotes long-term insulin independence after islet transplantation in type 1 diabetes. Am J Transplant 12:1576–1583

Bennet W, Sundberg B, Groth CG, Brendel MD, Brandhorst D, Brandhorst H, Bretzel RG, Elgue G, Larsson R, Nilsson B et al (1999) Incompatibility between human blood and isolated islets of Langerhans: a finding with implications for clinical intraportal islet transplantation? Diabetes 48:1907–1914

Berman DM, Willman MA, Han D, Kleiner G, Kenyon NM, Cabrera O, Karl JA, Wiseman RW, O'Connor DH, Bartholomew AM et al (2010) Mesenchymal stem cells enhance allogeneic islet engraftment in nonhuman primates. Diabetes 59:2558–2568

Bonner-Weir S, Taneja M, Weir GC, Tatarkiewicz K, Song KH, Sharma A, O'Neil JJ (2000) In vitro cultivation of human islets from expanded ductal tissue. Proc Natl Acad Sci USA 97:7999–8004

Borg DJ, Bonifacio E (2011) The use of biomaterials in islet transplantation. Curr Diab Rep 11:434–444

Brandhorst H, Muehling B, Yamaya H, Henriksnaes J, Carlsson PO, Korsgren O, Brandhorst D (2008) New class of oxygen carriers improves islet isolation from long-term stored rat pancreata. Transpl Proc 40:393–394

Brauker J, Martinson LA, Young SK, Johnson RC (1996) Local inflammatory response around diffusion chambers containing xenografts. Nonspecific destruction of tissues and decreased local vascularization. Transplantation 61:1671–1677

Brauker JH, Carr-Brendel VE, Martinson LA, Crudele J, Johnston WD, Johnson RC (1995) Neovascularization of synthetic membranes directed by membrane microarchitecture. J Biomed Mater Res 29:1517–1524

Brelje TC, Scharp DW, Lacy PE, Ogren L, Talamantes F, Robertson M, Friesen HG, Sorenson RL (1993) Effect of homologous placental lactogens, prolactins, and growth hormones on islet B-cell division and insulin secretion in rat, mouse, and human islets: implication for placental lactogen regulation of islet function during pregnancy. Endocrinology 132:879–887

Brissova M, Lacik I, Powers AC, Anilkumar AV, Wang T (1998) Control and measurement of permeability for design of microcapsule cell delivery system. J Biomed Mater Res 39:61–70

Broadhead KW, Biran R, Tresco PA (2002) Hollow fiber membrane diffusive permeability regulates encapsulated cell line biomass, proliferation, and small molecule release. Biomaterials 23:4689–4699

Bruni A, Gala-Lopez B, Pepper AR, Abualhassan NS, Shapiro AJ (2014) Islet cell transplantation for the treatment of type 1 diabetes: recent advances and future challenges. Diabetes Metab Syndr Obes: Targets Ther 7:211–223

Cantley J, Grey ST, Maxwell PH, Withers DJ (2010) The hypoxia response pathway and beta-cell function. Diabetes Obes Metab 12(Suppl 2):159–167

Cardona K, Korbutt GS, Milas Z, Lyon J, Cano J, Jiang W, Bello-Laborn H, Hacquoil B, Strobert E, Gangappa S et al (2006) Long-term survival of neonatal porcine islets in nonhuman primates by targeting costimulation pathways. Nat Med 12:304–306

Carlson DF, Tan W, Lillico SG, Stverakova D, Proudfoot C, Christian M, Voytas DF, Long CR, Whitelaw CB, Fahrenkrug SC (2012) Efficient TALEN-mediated gene knockout in livestock. Proc Natl Acad Sci USA 109:17382–17387

Chatenoud L (2003) CD3-specific antibody-induced active tolerance: from bench to bedside. Nat Rev Immunol 3:123–132

Chee J, Ko HJ, Skowera A, Jhala G, Catterall T, Graham KL, Sutherland RM, Thomas HE, Lew AM, Peakman M et al (2014) Effector-memory T cells develop in islets and report islet pathology in type 1 diabetes. J Immunol 192:572–580

Cheng K, Ho K, Stokes R, Scott C, Lau SM, Hawthorne WJ, O'Connell PJ, Loudovaris T, Kay TW, Kulkarni RN et al (2010) Hypoxia-inducible factor-1alpha regulates beta cell function in mouse and human islets. J Clin Investig 120:2171–2183

Citro A, Cantarelli E, Maffi P, Nano R, Melzi R, Mercalli A, Dugnani E, Sordi V, Magistretti P, Daffonchio L et al (2012) CXCR1/2 inhibition enhances pancreatic islet survival after transplantation. J Clin Investig 122:3647–3651

Collier TO, Anderson JM (2002) Protein and surface effects on monocyte and macrophage adhesion, maturation, and survival. J Biomed Mater Res 60:487–496

Contreras JL, Eckstein C, Smyth CA, Bilbao G, Vilatoba M, Ringland SE, Young C, Thompson JA, Fernandez JA, Griffin JH et al (2004) Activated protein C preserves functional islet mass after intraportal transplantation: a novel link between endothelial cell activation, thrombosis, inflammation, and islet cell death. Diabetes 53:2804–2814

Cooper DK, Hara H, Ezzelarab M, Bottino R, Trucco M, Phelps C, Ayares D, Dai Y (2013) The potential of genetically-engineered pigs in providing an alternative source of organs and cells for transplantation. J Biomed Res 27:249–253

Coronel MM, Stabler CL (2013) Engineering a local microenvironment for pancreatic islet replacement. Curr Opin Biotechnol 24:900–908

Dalvi MP, Umrani MR, Joglekar MV, Hardikar AA (2009) Human pancreatic islet progenitor cells demonstrate phenotypic plasticity in vitro. J Biosci 34:523–528

De Vos P, Vegter D, De Haan BJ, Strubbe JH, Bruggink JE, Van Schilfgaarde R (1996) Kinetics of intraperitoneally infused insulin in rats. Functional implications for the bioartificial pancreas. Diabetes 45:1102–1107

Denis MC, Mahmood U, Benoist C, Mathis D, Weissleder R (2004) Imaging inflammation of the pancreatic islets in type 1 diabetes. Proc Natl Acad Sci USA 101:12634–12639

Di Nicola M, Carlo-Stella C, Magni M, Milanesi M, Longoni PD, Matteucci P, Grisanti S, Gianni AM (2002) Human bone marrow stromal cells suppress T-lymphocyte proliferation induced by cellular or nonspecific mitogenic stimuli. Blood 99:3838–3843

Ding Y, Xu D, Feng G, Bushell A, Muschel RJ, Wood KJ (2009) Mesenchymal stem cells prevent the rejection of fully allogenic islet grafts by the immunosuppressive activity of matrix metalloproteinase-2 and -9. Diabetes 58:1797–1806

Dionne KE, Colton CK, Yarmush ML (1989) Effect of oxygen on isolated pancreatic tissue. ASAIO Trans 35:739–741

Dolgin E (2014) Encapsulate this. Nat Med 20:9–11

Dor Y, Brown J, Martinez OI, Melton DA (2004) Adult pancreatic beta-cells are formed by self-duplication rather than stem-cell differentiation. Nature 429:41–46

Dubois S, Madec AM, Mesnier A, Armanet M, Chikh K, Berney T, Thivolet C (2010) Glucose inhibits angiogenesis of isolated human pancreatic islets. J Mol Endocrinol 45:99–105

Durrbach A, Pestana JM, Pearson T, Vincenti F, Garcia VD, Campistol J, Rial Mdel C, Florman S, Block A, Di Russo G et al (2010) A phase III study of belatacept versus cyclosporine in kidney transplants from extended criteria donors (BENEFIT-EXT study). Am J Transplant 10:547–557

Dwyer KM, Mysore TB, Crikis S, Robson SC, Nandurkar H, Cowan PJ, D'Apice AJ (2006) The transgenic expression of human CD39 on murine islets inhibits clotting of human blood. Transplantation 82:428–432

Eriksson O, Eich T, Sundin A, Tibell A, Tufveson G, Andersson H, Felldin M, Foss A, Kyllonen L, Langstrom B et al (2009) Positron emission tomography in clinical islet transplantation. Am J Transplant 9:2816–2824

Faye A, Van Den Abeele T, Peuchmaur M, Mathieu-Boue A, Vilmer E (1998) Anti-CD20 monoclonal antibody for post-transplant lymphoproliferative disorders. Lancet 352:1285

Figliuzzi M, Bonandrini B, Silvani S, Remuzzi A (2014) Mesenchymal stem cells help pancreatic islet transplantation to control type 1 diabetes. World J Stem Cells 6:163–172

Figliuzzi M, Cornolti R, Perico N, Rota C, Morigi M, Remuzzi G, Remuzzi A, Benigni A (2009) Bone marrow-derived mesenchymal stem cells improve islet graft function in diabetic rats. Transpl Proc 41:1797–1800

Fisicaro N, Londrigan SL, Brady JL, Salvaris E, Nottle MB, O'Connell PJ, Robson SC, d'Apice AJ, Lew AM, Cowan PJ (2011) Versatile co-expression of graft-protective proteins using 2A-linked cassettes. Xenotransplantation 18:121–130

Fritschy WM, Strubbe JH, Wolters GH, van Schilfgaarde R (1991) Glucose tolerance and plasma insulin response to intravenous glucose infusion and test meal in rats with microencapsulated islet allografts. Diabetologia 34:542–547

Gabardi S, Martin ST, Roberts KL, Grafals M (2011) Induction immunosuppressive therapies in renal transplantation. Am J Health Syst Pharm: AJHP 68:211–218

Gala-Lopez B, Pepper AR, Shapiro AM (2013) Biologic agents in islet transplantation. Curr Diab Rep 13:713–722

Gill RG, Wolf L, Daniel D, Coulombe M (1994) $CD4^+$ T cells are both necessary and sufficient for islet xenograft rejection. Transpl Proc 26:1203

Goto M, Tjernberg J, Dufrane D, Elgue G, Brandhorst D, Ekdahl KN, Brandhorst H, Wennberg L, Kurokawa Y, Satomi S et al (2008) Dissecting the instant blood-mediated inflammatory reaction in islet xenotransplantation. Xenotransplantation 15:225–234

Gruessner AC (2011) 2011 update on pancreas transplantation: comprehensive trend analysis of 25,000 cases followed up over the course of twenty-four years at the International Pancreas Transplant Registry (IPTR). Rev Diabet Stud: RDS 8:6–16

Gruessner AC, Sutherland DE, Gruessner RW (2012) Long-term outcome after pancreas transplantation. Curr Opin Organ Transplant 17:100–105

Halloran PF (2004) Immunosuppressive drugs for kidney transplantation. N Engl J Med 351:2715–2729

Hart NJ, Chung WJ, Weber C, Ananthakrishnan K, Anderson M, Patek R, Zhang Z, Limesand SW, Vagner J, Lynch RM (2014) Hetero-bivalent GLP-1/glibenclamide for targeting pancreatic beta-cells. ChemBioChem 15:135–145

Hering BJ, Kandaswamy R, Ansite JD, Eckman PM, Nakano M, Sawada T, Matsumoto I, Ihm SH, Zhang HJ, Parkey J et al (2005) Single-donor, marginal-dose islet transplantation in patients with type 1 diabetes. JAMA 293:830–835

Hering BJ, Wijkstrom M, Graham ML, Hardstedt M, Aasheim TC, Jie T, Ansite JD, Nakano M, Cheng J, Li W et al (2006) Prolonged diabetes reversal after intraportal xenotransplantation of wild-type porcine islets in immunosuppressed nonhuman primates. Nat Med 12:301–303

Hilderink J, Otto C, Slump C, Lenferink A, Engelse M, van Blitterswijk C, de Koning E, Karperien M, van Apeldoorn A (2013) Label-free detection of insulin and glucagon within human islets of Langerhans using Raman spectroscopy. PLoS ONE 8:e78148

Huurman VA, Hilbrands R, Pinkse GG, Gillard P, Duinkerken G, van de Linde P, van der Meer-Prins PM, der Voort Versteeg-van, Maarschalk MF, Verbeeck K, Alizadeh BZ et al (2008) Cellular islet autoimmunity associates with clinical outcome of islet cell transplantation. PLoS ONE 3:e2435

Ito T, Itakura S, Todorov I, Rawson J, Asari S, Shintaku J, Nair I, Ferreri K, Kandeel F, Mullen Y (2010) Mesenchymal stem cell and islet co-transplantation promotes graft revascularization and function. Transplantation 89:1438–1445

Jansson L, Carlsson PO (2002) Graft vascular function after transplantation of pancreatic islets. Diabetologia 45:749–763

Ji M, Yi S, Smith-Hurst H, Phillips P, Wu J, Hawthorne W, O'Connell P (2011) The importance of tissue factor expression by porcine NICC in triggering IBMIR in the xenograft setting. Transplantation 91:841–846

Johansson H, Lukinius A, Moberg L, Lundgren T, Berne C, Foss A, Felldin M, Kallen R, Salmela K, Tibell A et al (2005) Tissue factor produced by the endocrine cells of the islets of Langerhans is associated with a negative outcome of clinical islet transplantation. Diabetes 54:1755–1762

Karnik SK, Chen H, McLean GW, Heit JJ, Gu X, Zhang AY, Fontaine M, Yen MH, Kim SK (2007) Menin controls growth of pancreatic beta-cells in pregnant mice and promotes gestational diabetes mellitus. Science 318:806–809

Keenan HA, Sun JK, Levine J, Doria A, Aiello LP, Eisenbarth G, Bonner-Weir S, King GL (2010) Residual insulin production and pancreatic ss-cell turnover after 50 years of diabetes: Joslin Medalist Study. Diabetes 59:2846–2853

Keymeulen B, Vandemeulebroucke E, Ziegler AG, Mathieu C, Kaufman L, Hale G, Gorus F, Goldman M, Walter M, Candon S et al (2005) Insulin needs after CD3-antibody therapy in new-onset type 1 diabetes. N Engl J Med 352:2598–2608

Kheradmand T, Wang S, Gibly RF, Zhang X, Holland S, Tasch J, Graham JG, Kaufman DB, Miller SD, Shea LD et al (2011) Permanent protection of PLG scaffold transplanted allogeneic islet grafts in diabetic mice treated with ECDI-fixed donor splenocyte infusions. Biomaterials 32:4517–4524

Kim JH, Jin SM, Oh SH, Lee S, Oh BJ, Kim SK, Suh S, Lee JH, Jung HS, Lee MS et al (2012) Counting small hypointense spots confounds the quantification of functional islet mass based on islet MRI. Am J Transplant 12:1303–1312

Kim SJ, Doudet DJ, Studenov AR, Nian C, Ruth TJ, Gambhir SS, McIntosh CH (2006) Quantitative micro positron emission tomography (PET) imaging for the in vivo determination of pancreatic islet graft survival. Nat Med 12:1423–1428

Koh A, Senior P, Salam A, Kin T, Imes S, Dinyari P, Malcolm A, Toso C, Nilsson B, Korsgren O et al (2010) Insulin-heparin infusions peritransplant substantially improve single-donor clinical islet transplant success. Transplantation 89:465–471

Komatsu N, Okamoto K, Sawa S, Nakashima T, Oh-Hora M, Kodama T, Tanaka S, Bluestone JA, Takayanagi H (2014) Pathogenic conversion of Foxp3(+) T cells into TH17 cells in autoimmune arthritis. Nat Med 20:62–68

Koulmanda M, Laufer TM, Auchincloss H Jr, Smith RN (2004) Prolonged survival of fetal pig islet xenografts in mice lacking the capacity for an indirect response. Xenotransplantation 11:525–530

Kriz J, Jirak D, Girman P, Berkova Z, Zacharovova K, Honsova E, Lodererova A, Hajek M, Saudek F (2005) Magnetic resonance imaging of pancreatic islets in tolerance and rejection. Transplantation 80:1596–1603

Kroon E, Martinson LA, Kadoya K, Bang AG, Kelly OG, Eliazer S, Young H, Richardson M, Smart NG, Cunningham J et al (2008) Pancreatic endoderm derived from human embryonic stem cells generates glucose-responsive insulin-secreting cells in vivo. Nat Biotechnol 26:443–452

Lanza RP, Butler DH, Borland KM, Staruk JE, Faustman DL, Solomon BA, Muller TE, Rupp RG, Maki T, Monaco AP et al (1991) Xenotransplantation of canine, bovine, and porcine islets in diabetic rats without immunosuppression. Proc Natl Acad Sci USA 88:11100–11104

Lee DY, Nam JH, Byun Y (2004) Effect of polyethylene glycol grafted onto islet capsules on prevention of splenocyte and cytokine attacks. J Biomater Sci Polym Ed 15:753–766

Li Y, Li XC, Zheng XX, Wells AD, Turka LA, Strom TB (1999) Blocking both signal 1 and signal 2 of T-cell activation prevents apoptosis of alloreactive T cells and induction of peripheral allograft tolerance. Nat Med 5:1298–1302

Lin CC, Anseth KS (2011) Cell-cell communication mimicry with poly(ethylene glycol) hydrogels for enhancing beta-cell function. Proc Natl Acad Sci USA 108:6380–6385

Loudovaris T, Mandel TE, Charlton B (1996) $CD4^+$ T cell mediated destruction of xenografts within cell-impermeable membranes in the absence of $CD8^+$ T cells and B cells. Transplantation 61:1678–1684

Lowe MC, Badell IR, Turner AP, Thompson PW, Leopardi FV, Strobert EA, Larsen CP, Kirk AD (2013) Belatacept and sirolimus prolong nonhuman primate islet allograft survival: adverse consequences of concomitant alefacept therapy. Am J Transplant 13:312–319

Ludwig B, Rotem A, Schmid J, Weir GC, Colton CK, Brendel MD, Neufeld T, Block NL, Yavriyants K, Steffen A et al (2012) Improvement of islet function in a bioartificial pancreas by enhanced oxygen supply and growth hormone releasing hormone agonist. Proc Natl Acad Sci USA 109:5022–5027

Ludwig B, Zimerman B, Steffen A, Yavriants K, Azarov D, Reichel A, Vardi P, German T, Shabtay N, Rotem A et al (2010) A novel device for islet transplantation providing immune protection and oxygen supply. Horm Metab Res 42:918–922

Lumelsky N, Blondel O, Laeng P, Velasco I, Ravin R, McKay R (2001) Differentiation of embryonic stem cells to insulin-secreting structures similar to pancreatic islets. Science 292:1389–1394

Lutz AJ, Li P, Estrada JL, Sidner RA, Chihara RK, Downey SM, Burlak C, Wang ZY, Reyes LM, Ivary B et al (2013) Double knockout pigs deficient in *N*-glycolylneuraminic acid and galactose alpha-1,3-galactose reduce the humoral barrier to xenotransplantation. Xenotransplantation 20:27–35

Maki T, Ubhi CS, Sanchez-Farpon H, Sullivan SJ, Borland K, Muller TE, Solomon BA, Chick WL, Monaco AP (1991) Successful treatment of diabetes with the biohybrid artificial pancreas in dogs. Transplantation 51:43–51

Marino E, Walters SN, Villanueva JE, Richards JL, Mackay CR, Grey ST (2014) BAFF regulates activation of self-reactive T cells through B-cell dependent mechanisms and mediates protection in NOD mice. Eur J Immunol 44:983–993

Mattsson G, Jansson L, Carlsson PO (2002) Decreased vascular density in mouse pancreatic islets after transplantation. Diabetes 51:1362–1366

Moberg L, Johansson H, Lukinius A, Berne C, Foss A, Kallen R, Ostraat O, Salmela K, Tibell A, Tufveson G et al (2002) Production of tissue factor by pancreatic islet cells as a trigger of detrimental thrombotic reactions in clinical islet transplantation. Lancet 360:2039–2045

Moberg L, Korsgren O, Nilsson B (2005) Neutrophilic granulocytes are the predominant cell type infiltrating pancreatic islets in contact with ABO-compatible blood. Clin Exp Immunol 142:125–131

Monti P, Scirpoli M, Maffi P, Ghidoli N, De Taddeo F, Bertuzzi F, Piemonti L, Falcone M, Secchi A, Bonifacio E (2008) Islet transplantation in patients with autoimmune diabetes induces homeostatic cytokines that expand autoreactive memory T cells. J Clin Investig 118:1806–1814

Moritz W, Meier F, Stroka DM, Giuliani M, Kugelmeier P, Nett PC, Lehmann R, Candinas D, Gassmann M, Weber M (2002) Apoptosis in hypoxic human pancreatic islets correlates with HIF-1alpha expression. FASEB J 16:745–747

Morris PJ, Russell NK (2006) Alemtuzumab (Campath-1H): a systematic review in organ transplantation. Transplantation 81:1361–1367

Nauta AJ, Kruisselbrink AB, Lurvink E, Willemze R, Fibbe WE (2006) Mesenchymal stem cells inhibit generation and function of both $CD34^{+}$-derived and monocyte-derived dendritic cells. J Immunol 177:2080–2087

Naziruddin B, Iwahashi S, Kanak MA, Takita M, Itoh T, Levy MF (2014) Evidence for instant blood-mediated inflammatory reaction in clinical autologous islet transplantation. Am J Transplant 14:428–437

Niclauss N, Morel P, Berney T (2014) Has the gap between pancreas and islet transplantation closed? Transplantation 98:593–599

Nilsson B, Ekdahl KN, Korsgren O (2011) Control of instant blood-mediated inflammatory reaction to improve islets of Langerhans engraftment. Curr Opin Organ Transplant 16:620–626

O'Connell PJ, Cowan PJ, Hawthorne WJ, Yi S, Lew AM (2013a) Transplantation of xenogeneic islets: are we there yet? Curr Diab Rep 13:687–694

O'Connell PJ, Holmes-Walker DJ, Goodman D, Hawthorne WJ, Loudovaris T, Gunton JE, Thomas HE, Grey ST, Drogemuller CJ, Ward GM et al (2013b) Multicenter Australian trial of islet transplantation: improving accessibility and outcomes. Am J Transplant 13:1850–1858

O'Shea GM, Sun AM (1986) Encapsulation of rat islets of Langerhans prolongs xenograft survival in diabetic mice. Diabetes 35:943–946

Olack BJ, Jaramillo A, Benshoff ND, Kaleem Z, Swanson CJ, Lowell JA, Mohanakumar T (2002) Rejection of porcine islet xenografts mediated by $CD4^{+}$ T cells activated through the indirect antigen recognition pathway. Xenotransplantation 9:393–401

Ozmen L, Ekdahl KN, Elgue G, Larsson R, Korsgren O, Nilsson B (2002) Inhibition of thrombin abrogates the instant blood-mediated inflammatory reaction triggered by isolated human islets: possible application of the thrombin inhibitor melagatran in clinical islet transplantation. Diabetes 51:1779–1784

Padera RF, Colton CK (1996) Time course of membrane microarchitecture-driven neovascularization. Biomaterials 17:277–284

Pagliuca FW, Millman JR, Gurtler M, Segel M, Van Dervort A, Ryu JH, Peterson QP, Greiner D, Melton DA (2014) Generation of functional human pancreatic beta cells in vitro. Cell 159:428–439

Pedraza E, Coronel MM, Fraker CA, Ricordi C, Stabler CL (2012) Preventing hypoxia-induced cell death in beta cells and islets via hydrolytically activated, oxygen-generating biomaterials. Proc Natl Acad Sci USA 109:4245–4250

Piemonti L, Everly MJ, Maffi P, Scavini M, Poli F, Nano R, Cardillo M, Melzi R, Mercalli A, Sordi V et al (2013) Alloantibody and autoantibody monitoring predicts islet transplantation outcome in human type 1 diabetes. Diabetes 62:1656–1664

Porat S, Weinberg-Corem N, Tornovsky-Babaey S, Schyr-Ben-Haroush R, Hija A, Stolovich-Rain M, Dadon D, Granot Z, Ben-Hur V, White P et al (2011) Control of pancreatic beta cell regeneration by glucose metabolism. Cell Metab 13:440–449

Posselt AM, Szot GL, Frassetto LA, Masharani U, Tavakol M, Amin R, McElroy J, Ramos MD, Kerlan RK, Fong L et al (2010) Islet transplantation in type 1 diabetic patients using calcineurin inhibitor-free immunosuppressive protocols based on T-cell adhesion or costimulation blockade. Transplantation 90:1595–1601

Putnam AL, Safinia N, Medvec A, Laszkowska M, Wray M, Mintz MA, Trotta E, Szot GL, Liu W, Lares A et al (2013) Clinical grade manufacturing of human alloantigen-reactive regulatory T cells for use in transplantation. Am J Transplant 13:3010–3020

Qi M, Kinzer K, Danielson KK, Martellotto J, Barbaro B, Wang Y, Bui JT, Gaba RC, Knuttinen G, Garcia-Roca R et al (2014) Five-year follow-up of patients with type 1 diabetes transplanted with allogeneic islets: the UIC experience. Acta Diabetol 51:833–843

Rezania A, Bruin JE, Arora P, Rubin A, Batushansky I, Asadi A, O'Dwyer S, Quiskamp N, Mojibian M, Albrecht T et al (2014) Reversal of diabetes with insulin-producing cells derived in vitro from human pluripotent stem cells. Nat Biotechnol 32:1121–1133

Rickels MR, Liu C, Shlansky-Goldberg RD, Soleimanpour SA, Vivek K, Kamoun M, Min Z, Markmann E, Palangian M, Dalton-Bakes C et al (2013) Improvement in beta-cell secretory capacity after human islet transplantation according to the CIT07 protocol. Diabetes 62:2890–2897

Risbud MV, Bhargava S, Bhonde RR (2003) In vivo biocompatibility evaluation of cellulose macrocapsules for islet immunoisolation: implications of low molecular weight cut-off. J Biomed Mater Res, Part A 66:86–92

Rothstein DM, Sayegh MH (2003) T-cell costimulatory pathways in allograft rejection and tolerance. Immunol Rev 196:85–108

Sakata N, Yoshimatsu G, Tsuchiya H, Aoki T, Mizuma M, Motoi F, Katayose Y, Kodama T, Egawa S, Unno M (2013) Imaging of transplanted islets by positron emission tomography, magnetic resonance imaging, and ultrasonography. Islets 5:179–187

Salthouse TN (1984) Some aspects of macrophage behavior at the implant interface. J Biomed Mater Res 18:395–401

Salvay DM, Rives CB, Zhang X, Chen F, Kaufman DB, Lowe WL Jr, Shea LD (2008) Extracellular matrix protein-coated scaffolds promote the reversal of diabetes after extrahepatic islet transplantation. Transplantation 85:1456–1464

Samy KP, Martin BM, Turgeon NA, Kirk AD (2014) Islet cell xenotransplantation: a serious look toward the clinic. Xenotransplantation 21:221–229

Sato Y, Endo H, Okuyama H, Takeda T, Iwahashi H, Imagawa A, Yamagata K, Shimomura I, Inoue M (2011) Cellular hypoxia of pancreatic beta-cells due to high levels of oxygen consumption for insulin secretion in vitro. J Biol Chem 286:12524–12532

Scharp DW, Marchetti P (2014) Encapsulated islets for diabetes therapy: history, current progress, and critical issues requiring solution. Adv Drug Deliv Rev 67–68:35–73

Shapiro AM, Lakey JR, Ryan EA, Korbutt GS, Toth E, Warnock GL, Kneteman NM, Rajotte RV (2000) Islet transplantation in seven patients with type 1 diabetes mellitus using a glucocorticoid-free immunosuppressive regimen. N Engl J Med 343:230–238

Shapiro AM, Ricordi C (2004) Unraveling the secrets of single donor success in islet transplantation. Am J Transplant 4:295–298

Shimoda M, Chen S, Noguchi H, Matsumoto S, Grayburn PA (2010) In vivo non-viral gene delivery of human vascular endothelial growth factor improves revascularisation and restoration of euglycaemia after human islet transplantation into mouse liver. Diabetologia 53:1669–1679

Soon-Shiong P, Heintz RE, Merideth N, Yao QX, Yao Z, Zheng T, Murphy M, Moloney MK, Schmehl M, Harris M et al (1994) Insulin independence in a type 1 diabetic patient after encapsulated islet transplantation. Lancet 343:950–951

Sorenson RL, Brelje TC, Roth C (1993) Effects of steroid and lactogenic hormones on islets of Langerhans: a new hypothesis for the role of pregnancy steroids in the adaptation of islets to pregnancy. Endocrinology 133:2227–2234

Stendahl JC, Kaufman DB, Stupp SI (2009) Extracellular matrix in pancreatic islets: relevance to scaffold design and transplantation. Cell Transplant 18:1–12

Sullivan SJ, Maki T, Borland KM, Mahoney MD, Solomon BA, Muller TE, Monaco AP, Chick WL (1991) Biohybrid artificial pancreas: long-term implantation studies in diabetic, pancreatectomized dogs. Science 252:718–721

Suszynski TM, Rizzari MD, Scott WE, Eckman PM, Fonger JD, John R, Chronos N, Tempelman LA, Sutherland DE, Papas KK (2013) Persufflation (gaseous oxygen perfusion) as a method of heart preservation. J Cardiothorac Surg 8:105

Sutherland DE, Sibley R, Xu XZ, Michael A, Srikanta AM, Taub F, Najarian J, Goetz FC (1984) Twin-to-twin pancreas transplantation: reversal and reenactment of the pathogenesis of type I diabetes. Trans Assoc Am Physicians 97:80–87

Teo AK, Wagers AJ, Kulkarni RN (2013) New opportunities: harnessing induced pluripotency for discovery in diabetes and metabolism. Cell Metab 18:775–791

Thompson P, Badell IR, Lowe M, Cano J, Song M, Leopardi F, Avila J, Ruhil R, Strobert E, Korbutt G et al (2011) Islet xenotransplantation using gal-deficient neonatal donors improves engraftment and function. Am J Transplant 11:2593–2602

Tibell A, Rafael E, Wennberg L, Nordenstrom J, Bergstrom M, Geller RL, Loudovaris T, Johnson RC, Brauker JH, Neuenfeldt S et al (2001) Survival of macroencapsulated allogeneic parathyroid tissue one year after transplantation in nonimmunosuppressed humans. Cell Transplant 10:591–599

Toso C, Isse K, Demetris AJ, Dinyari P, Koh A, Imes S, Kin T, Emamaullee J, Senior P, Shapiro AM (2009) Histologic graft assessment after clinical islet transplantation. Transplantation 88:1286–1293

Turgeon NA, Avila JG, Cano JA, Hutchinson JJ, Badell IR, Page AJ, Adams AB, Sears MH, Bowen PH, Kirk AD et al (2010) Experience with a novel efalizumab-based immunosuppressive regimen to facilitate single donor islet cell transplantation. Am J Transplant 10:2082–2091

van der Windt DJ, Bottino R, Casu A, Campanile N, Smetanka C, He J, Murase N, Hara H, Ball S, Loveland BE et al (2009) Long-term controlled normoglycemia in diabetic non-human primates after transplantation with hCD46 transgenic porcine islets. Am J Transplant 9:2716–2726

Vantyghem MC, Defrance F, Quintin D, Leroy C, Raverdi V, Prevost G, Caiazzo R, Kerr-Conte J, Glowacki F, Hazzan M et al (2014) Treating diabetes with islet transplantation: lessons from the past decade in Lille. Diabetes Metab 40:108–119

Vendrame F, Pileggi A, Laughlin E, Allende G, Martin-Pagola A, Molano RD, Diamantopoulos S, Standifer N, Geubtner K, Falk BA et al (2010) Recurrence of type 1 diabetes after simultaneous pancreas-kidney transplantation, despite immunosuppression, is associated with autoantibodies and pathogenic autoreactive CD4 T-cells. Diabetes 59:947–957

Veriter S, Gianello P, Dufrane D (2013) Bioengineered sites for islet cell transplantation. Curr Diab Rep 13:745–755

Walsh PT, Taylor DK, Turka LA (2004) Tregs and transplantation tolerance. J Clin Investig 114:1398–1403

Wang P, Yigit MV, Medarova Z, Wei L, Dai G, Schuetz C, Moore A (2011) Combined small interfering RNA therapy and in vivo magnetic resonance imaging in islet transplantation. Diabetes 60:565–571

Wekerle T, Kurtz J, Bigenzahn S, Takeuchi Y, Sykes M (2002) Mechanisms of transplant tolerance induction using costimulatory blockade. Curr Opin Immunol 14:592–600

Wu DC, Hester J, Nadig SN, Zhang W, Trzonkowski P, Gray D, Hughes S, Johnson P, Wood KJ (2013) Ex vivo expanded human regulatory T cells can prolong survival of a human islet allograft in a humanized mouse model. Transplantation 96:707–716

Xie D, Smyth CA, Eckstein C, Bilbao G, Mays J, Eckhoff DE, Contreras JL (2005) Cytoprotection of PEG-modified adult porcine pancreatic islets for improved xenotransplantation. Biomaterials 26:403–412

Xu X, D'Hoker J, Stange G, Bonne S, De Leu N, Xiao X, Van de Casteele M, Mellitzer G, Ling Z, Pipeleers D et al (2008) Beta cells can be generated from endogenous progenitors in injured adult mouse pancreas. Cell 132:197–207

Yamasaki T, Inoue K, Hayashi H, Gu Y, Setoyama H, Ida J, Cui W, Kawakami Y, Kogire M, Imamura M (1998) Effect of a new immunosuppressive agent, FTY720, on survival of islet allografts. Cell Transplant 7:403–406

Yazaki S, Iwamoto M, Onishi A, Miwa Y, Hashimoto M, Oishi T, Suzuki S, Fuchimoto D, Sembon S, Furusawa T et al (2012) Production of cloned pigs expressing human thrombomodulin in endothelial cells. Xenotransplantation 19:82–91

Yi S, Feng X, Hawthorne WJ, Patel AT, Walters SN, O'Connell PJ (2002) $CD4^+$ T cells initiate pancreatic islet xenograft rejection via an interferon-gamma-dependent recruitment of macrophages and natural killer cells. Transplantation 73:437–446

Zhang N, Richter A, Suriawinata J, Harbaran S, Altomonte J, Cong L, Zhang H, Song K, Meseck M, Bromberg J et al (2004) Elevated vascular endothelial growth factor production in islets improves islet graft vascularization. Diabetes 53:963–970

Chapter 12
Beta Cell Therapies for Type 1 Diabetes

Wayne John Hawthorne

Introduction

As discussed in preceding chapters in this book, it is widely accepted that type 1 diabetes mellitus (T1D) is an autoimmune disorder that presents clinically with hyperglycaemia. The root cause is insulin deficiency (Rafeullah et al. 2012), as a result of islet β-cell destruction (Waldron-Lynch and Herold 2011). If left untreated, patients will develop severe hyperglycaemia and ketoacidosis, becoming so metabolically unwell that they would require intervention, hospitalisation, and treatment. If untreated, they can potentially die from their severe metabolic state (Gubitosi-Klug 2014). Even when diagnosed and treated with ongoing insulin, if patients maintain poorly controlled blood glucose levels (BGLs), their ongoing hyperglycaemia and unstable insulin levels can contribute to end-stage micro- and macro-vascular damage and even contribute to organ failure such as neuropathy, nephropathy, retinopathy, and peripheral vascular disease associated with morbidity and mortality (Katagi et al. 2014; Babizhayev et al. 2013). Type 1 diabetes can present at different ages and forms that can display a myriad of clinical presentations and features that will not be discussed in this chapter. The treatments outlined in this chapter revolve around the potential prevention of, or the treatment for, type 1 diabetes (~10 % of diabetes). The greater proportion (~90 %) of cases

W.J. Hawthorne (✉)
Department of Surgery, Westmead Clinical School, University of Sydney, Westmead Hospital, Westmead, NSW 2145, Australia
e-mail: wayneh@med.usyd.edu.au

W.J. Hawthorne
National Pancreas and Islet Transplant Laboratories, The Westmead Institute for MedicalResearch, Westmead, NSW 2145, Australia

A.A. Hardikar (ed.), *Pancreatic Islet Biology*, Stem Cell Biology and Regenerative Medicine, DOI 10.1007/978-3-319-45307-1_12

of diabetes are of those with type 2 diabetes (T2D), while a small number of cases being made up of and including gestational diabetes, endocrinopathies, and postviral infection-induced diabetes (McCulloch 2007; Boitard 2012), which will not be discussed here. Although hyperglycaemia is an indicator for both T1D and T2D, the clinical features and pathophysiology between the two disorders are quite different (Glaser 2007). The focus of this chapter, however, will be to provide an outline of the current cutting edge and novel technologies that are being investigated for the treatment for individuals with type 1 diabetes. These new and potential clinical therapies are already under investigation by researchers and clinicians worldwide in the hope to provide newer options to treat individuals with type 1 diabetes.

Gold Standard Treatments for Type 1 Diabetes

Insulin Therapies

Current gold standard treatments for type 1 diabetes remain anchored to insulin administration; extracted insulin, the mainstay of treatment since its discovery in 1922 by Banting et al. (1922), has been superseded by recombinant human insulin as replacement therapy for type 1 diabetes (Karamitsos 2011). Despite standard insulin therapy preventing acute ketoacidosis and associated morbidity and mortality, it remains a heavy burden to lifestyle that neither alleviates all the problems of diabetic individuals nor prevents the serious long-term causality to vital organs such as the kidney, eye, and the heart as well as the entire vascular system (Karamitsos 2011). Intensive insulin therapy can also inadvertently result in hypoglycaemic episodes with minor (e.g. blurred vision and lethargy) to extreme manifestations (coma and death) (Gubitosi-Klug 2014). There is significant evidence that the use of insulin pumps with a global approach to improvement to lifestyle by exercising regularly, maintaining a healthy weight, and eating healthily (Coyle et al. 2013) have not reduced severe hypoglycaemic episodes in at-risk patients (Bergenstal et al. 2010).

Current Transplantation Options

Major technological advances and changes in the development of newer generation immunosuppressive agents have heralded the advent of transplantation to treat individuals with the more problematic and complicated cases of type 1 diabetes. We can now offer patients with type 1 diabetes and severe hypoglycaemic unawareness the option of treatment with a clinical allo-islet cell transplant as a means to cure their diabetes (O'Connell et al. 2013; O'Connell et al. 2006). This cutting edge technology has been available and utilised for over a decade with variable success (Balamurugan et al. 2014) and is discussed in an accompanying chapter by Helen Thomas and colleagues. We have demonstrated advances in both techniques for isolation and transplantation of islet cells, but also with the advent of newer immunosuppressives (Shapiro et al. 2006) provided significant improvements in transplantation results and an overall increase to long-term outcomes, with insulin independence rates up to 5 years post-transplant with minimal complications (Bruni et al. 2014). However, this option has limited application to the broader population of individuals with type 1 diabetes due to its focus on treatment for patients with severe hypoglycaemic unawareness and its reliance on the availability of cadaveric donor availability and selection, engraftment, and the side effects of immunosuppression remaining as existing obstacles to be addressed to further improve this therapy (Bruni et al. 2014). Underpinning transplantation of patients with type 1 diabetes has been the major surgical procedure of simultaneous transplantation of a pancreas and a kidney transplant (SPK). This type of transplant has been utilised to treat the subpopulation of individuals with diabetes and secondary renal failure. This has been the mainstay of treatment for these patients for a number of decades with extremely good results worldwide for those individuals with diabetes and renal failure (Thwaites et al. 2012). Overall, our own experience has shown that pancreas transplantation is currently the only proven option to achieve insulin independence over periods of more than a decade (Lam et al. 2010). However, a pancreas allograft alone or simultaneous pancreas and kidney transplant requires major surgery, and most patients only receive such when they also require treatment for renal failure, thus necessitating the kidney allograft. The advantage of transplanting the pancreas at the same time provides a great benefit to the patient in a cure for their diabetes but also providing significant protective advantage to the simultaneous kidney graft (Allen et al. 2001). A significant advantage of islet allograft transplantation is that it is a relatively minor procedure, although it usually requires two or more transplants to achieve insulin independence (O'Connell et al. 2006). It has been very successful in reducing the underlying issue in these patients of their rather severe hypoglycaemic episodes (Tiwari et al. 2012).

Both islet cell and simultaneous pancreas transplants have good success rates with excellent long-term outcomes being shown to prevent ongoing progression of the other secondary complications of diabetes (Thompson et al. 2011), and specific studies have shown significant improvements in retinopathy (Chow et al. 1999) and neuropathy (Lam et al. 2010; Allen et al. 1997; Nankivell et al. 1997). The major hurdle to increasing the number of potential recipients is the ongoing reliance on the generous altruistic but somewhat low donation rate from cadaveric organ donors and issues associated with the requirement for lifelong immunosuppression (Tuduri et al. 2012). Ultimately, these forms of treatment offer extremely good outcomes for patients in these subpopulations that can receive them and the use of these current transplantation technologies. The current treatment rates for patients with type 1 diabetes by these methods remain low and to be able to increase treatment rates for a greater proportion of patients we rely on the development of newer methods to increase the low organ donor rates (Shahrestani et al. 2016). However, even with improvements made to increase organ donor rates by such methods as the use of suboptimal organ donors, we will unlikely be able to transplant the ever-increasing number of patients who require treatment for their type 1 diabetes and development of their secondary complications such as renal failure (Morrissey and Monaco 2014).

Peptides, such as cyclosporine and core peptide (CP), prevent or suppress IL-2 production in T cells, and their immunosuppressive properties may be used in conjunction with simultaneous pancreas and kidney or islet transplantation to prevent rejection (Manolios et al. 1997; Sobel et al. 2010). To remove the need for daily administration of insulin or immunosuppressive drugs following transplantation, gene therapy approaches for type 1 diabetes treatment are being widely researched. Various targets of immune intervention may also be coupled with gene therapy to provide a synergistic effect, some of which will be discussed in the following sections of this chapter.

New Frontiers in β-Islet Cell Replacement

Genetic Engineering and Modification

Currently used genetic engineering methods are performed by using germ line or somatic manipulation. In germ line manipulation, we can transfer genes to the target individual as well as their offspring, whereas somatic genetic manipulation only effects the individual into which the transgene is introduced (Wirth et al. 2013). Gene transfer can be divided into in vivo or in vitro transfer. For successful in vivo delivery, the vehicle for the transgene must be appropriately directed to the target cells and the gene product must be protected from immune attack. Manipulating cells genetically in vitro is less invasive than in vivo techniques; however, target cells are required to be easily removed and transplanted back into the host.

In type 1 diabetes, islets are the target for auto-reactive T cell destruction. Gene therapy is a useful technique to treat type 1 diabetes as it can be applied from many different directions and against very specific targets. These can occur at the earlier more interventional time point where the insulin gene can be replaced in a host, or the auto-reactive T cells suppressed. Later in the course of the disease, alternative treatments can be offered by using genetically modified donor pigs as potential donors for xenotransplantation. These different methods of gene transfer are discussed in the following sections of this chapter.

Potential Non-viral Gene Transfer Treatments

We have a number of non-viral ways in which we can quite simply and relatively non-expensively genetically modify cells including the very fragile islet cells. One such method utilises calcium phosphate co-precipitation where the DNA of the target molecule is added to buffered saline/phosphate solution and a precipitate forms. Cells can endocytose or phagocytose the DNA-containing precipitate. This method has been tested in a variety of cell types and can produce either transiently transfected cells or cells that are able to stably express the transgene. Liposomes have also been used as high efficiency transfection agents of cells both in vivo and in vitro, unlike calcium phosphate co-precipitation, which is conducted in vitro. The advantage of in vivo lipofection is that the liposomes may be injected into the bloodstream and is less invasive than other treatments, such as transplantation. Liposomes containing DNA have minimal positive charges which improve their interaction with target cells and the consequent transfection efficiency (Torchilin 2006). Directly injecting DNA into cells is an effective method for transfecting cells. However, as each cell needs to be targeted individually, this is a labour-intensive technique and is not suited for the targeting of large cell numbers. Electroporation creates permeable membranes for gene transfer by applying high voltages to cells and, in many cases, causes cell death. Electroporation is a good technique to enhance plasmid-based gene delivery, but is hampered by the fact that efficient gene transfer into β-cells requires dissociation of the tightly clustered sac of cells (islets) into single-cell suspensions (Prud'homme et al. 2007). Without the maintenance of their morphology, dissociated islets do not respond well to glucose (Callewaert et al. 2007). In comparison with both lipofection and calcium phosphate co-precipitation, biolistic transfection produces higher transfection efficiencies. Biolistics is the use of a 'gene gun' to transfect cells with a transgene (Xia et al. 2011). The 'gene gun' rapidly discharges DNA—microprojectiles into cells. Non-viral methods are cost-effective and have low host immunogenicity, but gene delivery by viral vectors is more efficient (Nayerossadat et al. 2012), which makes viral vectors a more attractive gene therapy option, and is further discussed.

Potential Use of Viral Vectors

Obviously, the use of the new cutting edge tools takes significant new skills and a degree of technical ability, but the choice of the tool also requires rather complex and intricate knowledge of the tools to be used. In the case of using viral vectors, these are the tools that we can use to deliver genetic material into cells. The process can be performed either in vitro or in vivo both with inherent advantages and disadvantages. Using a particular viral vector to harness the viruses' specialised molecular machinery to transport and insert their payload into the target cells (transduction) has been done reliably since the 1970s (Cole et al. 1979). However, in order to be successful, vectors need to be simple to manufacture in large numbers, have the ability to be targeted to a specific site, be able to transduce both dividing and non-dividing cells, result in high transduction efficiency, not elicit a strong immune response, and allow for long-term expression of the transgene (Lu 2004). For transgene delivery into islets, the vector is required to pass through the islet membrane and transduce the cells within. Experiments by Leibowitz et al. (1999) have previously shown that successful transduction of the cells within islets only occurs at the periphery of the islet (approximately 10 % of cells), and cells in the core of the islet are not transduced. The main disadvantage of retroviral transduction is that they are only able to transduce cells that are currently dividing—non-dividing islets cannot be transduced by retroviral vectors (Liu and Berkhout 2014). There may also be random integration of the transgene into the host genome, resulting in insertional mutagenesis (Bobisse et al. 2007; Naldini 2011; Schambach et al. 2013).

Adenoviral vectors have the advantage over retroviral vectors in that they are able to transduce both dividing and non-dividing cells and can be prepared in high titres (Franceschi and Ge 2008). Adenoviruses can therefore infect non-dividing insulin-secreting cells and have been shown to be able to transduce rodent islets (Becker et al. 1994; Csete et al. 1995; Sigalla et al. 1997). Barbu et al. (2006) have shown by confocal sectioning of intact islets transduced with GFP that expression on the cells was in fact only on the periphery of the islets and as such transduction efficiencies are approximately only 30 %.

The weaknesses of this type of gene transfer are that the vector antigens elicit potent immune responses and the inserted DNA is episomal, resulting in short-term transgene expression (Kawabata et al. 2010).

Lentiviral vectors have similar characteristics to both retroviral vectors and adenoviral vectors. The retroviral characteristics are the ability to integrate the transgene into host chromosomal DNA and to alter the surface envelope proteins. Lentiviral vectors are able to transduce primary and post-mitotic cells such as neurons, liver, muscle cells, primary endothelial cells, and islets (Leibowitz et al. 1999; He et al. 2006), and to transduce dividing and non-dividing cells without the potent immune responses that adenoviral vectors elicit (Liu and Berkhout 2014).

Recently, lentiviral vectors have been produced where the genes unnecessary for transduction are removed, decreasing the possibility of active recombinant virus production (Bobisse et al. 2007). Self-inactivating vectors (SINs) are unable to produce a full-length vector (Bobisse et al. 2007) because there is a 400-base pair (bp) deletion within the U3 region of the 3' long terminal repeat (LTR) (Fernandes et al. 2004). It has also been noted that the insertion of a 178-bp fragment amplified from HIV-1 strain.

NL4-3 between the *rev* response element and the internal cytomegalovirus (CMV) promoter resulted in enhanced transduction efficiency (Fernandes et al. 2004). A number of other groups have developed methods to transduce pig islet cells such as neonatal islet cell clusters as can be seen in Fig. 12.1 that can be transduced and be used as a potential future donor source for xenotransplantation. Ginn et al. (2004) demonstrated the efficient use of lentivirus vector-mediated transduction of neonatal porcine islet cell clusters as a means to transfer or confer protection of these transplanted cells for future use in xenotransplantation.

After dissociation and enzymatic digestion, NICC cells were exposed to a lentiviral vector, containing a CMV-enhanced green fluorescent protein (EGFP) expression cassette, at multiplicities of infection (MOI) ranging between 10 and 1000. While a high proportion of NICCs could be transduced at relatively low MOI, higher MOI resulted in more intense EGFP expression. An MOI of approximately 200 was therefore chosen for subsequent experimentation. Following 6 days in culture, 20,000 transduced NICCs were transplanted beneath the kidney capsule

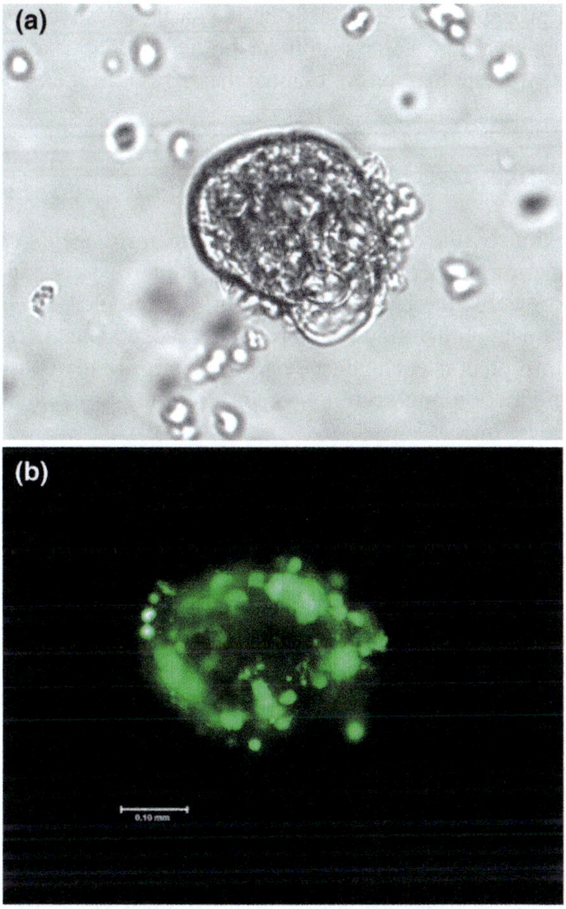

Fig. 12.1 Photomicrograph demonstrating neonatal islet cell clusters after being treated with lentiviral vector containing green fluorescent protein (GFP) the first showing a bright-field image of a neonatal islet cell cluster after being treated with GFP demonstrating intact cell wall and normal morphology (**a**). The second showing a neonatal islet cell cluster following being treated with lentiviral vector containing GFP for three days demonstrating its GFP expression (**b**). Magnification ×200

of diabetic SCID mice. Transplanted kidneys were removed on days 1, 7, 21, and 120, and processed for frozen section and histology to demonstrate the presence of the EGFP and also the intact NICC. Macroscopically, SCID mice demonstrated the continued presence of transplanted NICCs; furthermore, histological analyses from early biopsy time points showed persistence of EGFP expression in the transplanted NICC tissue. Constructing lentiviral vectors encoding molecules with the potential to prevent acute graft rejection or other agents to help prevent inflammation or thrombosis has great potential. Candidate molecules include CD39, which regulates vascular inflammation and thrombosis, and tissue factor pathway inhibitor and thrombomodulin, which are important endothelial cell regulators of the coagulation process. Islets and other cells can be transduced by lentivirus at high efficiency making this vector system an attractive tool for xenotransplantation research and potential application for clinical therapy (Ginn et al. 2004).

Choosing the Correct Target to Prevent Diabetes

As outlined previously, the prevention or treatment for diabetes can be achieved by using a number of genetic manipulations, including tolerance induction, use of immunoregulatory cells, the inhibition of apoptosis, ectopic gene expression, and transplantation along with immunosuppression. These gene therapy approaches are used to either protect islets against destruction prior to the onset of diabetes or reproduce the production and secretion of insulin after onset of hyperglycaemia, and are discussed below.

Protection of Islet Cells Prior to Destruction: A number of programmes have been undertaken to identify people who are at a high risk of developing type 1 diabetes. This has included genome-wide scanning for genetically susceptible individuals, which has shown that the human leucocyte antigen (HLA) locus on chromosome 6q21.3 induces the highest susceptibility to type 1 diabetes (Holm et al. 2004). More specifically, the reports by Golden et al. have suggested that the HLA haplotypes DR3-DQB1*0201 and DR4-DQB1*0302 increase susceptibility to type 1 diabetes. DR3 and DQB1*0302 are thought to be the primary alleles linked with increased susceptibility, and that DQB1*0201 and DR4 genes play secondary roles (Golden et al. 2005). Obviously, armed with this knowledge, we can take steps to pre-emptively target these individuals in order to potentially provide interventional treatment strategies prior to the onset of their type 1 diabetes.

Targeting Mechanisms for Induction of Immune Tolerance One such avenue is to target the various mechanisms that either fail or are attacked and provide support to these. One approach for type 1 diabetes therapy is to induce tolerance to islet antigens by the transduction of stem cells. Autologous stem cells, transduced with a retrovirus encoding major histocompatibility complex (MHC) class II I-A β-chain molecules and transplanted into diabetic mice, protected mice against type 1 diabetestype 1 diabetes (Tian et al. 2007). Diabetes was precipitated in mice by blocking the programmed death-1-programmed death ligand-1 (PD-1-PD-L1) pathway (Ansari et al. 2003). BGLs of mice treated with the transduced stem cells remained normal for more than 180 days (end of study), suggesting that MHC class II I-A β-chain molecules can induce self-tolerance to islet auto-antigens and resistance to diabetes.

Targeting Immunoregulatory Cytokines Another potential target is the use of immunoregulatory cytokines. Cytokines, such as IL-4 or IL-10, have immunosuppressive properties while others, such as transforming growth factor (TGF)-β, can inhibit natural killer (NK) cell function, thymocyte proliferation, antibody production, and T cell activity (Hu et al. 1994; Mia et al. 2014). Kodama et al. found that there was a gradual decrease in IL-4 in the pancreatic lymph nodes of older NOD mice compared with IL-4 levels in pancreatic lymph nodes in diabetes-resistant

NOD.B10 mice (Kodama et al. 2008). Using this information, Creusot et al. reintroduced IL-4 using lentivirally transduced dendritic cells (DCs) into diabetes-susceptible NOD mice and monitored their BGLs for the development of diabetes (Creusot et al. 2008). DCs with a lentivirus expressing IL-4, under the control of a CMV promoter, were transduced and injected into the tail vein on 12-week-old non-diabetic NOD mice. The progress of these mice was compared with mice injected with phosphate-buffered saline (PBS) into the tail vein. There was delayed onset of type 1 diabetes by 4–8 weeks and reduced proportion of mice developing diabetes when compared to control (30 vs. 80 %). It was also shown that MHC expression played a major role in the protective effect of IL-4 in diabetes. The BGLs of normal NOD mice were compared to the BGLs of MHC-deficient (C2ta$^{-/-}$β2m$^{-/-}$) mice when both groups received the same DC/IL-4 treatment. In the MHC-deficient group, more than 70 % of mice developed diabetes by 29 weeks of age, whereas only 20 % of mice that exhibited normal MHC expression developed diabetes. These data indicate that IL-4 expression is able to protect NOD mice from diabetes when administered prophylactically, but the co-expression of MHC by DCs is also required.

To reverse diabetes with IL-4 expression, the presence of intrapancreatic CC chemokine ligand (CCL) 4 was required (Meagher et al. 2007) as well as MHC co-expression, which played a major role in the protective effect against diabetes, reducing the incidence of type 1 diabetes when compared to MHC-deficient mice (Creusot et al. 2008). T cells from the spleen of IL-4- and CCL4-treated NOD mice were transferred to NOD.SCID mice. The NOD.SCID mice displayed reduced type 1 diabetes incidence (4/10 mice) and delayed onset (98 days) compared with control NOD.SCID mice (78 days). To show that CCL4 played an important role in this treatment, it was sequestered from nine of the NOD.SCID mice, which developed type 1 diabetes within 66 days. When removed, CCL4 enabled the T cells from IL-4-treated mice to regain their pathogenic ability and reduced the time of onset of type 1 diabetes. CCL4 was then cloned into a plasmid and was given to mice intradermally weekly from 3 weeks of age until destructive insulitis (14 weeks). In the control group, insulitis progressively worsened from 14 weeks until 33 weeks. By contrast, mice treated with CCL4 exhibited reduced insulitis and incidence of type 1 diabetes (reduced from 72 to 27 %) by 35 weeks of age. These data suggest that CCL4 potentially can be used as a treatment for type 1 diabetes.

Islet auto-antigens, such as glutamic acid decarboxylase (GAD) and insulin (INS), conjugated with cholera toxin B (CTB) subunit have been shown to suppress type 1 diabetes (Bregenholt et al. 2003; Sadeghi et al. 2002; Sun et al. 1994). When recombinant vaccinia viruses, encoding the conjugates (CTB::GAD and CTB::INS) and IL-10, were injected into 4-week-old NOD mice, diabetes was reversed (Denes et al. 2006). In the control group, by 29 weeks of age, 70 % had developed diabetes. Comparatively only 30 % of CTB::GAD-treated mice and 40 % of IL-10-treated mice developed diabetes. CTB::INS appeared to be an improved treatment option compared to CTB::GAD or IL-10, with only 20 % of mice developing diabetes by 31 weeks of age. There was also delayed onset of diabetes when comparing treatment in 11-week-old NOD mice and untreated controls showing that CTB::INS delivered by recombinant vaccinia virus may be a good treatment for type 1 diabetes.

Low-dose IL-2 has been shown to induce long-lasting remission in diabetic NOD mice by decreasing IFN-γ production and increasing Foxp3$^+$ regulatory T cells in the pancreas (Grinberg-Bleyer et al. 2010). In a recent study taking advantage of this finding, IL-2 expression was induced islets of NOD mice by adeno-associated virus (AAV) vector gene delivery, driven by a mouse insulin promoter (Johnson et al. 2013). Mice received a single dose of the double-stranded AAV vector encoding IL-2 at either 12 or 16 weeks of age and displayed increased Foxp3$^+$ regulatory T cells and an increase in the anti-apoptotic Bcl-2 molecule. None of the mice thus treated developed diabetes. The regulatory T cells were found to be increased only around the infected islets and not in the draining pancreatic lymph nodes, which showed that IL-2 expression was localised, avoiding potential toxic effects of systemic IL-2. In this study, islets were genetically modified to express IL-2-induced expansion of regulatory T cells, which successfully suppressed type 1 diabetes.

Adenovirus-mediated IL-10 transfer into a rat β-cell line resulted in increased insulin secretion in response to glucose as well as a reduction of IL-1β-mediated Fas-mediated apoptosis (Xu et al. 2010). The number of IL-10-transfected islets expressing Fas was significantly lower than control islets (24.6 ± 1.0 vs. 32.6 ± 1.1 %). This finding was supported by data that showed a reduction of apoptosis in IL-10-transfected islets compared to control islets (9.4 ± 1.1 vs. 20.6 ± 2.3 %, $P < 0.05$), suggesting that using gene therapy to deliver IL-10 to islets can promote islet survival in diabetic patients.

Targeting Cytotoxic-Induced Apoptosis One very interesting avenue that has been targeted is the protection of islets against cytotoxic-induced apoptosis. In a number of studies, islets have been genetically modified to express anti-apoptotic molecules. These molecules have included the use of the anti-apoptosis gene—Bcl-2 (Contreras et al. 2002; Dupraz et al. 1999; Rabinovitch et al. 1999); anti-Fas molecules (Klein et al. 2000); and anti-apoptotic proteins—A20 (Grey et al. 1999), insulin-like growth factor I (IGF-I) (Giannoukakis et al. 2000), and the X-linked inhibitor of apoptosis protein (XIAP) (Plesner et al. 2010). Bcl-2 was expressed in various insulin-expressing cells by lentivirus (Dupraz et al. 1999), a replication-defective herpes simplex virus amplicon vector (Rabinovitch et al. 1999), and adenovirus (Contreras et al. 2002). Bcl-2-expressing βTC-tet cells, immortalised pancreatic β-cells, were protected against cytokines and staurosporine-induced apoptosis, as well as hypoxia-induced apoptosis (Dupraz et al. 1999). In vivo models using STZ-diabetic C3H mice showed that Bcl-2-expressing βTC-tet cells promoted lower postprandial basal BGLs when transplanted under the kidney capsule, reversing hyperglycaemia for several months. Similarly, when β-cells were transfected with Bcl-2, they survived a challenge with pro-apoptotic cytokines (IL-1β, TNF-α, IFN-γ) (Rabinovitch et al. 1999). In another study, Bcl-2 expression was induced in islets by a CMV-driven adenovirus. When Bcl-2-expressing islets were transplanted into STZ-induced diabetic SCID mice, diabetes was reversed. By contrast, mice that received control islets remained hyperglycaemic.

A20, a zinc finger protein shown to be an anti-apoptotic TNF-α-induced gene in endothelial cells, protects islets against IL-1β- and IFN-γ-induced apoptosis (Grey et al. 1999). Rodent islets, transduced with adenoviruses encoding A20, exhibited reduced levels of apoptosis, compared to untransduced islets. In a similar study aiming to prevent Fas-mediated apoptosis in islets, Klein et al. (2000) transduced an anti-Fas tRNA-ribozyme into a βTC-3 insulinoma cell line, NIT-1 cells and dissociated islets. The results showed that the tRNA-ribozyme reduced Fas expression by 80 %, when compared to mock transduced cells, and protected against apoptosis in all three-cell types. Human IGF-I is another molecule that prevents islet apoptosis (Giannoukakis et al. 2000). An adenoviral vector under the control of a CMV promoter was used to transfer IGF-I to islets. Caspase-3 activity, i.e. apoptosis, was reduced in transduced islets (23 ± 5 %) compared to control islets. In a more recent study, a chicken β-actin promoter-driven recombinant adenovirus encoding another anti-apoptotic protein, XIAP, was transduced into islets (Plesner et al. 2010). The survival of these XIAP-expressing islets, when transplanted into STZ-induced diabetic mice, was enhanced compared to controls. Taken together, these studies show that anti-apoptotic molecules can be protective against diabetes.

Treating Diabetes Following Islet Destruction: Unfortunately, there will be patients who are unable to be treated prior to the development of type 1 diabetes, and these patients will then require interventional treatments that utilise a number of methods to provide surrogates to the destroyed β-cells. These assorted methods include those of viral gene delivery. Additionally, some groups are investigating non-viral methods to target cells because they offer potentially safer gene delivery options. Chitosan nanoparticles have been used to wrap an expression plasmid encoding human insulin before administration to diabetic rats via the gastrointestinal tract (Niu et al. 2008). Following treatment, the BGLs of rats decreased from 22.12 ± 1.31 to 5.63 ± 0.48 and 5.07 ± 0.37 mmol/L in the lavage and coloclysis groups, respectively. In another study, glucagon-like peptide-1 (GLP-1)/IgG-Fc fusion protein was transferred to muscle cells of STZ-induced diabetic mice by electroporation (Soltani et al. 2007). GLP-1/IgG-Fc was subcloned into a vector driven by a CMV immediate-early enhancer–promoter. Mice received the vector through intramuscular DNA injections followed by electrical current locally applied to the skin. Compared with the control group that received IgG-Fc alone, GLP-1/IgG-Fc-treated mice exhibited an approximate 50 % decrease in the incidence of type 1 diabetes. The treatment increased insulin levels and was shown to improve β-cell numbers following depletion by STZ injections. Although both of these non-viral treatments successfully reduced BGLs of diabetic rodents, the highly efficient viral methods are still more widely used as discussed in this chapter.

Producing Surrogate β-Cells Ectopic gene expression is the expression of the target gene in an abnormal place in another cell. These cells would not normally produce insulin or other hormones that the islet cell produces, but this surrogacy is one potential way to treat type 1 diabetes. Ideally, we can place β-cell properties in cell types not targeted by the immune system. It is a widely used technique, which does not require immunosuppressive measures because the target cell originates from the graft recipient, offering an unlimited source of autologous cells for genetic manipulation. A number of suitable alternatives to β-cells for manipulation into insulin-producing cells include hepatocytes (Fodor et al. 2007; Gerace et al. 2013; Halban et al. 2001; Hsu et al. 2008; Koizumi et al. 2006; Olson et al. 2008; Ren et al. 2013; Shternhall-Ron et al. 2007; Tang et al. 2013), stem cells (Rahmati et al. 2013; Talebi et al. 2012; Xu et al. 2007), neurons (Kojima et al. 2009), fibroblasts (Selden et al. 1987), muscle, keratinocytes (Lei et al. 2007; Tian et al. 2008), neuroendocrine cells, and many other endocrine cells (Zhang et al. 2008).

Ectopic insulin expression has been trialed in cells such as mouse pituitary corticotroph (AtT20 cells) in 1983 (Moore et al. 1983) and fibroblasts in 1987 (Selden et al. 1987). AtT20 cells are similar to β-cells in that they are able to secrete proinsulin and also express the proconvertases, proprotein convertase (PPC)2 and PPC3, to convert proinsulin into mature insulin (Mitanchez et al. 1997). The expression of insulin from AtT20 cells confirmed that non-β-cells could be modified to secrete insulin. Important features of β-cells which are essential in the body's ability to regulate BGLs include the continuous biofeedback monitoring of active hormone levels, regulated transcription and translation of pro-insulin, regulated pro-insulin processing to mature insulin, regulated storage of mature insulin, and regulated secretion of mature insulin to a stimulus, such as glucose (Ren et al. 2013). For this finding to be put into use, target cells for genetic engineering must have a regulated secretory pathway. In a subsequent series of experiments, fibroblasts were engineered to express pro-insulin under a metallothionein promoter resulting in constitutive insulin production. The unregulated insulin release caused hypoglycaemia in animals transplanted with the modified fibroblasts and unfortunately subsequent hypoglycaemic coma and death. Following from these early experiments, many groups have attempted to recreate the β-cell in other cell types as a treatment for diabetes and these are discussed below.

(a) Epithelial cells as surrogate β-cells

Epidermal keratinocytes differ from islets in that they are self-renewing cells and are an unlimited source of autologous cells that can be used for genetic modification. In experiments to transduce keratinocytes to express insulin, furin-cleavable proinsulin was used because the proteases (prohormone convertases, PHC1/PHC3 and PHC2) required to produce mature insulin by cleaving the disulphide bonds linking C-peptide to the A-chain and B-chain to release C-peptide and mature insulin are only present in islets and not in cells such as keratinocytes and hepatocytes (Tian et al. 2008). By altering the human insulin gene to produce a recognition site for a propeptide endoprotease which is abundant in the Golgi apparatus (furin), liver cells and keratinocytes were able to process proinsulin into mature insulin (Tian et al. 2008).

Keratinocytes have been modified to ectopically express furin-cleavable proinsulin using a recombinant retrovirus system (Lei et al. 2007). Enzyme-linked immunosorbent assay (ELISA) was used to quantify proinsulin production from transduced cells. The amount of mature insulin produced by transduced cells was measured indirectly using ELISA by the quantification of C-peptide, the product released from the conversion of proinsulin to mature insulin. The levels of both proinsulin and C-peptide were high, suggesting that mature insulin was successfully produced. In another study, furin-cleavable proinsulin was again engineered into epidermal keratinocytes and was evaluated for insulin expression in vivo by transplantation into athymic streptozotocin (STZ)-induced diabetic mice (Tian et al. 2008). The expression of insulin was successfully detected in the plasma of transplanted mice, compared to control mice that did not receive transduced cells and did not express insulin in the plasma, confirming that keratinocytes are good candidates for ectopic gene expression of insulin.

K cells are a different type of enteroendocrine epithelial cells that are located in the intestinal lining sharing similar characteristics to β-cells and making them a good candidate for genetic manipulation. A murine enteroendocrine K cell-derived cell line (STC-1) was genetically engineered to express insulin (Zhang et al. 2008). Transduced STC-1 cells were transplanted into STZ-induced diabetic nude mice. BGLs in all mice transplanted with engineered STC-1 cells returned to normal levels and remained so until the end of the study at 49 days post-transplant.

(b) Hepatocytes as surrogate β-cells

Hepatocytes are similar to β-cells in the sense that they can facilitate the removal of glucose from the bloodstream via glucokinase phosphorylation in response to hyperglycaemia (Cullen et al. 2014). They are good cellular candidates for genetic manipulation because they are the essential regulators of carbohydrate metabolism via insulin, are already equipped with the machinery to respond to fluctuating glucose levels, and can be modified to produce insulin as glucose levels increase. Various research groups have modified hepatocytes with either an insulin gene or a gene that will promote insulin expression. Olson et al. (2008) examined the use of an adenovirus vector encoding an insulin transgene under the control of a glucose- and insulin-responsive, liver-specific promoter. BBDR/Wor rats, made diabetic by intraperitoneal injections of polyinosinic–polycytidylic acid (poly-I:C), were administered virus via the jugular vein. When virus-treated rats were compared to control, the treated rats were euglycaemic, had normal metabolic switching, and reduced intra-abdominal fat deposits. By contrast, control rats remained severely hyperglycaemic.

Human pancreatic duodenal homeobox 1 (PDX-1) is a pancreatic transcription factor that is essential for pancreatic development and drives the production of insulin (Koizumi et al. 2006). The PDX-1 gene has been transduced into hepatocytes in different studies and tested for its ability to reverse type 1 diabetes (Fodor et al. 2007; Tang et al. 2013). Fodor et al. (2007) investigated the outcome of transplanting PDX-1-transduced hepatocytes into STZ-induced diabetic SCID mice. Following the transplant, the mice were challenged with an intraperitoneal injection of glucose and BGLs returned to normal levels within 8 weeks. The successful return of euglycaemia was attributed to the presence of the transplanted hepatocytes, as removal of the graft resulted in recurrence of hyperglycaemia. In another study, also studying the effects of PDX-1 on hepatocytes, adeno-associated virus (AAV) serotype 2 vector-expressing PDX-1 and Ngn3 were injected into the portal vein of STZ-induced diabetic NOD.SCID mice (Tang et al. 2013). Livers treated with PDX-1/Ngn3 or PDX-1 alone responded to a glucose challenge by secreting insulin, and both insulin and glucagon were detected in hepatocytes located along the edge of central veins of the liver. When the vectors PDX-1/Ngn3 or PDX-1 alone were injected into diabetic mice, the BGLs returned to normal levels in contrast to diabetic mice that received Ngn3 alone and remained hyperglycaemic. The expression of the PDX-1 gene in hepatocytes has been shown by these studies to be a promising treatment for type 1 diabetes.

Furin-cleavable insulin, which has previously been used in epithelial cells, has been delivered into hepatocytes as a treatment for type 1 diabetes (Gerace et al. 2013; Hsu et al. 2008; Ren et al. 2013; Ren et al. 2007). Ren et al. (2007) transduced hepatocytes to express insulin with furin-recognition sequences.

Transduced cells were administered by intervallic infusion in full flow occlusion to the livers of STZ-induced diabetic rats. Following transduction, an immediate reduction of BGLs was seen in the diabetic rats, and within 5 days, BGLs of diabetic rats were comparable to non-diabetic healthy rats. This study was the first to demonstrate long-term normalisation of BGLs of STZ-induced diabetes with the use of a lentiviral delivery system of furin-cleavable insulin. The same group also studied the effect of furin-mediated insulin on diabetic NOD mice livers (Ren et al. 2013). By removing the liver from the circulation during lentiviral delivery, liver-to-pancreas transdifferentiation was achieved as can be seen in Fig. 12.2. Pancreatic hormones (insulin, glucagon, and somatostatin), synaptophysin (major membrane glycoprotein located on pancreatic endocrine cells), and pancreatic transcription factors (PDX-1, Neurod1, Neurog3, Nkx2-2, Pax4) were detected in transduced cells but not in untreated livers. Diabetes was successfully reversed in lentivirus-treated NOD mice until the experimental end point, at 150 days, which contrasted with mice treated with empty vector and remained hyperglycaemic.

Most impressively in a follow-up to this work in a large animal study, Gerace et al. (2013) investigated the transdifferentiation of liver cells resulting in similar characteristics to β-cells in the sense that they were able to secrete insulin in response to glucose challenge in Westran pigs, again using furin-cleavable human insulin. Successful reversal of diabetes was detected in one of the Westran pigs, which achieved a normal intravenous glucose tolerance test at day 30 post-transduction and retained normal BGLs until the experimental end point. Pancreatic hormones (insulin, glucagon, and somatostatin) and β-cell transcription factors (PDX-1, Neurod1, and Nkx2-2) were detected in the transduced liver. These can be seen in Fig. 12.3.

The redirection of hepatocytes into cells displaying pancreatic cell attributes can be accomplished by systemic administration of a recombinant adenovirus encoding for PDX-1 (Shternhall-Ron et al. 2007). This study, performed in cyclophosphamide-induced diabetic NOD mice, assessed whether liver-to-pancreas redirection could occur when the pancreas was under autoimmune attack.

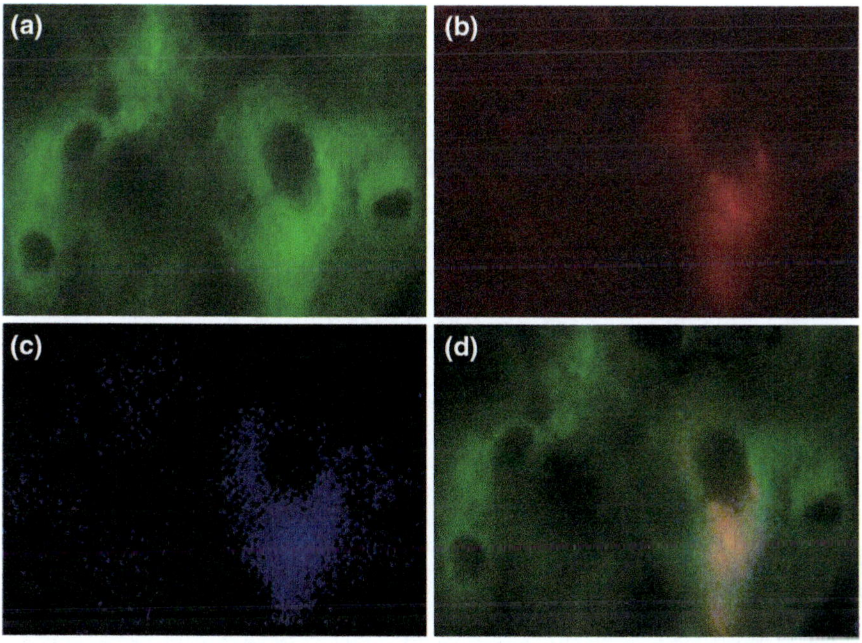

Fig. 12.2 Expression of pancreatic hormones following reversal of spontaneous hyperglycaemia in non-obese diabetic mice that had been treated with lentiviral vector (HMD) containing furin-cleavable human insulin. Photomicrographs of mouse liver at 5 months for anti-insulin (**a**), anti-somatostatin (**b**), anti-glucagon (**c**), and merged image of all three hormones (**d**). Magnification ×1000

Successful redirection of hepatocytes to islet-like functions was confirmed by the detection of endocrine hormones (insulin, glucagon, and somatostatin) in more than 80 % of PDX-1-treated mice. When compared to control mice that all remained hyperglycaemic, the treated mice displayed 55-fold higher levels of insulin and a 43 % reversal of diabetes.

Polyethylenimine (PEI) is an agent that enhances insulin expression and was used in combination with recombinant adeno-associated virus (rAAV) infection in animal studies to increase the expression of furin-cleavable human insulin in hepatoma cells (Hsu et al. 2008). In STZ-induced diabetic C57BL/6J mice treated with rAAV–insulin–PEI complexes, euglycaemia was achieved. It was also noted that there was a 30-min delay in insulin release in response to glucose compared to healthy non-diabetic mice, possibly caused by delayed regulation of insulin transcription. For this approach to be feasible in humans, the delay in insulin release by hepatocytes needs to be reduced to mimic the speed of natural insulin secretion, where islets release insulin that was previously produced and stored. These results suggest that the presence of insulin in the liver can cause hepatocytes to transdifferentiate into pancreatic cells, and that this approach can potentially be a treatment for type 1 diabetes.

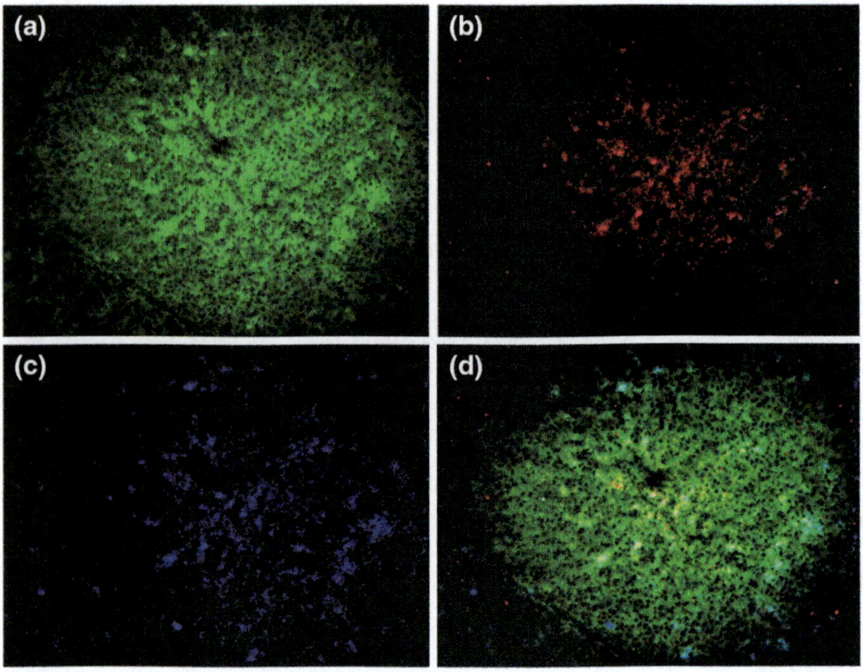

Fig. 12.3 Expression of pancreatic hormones following reversal of hyperglycaemia in a diabetic pig at 44 days, following treatment with furin-cleavable human insulin. Photomicrographs of liver tissue showing anti-insulin (**a**), anti-somatostatin (**b**), anti-glucagon (**c**), and merged image (**d**). Magnification ×200

(c) Neurons as surrogate β-cells

Like the hepatocytes discussed previously, neurons are able to sustain transferred genes for the duration of their lifespan, and Kojima et al. (2009) took advantage of this property to induce long-term stable gene expression. Leptin is a hormone that normally regulates energy homeostasis and has been shown, in the hypothalamus, to decrease tonic episodic insulin secretion and to reduce BGLs (Kojima et al. 2009). Ependymal cells, transduced with a rAAV encoding for leptin, were injected into the cerebroventricular cavity of STZ-induced diabetic mice. Mice injected with the control vector remained hyperglycaemic and died within 6 weeks. By contrast, the BGLs of treated mice reduced to levels comparable to naïve mice. This study suggests that leptin treatment in neurons may be a possible avenue for type 1 diabetes reversal.

(d) Stem cells as surrogate β-cells

A number of very novel investigational therapies involving stem cell transplantation and their ability to differentiate into insulin-producing or immunomodulating cells have also been studied (Couri et al. 2009; Voltarelli et al. 2007). Stem cells have many characteristics that make them attractive for use as surrogate β-cells. They have the ability to multiply in culture, differentiate, are immune-privileged and capable of circumventing the autoimmune response that destroys islets (Xu et al. 2007). Xu et al. studied the effect of transduced murine mesenchymal stem cells on the BGLs of STZ-induced diabetic mice. The stem cells were transduced with a recombinant retroviral plasmid encoding an insulin gene, and mice treated with these cells had lowered BGLs by the end of the study at 42 days. These results suggest that murine mesenchymal stem cells can be genetically modified to express human insulin and maintain normoglycaemia for at least 42 days.

Transdifferentiation of mouse mesenchymal stem cells into islet-like structures can be achieved by lentiviral transduction encoding the PDX-1 gene (Rahmati et al. 2013). Transduced cells began to form a ball-like appearance or three-dimensional spherical or grape-like clusters, which were compared to non-transduced cells that remained adherent spindle and fibroblast-like cells. Expression of three genes unique to the pancreas (insulin 1, insulin 2, and PDX-1) confirmed the pancreatic transdifferentiation by the stem cells, which secreted insulin in response to high levels of glucose. In a similar study, Talebi et al. (2012) also transduced mesenchymal stem cells with a lentivirus encoding PDX-1. Endocrine pancreatic marker genes (insulin 1, glucagon, Glut-2, PDX-1, and Ngn3) and insulin secretion were detected in transduced stem cells, which were then transplanted into alloxan-induced diabetic rats. Euglycaemia was achieved after 3 days in transplanted rats, contrasting with diabetic rats that were not transplanted with transduced stem cells and remained hyperglycaemic. These studies highlight the advantages of using mesenchymal stem cells as surrogate β-cells.

Protection of the Transplanted β-Cell

One presently very successful avenue for treating patients in several subpopulations that have type 1 diabetes is to transplant them with β-cells in one of the various forms of transplants as discussed. However, in order to be able to transplant foreign organs, tissue, or cells, we need to provide comprehensive concurrent immunosuppression to prevent graft rejection. This protection can be provided by a number of methods. Methods include grafting into immunologically privileged sites (Vagesjo et al. 2014); isolation from the immune system (encapsulation) (Jacobs-Tulleneers-Thevissen et al. 2013); inducing tolerance to grafts (Lee et al. 2014); or producing locally secreted immunosuppression (Carrington et al. 2010).

As discussed previously, the transplantation of human islets is one method for the treatment for diabetes but is dependent on the availability of cadaver donor organs. During islet isolation from the donor pancreas, much of the essential microvasculature for islet survival is stripped from the cells and the microvasculature destroyed (Panakanti and Mahato 2009). In a study aiming to repair some of the essential microvasculature, the routes of nutrient and oxygen delivery to islets, Panakanti and Mahato (2009), transduced islets with an adenovirus encoding hepatocyte growth factor (hHGF) and IL-1 receptor antagonist (hIL-1Ra), which increases the growth and proliferation of β-cells and prevents islet apoptosis, respectively. Transduced islets had a positive effect on the anti-apoptotic Bcl-2 gene and a negative effect on the pro-apoptotic Bax gene, resulting in reduction in apoptosis. When transduced islets were transplanted under the kidney capsule of STZ-induced diabetic NOD.SCID mice, BGLs were normalised for more than 28 days.

Rat islets genetically modified with type II membrane protein TNF-related apoptosis-inducing ligand (TRAIL) was transplanted into STZ-induced diabetic mice (Dirice et al. 2009). BGLs of mice that were transplanted with TRAIL-infected islets returned to normal levels by day 7 post-transplant and were maintained for 60 days. By contrast, BGLs of mice transplanted with mock-infected islets did not display prolonged euglycaemia and mice reverted to hyperglycaemia. Reversal of diabetes was confirmed to be due to the transplanted islets when mice again became hyperglycaemic after removal of the islet grafts. Immunohistochemistry revealed significant insulitis in mice transplanted with mock-infected islets contrasting with mice that received TRAIL-infected islets and displayed lower levels of insulitis. Evaluation of the data when combined with these data shows that adenoviral-mediated TRAIL gene transfer is a promising option for the treatment for diabetes.

Xenotransplantation of non-human islets, such as porcine islets, can overcome the limited availability of human islets, although the disadvantage of xenografts is that a greater immune responses need to be overcome when compared to the immune response to human cells (O'Connell et al. 2013). As part of the attempt to get these trials to the clinic, a number of groups have been investigating the possibility of using viral vectors carrying immunosuppressive agents to transduce porcine neonatal islet cells so that they can express these immunosuppressive agents from the surface of the islets. Once the transduced islets are transplanted, they can potentially produce localised immunosuppression in order to protect the islets without causing systemic immunosuppression of the recipient. Brady et al. eloquently demonstrated this in a novel study where humanised mice were transplanted with either control-transduced porcine neonatal islet cell clusters or diliximab (anti-CD2) Ab-transduced porcine neonatal islet cell clusters. The human T cells within grafts and spleens were enumerated by flow cytometry to demonstrate the differential in numbers caused by the local immunosuppression. They demonstrated that their diliximab Ab-transduced porcine neonatal islet cell clusters inhibited a human anti-pig xenogeneic response. The diliximab Ab bound human T cells, specifically depleting the human CD3(+) T cells in the humanised mouse model without inducing upregulation of activation markers or significant release of cytokines. More specifically, the humanised mice transplanted with diliximab Ab-transduced porcine neonatal islet cell clusters afforded depletion of CD3(+) T cells at the graft site leaving the peripheral immune system intact as can be seen in Fig. 12.4 (Brady et al. 2013).

Porcine islets have also been transfected with an adenovirus encoding for human decoy Fas antigen or membrane-bound human FasL and transplanted under diabetic

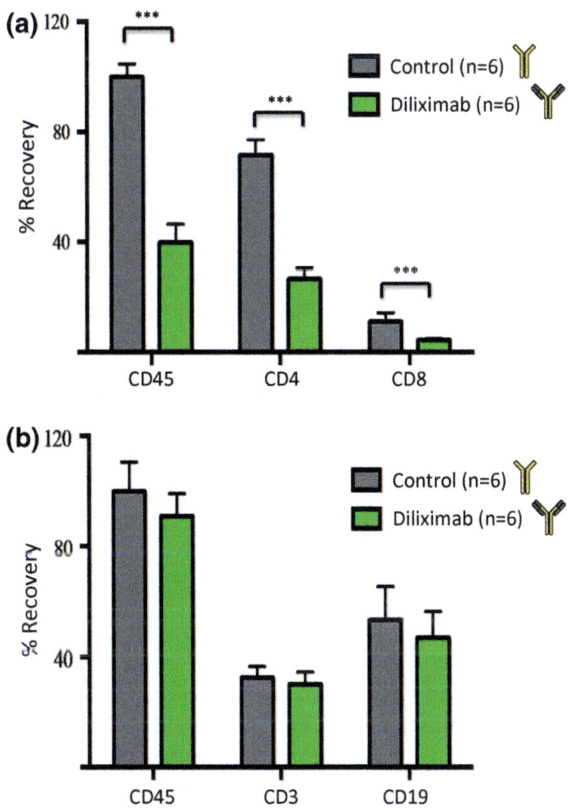

Fig. 12.4 Anti-CD2-producing pig xenografts affect localised depletion of human T cells in the graft site (**a**) and spleen (**b**) of treated huSCID mice

rat kidney capsules (Kawamoto et al. 2008). The idea behind inducing the expression of human decoy Fas antigen and membrane-bound human FasL in transplanted islets was to inhibit $CD8^+$ cytotoxic T-lymphocyte-mediated xeno-cytotoxicity. Compared to control animals that displayed graft rejection by day 3 post-transplant, transfected islet grafts survived past day 5 post-transplant, showing that the overexpression of either human decoy Fas or membrane-bound FasL effectively prevented islet graft destruction mediated by infiltrating $CD8^+$ cytotoxic T lymphocytes and macrophages.

β-cell lines are an inexhaustible source of β-cells that can be used for type 1 diabetes studies and xenotransplantation. Two murine cell lines that have been studied are the MIN and the NIT-1 cell lines. The MIN cell line was derived from transgenic mice that expressed the SV40 T-antigen gene from an 1867-bp human insulin promoter fragment (Miyazaki et al. 1990). MIN6 cells were capable of releasing insulin in response to glucose but could not transcribe insulin in response to glucose from the insulin gene in a controlled manner. The NIT-1 cell line is an insulinoma derived from NOD mice that expresses rat insulin promoter-SV40 T-antigen (RIP-Tag) transgene (Hamaguchi et al. 1991). These cells are not targeted by T cells because they do not express MHC on their cell surface in the absence of IFN-γ (Levine 1997). At low passage number, NIT-1 cells regulate release of insulin in response to glucose. However, as passage number increases, insulin release is constitutive (Levine 1997) and results in death of immune-compromised recipients (Georgiou et al. 1997). Recently, human β-cell lines, EndoC-βH1 and EndoC-βH2, were generated using targeted oncogenesis (Ravassard et al. 2011; Scharfmann et al. 2014). These cell lines expressed various β-cell-specific markers and secreted insulin. EndoC-βH1 and EndoC-βH2 were similar to adult β-cells in phenotype and responses to glucose, but with the additional capability to proliferate. Human β-cell lines are a limitless source of β-cells that, when transplanted, will elicit a reduced immune response compared to xenografts (O'Connell et al. 2013). To remove the possibility of tumour formation, the proliferative ability of EndoC-βH2 cells can be removed by excision of immortalising transgenes, which also increased β-cell-specific features and insulin production (Scharfmann et al. 2014). Potentially, human EndoC-βH2 cells could be used as surrogate cadaver islets in islet transplantation, especially if genetically modified with immunosuppressive molecules to enhance post-transplantation survival.

Xenotransplantation of the β-*Cell*

For a number of decades, there has been a concerted effort to use various forms of porcine β-cells as a replacement alternative to cadaveric human islet donors, and this has seen major successes and advances in the development of multitransgenic donor pigs to be used for transplantation. This has been achieved with cellular transplants leading the way using porcine neonatal islet cell transplants as a form of β-cell replacement to treat diabetic baboons (Hawthorne et al. 2011; Yi et al. 2012). They have the potential to offer an unlimited source of insulin-producing cells but have the limitation at present of requiring aggressive immunosuppression to prevent rejection of the xenotransplant. Techniques and methods to reduce the need for such heavy immunosuppression discussed previously are also addressed in the following sections dealing with the transgenic manipulation of the donor pig.

The pig has taken precedence as the most appropriate donor species, as it has a long-standing association with humans and a glucose physiology that is very similar to humans. The long-standing history of pig insulin use in humans substantiates its efficacy and overall ability for treating type 1 diabetes in the long term (Karamitsos 2011). Also ethically and providing social acceptance, the fact that we utilise pigs as both a food source and already for many years for prosthetic components in surgical procedures such as heart valves underpins their use as clinical substitutes.

Choice of Tissue to Be Transplanted

One of the major issues with β-cells being used as a replacement alternative to cadaveric human islet donors is in what form the tissue source to be transplanted will take. This has been a real issue due to the nature of the donor source, the pig, as, unfortunately, adult pig islets are difficult to isolate and primary non-function has been a long understood problem (Groth et al. 1998). For several decades, the islet isolation and purification process has been hampered by the lack of an outer capsule membrane, predisposing the islets to fragmentation (Falorni et al. 1996). As such, the isolation of adult pig islets requires quite mature pigs of more than 2 years of age (MacKenzie et al. 2003) and even since the description of the semi-automated isolation technique in the late 1980s, yields of pig islet isolations have not increased significantly (Toso et al. 2000). The production of a reliable supply of usable numbers of adult pig islets requires the most advanced pancreas procurement techniques including ductal preservation and selective arterial flush with cold preservation in which few groups can achieve good results in porcine islet isolation outcomes (Anazawa et al. 2010). A number of long-standing and dedicated islet isolation groups have even taken a very expensive approach to this process and have selected their own donor source herds of pigs for xenotransplantation and in order to obtain adult islets for transplantation (Sachs and Galli 2009). Some groups have even investigated such steps as pretreatment of donor pigs with various diets to alter islet isolation outcomes. Hering's group demonstrated that a diet rich in soybean oil could increase the yield of isolated islets in their own purpose-bred line of pigs (Loganathan et al. 2014). Despite the success of a small number of dedicated groups, the limitations to their success have forced the field to investigate alternative sources of donor tissues.

For these reasons, the field has investigated a number of alternative means to derive a reproducible and safe β-cell replacement alternative. Because foetal pigs are taken directly from the uterus, they are free of exogenous pathogens and foetal islet-like cells are easier to isolate using simple tissue culture techniques. Hence, for logistical and technical reasons, they have real potential as the future tissue source for clinical islet transplantation. Their main drawback is their relative immature state; the foetal pancreas contains pancreatic precursor cells that require several months of maturation before they develop into mature β-cells and provide adequate glucose control (Hawthorne et al. 2000).

Foetal pancreatic tissue has been demonstrated to reverse diabetes in rodent models. Allogeneic foetal rodent pancreatic transplantation with well-preserved islet architecture using immunosuppression with cyclosporine A and mycophenolate mofetil has been reported to successfully reverse streptozotocin-induced diabetes (du Toit et al. 1998). Cultured transgenic foetal pancreata were transplanted under the kidney capsule of wholly allogeneic CBA recipient mice, and CTLA4Ig expresser grafts showed enhanced survival at 6 weeks post-transplant (Sutherland et al. 2000). Foetal pig pancreatic tissue has also been shown to survive over

5 months in immunosuppressed NOD mice (Mandel and Koulmanda 1995) and in non-immunosuppressed nude mice (Korsgren et al. 1991) and NOD-SCID mice (Yi et al. 2003). FPPF have also been shown to have a number of distinct advantages for use as they have negligible amounts of cytokine-induced expression of inducible nitric oxide synthetase (iNOS) and production of nitric oxide (NO), these being potential mechanisms for beta cell destruction (Feng et al. 2000).

A number of large animal studies followed, with the long-term survival and function of foetal fragments transplanted into pigs by a number of methods and into a number of sites. These allowed the determination of the time course for maturation of the foetal porcine pancreatic fragments (FPPF) and production of the necessary hormones required to produce normoglycaemia in the clinical setting. The first of these studies demonstrated that FPPF required a period of at least 28 days to be able to produce endocrine-like nests and produce insulin, glucagon, and somatostatin. The FPPF also demonstrated they could survive, grow, and mature when transplanted into the thymus, spleen, liver, or kidney (Hawthorne et al. 2000). To further elaborate the therapeutic potential for FPPF use in the clinical setting and to establish their long-term function, a preclinical model of diabetes was established with subcapsular foetal pig pancreas fragments transplanted into pancreatectomised pigs. These studies demonstrated normal blood glucose control in a large animal model for more than 266 days post-transplantation, providing normal intravenous (IVGTT) and oral glucose tolerance tests (OGTT) as well as assessment of insulin and glucagon responses to arginine (Hawthorne et al. 2011). The histology from these grafts shown in Fig. 12.5 demonstrated endocrine-like nests that produced all of the hormones of normal β-cells.

To further demonstrate the utility of this type of graft, another preclinical large animal study demonstrated the resilience of the FPPF grafts to ischaemic injury by using a composite FPPF/kidney graft. The FPPF grafts were pre-transplanted under the capsule of a donor kidney. This kidney with the piggyback FPPF graft was then retranslated as a concomitant FPPF/kidney graft. This had a number of significant advantages for the potential transplantation into individuals with both type 1 diabetes and renal failure as a replacement for the simultaneous pancreas kidney allografts described earlier. These had the advantage that they simplified the combined surgical procedure and protected the islet graft from the immediate innate immune response. The foetal tissue survived the ischaemic insult associated with the concomitant transplant procedure, providing good long-term glucose handling and renal function from the composite FPPF/kidney transplant; the histopathology from these grafts is well demonstrated in Fig. 12.6 with all the β-cell-specific hormones staining positive in the graft (Hawthorne et al. 2011).

Despite the overwhelming utility and robustness of the FPPF transplants, current clinical practice dictates the use of a source of tissue that can be utilised in a similar fashion to currently transplanted islet cell transplants, being infused directly into the vasculature of the liver. Neonatal islet tissues can fulfil this role, and the source tissues are taken within a few days of birth allowing minimal time for the donor animal to be kept in specific pathogen-free facilities, thus reducing the potential logistical and financial costs (Korbutt et al. 1996). Additionally,

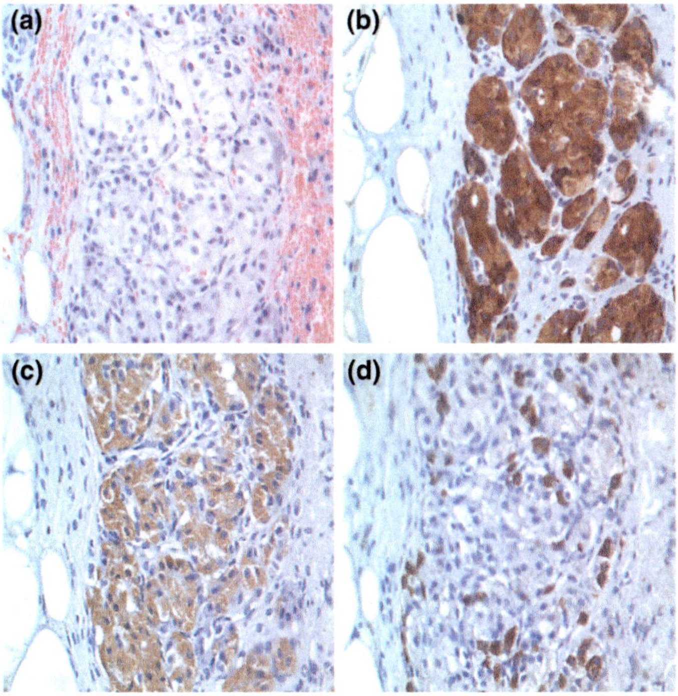

Fig. 12.5 Histopathology of foetal pig pancreas fragments (FPPF) that had been transplanted into the spleen of pancreatectomised (diabetic) Westran pigs for a period of 266 days. The FPPF grafts cured the induced diabetic state and demonstrated all the characteristics of islet cells with H&E **(a)** and positive staining for chromogranin A **(b)**, insulin **(c)**, and glucagon **(d)**. Magnification ×400

neonatal islet cell clusters (NICC) are relatively easy to isolate using simple tissue culture techniques and can be produced in extremely large numbers. Therefore, they have great potential as the future tissue source for clinical islet transplantation. Their inherent ability for proliferation and differentiation is a distinct advantage, and numerous studies have shown that neonatal pig donors yield islet cell clusters that consist primarily of duct cells and pancreatic endocrine cells. But first and foremost, NICC are capable of hormonal maturation and reversing hyperglycaemia in large animals and non-human primates (Korbutt et al. 1996; Hawthorne et al. 2014).

Although relatively minor, there are potential drawbacks to use of NICC as a tissue source. Neonatal pancreas tissue expresses antigens recognised by human natural antibodies that are not just limited to the alpha-Gal epitope (Vizzardelli et al. 2002). NICC do also require a month or so of maturation before they develop into mature β-cells and provide adequate glucose control (Hawthorne et al. 2014). Despite their potential issues, they have been shown to work well in a large animal study where they demonstrated reversal of diabetes in pancreatectomised pigs after transplantation of NICC (Kin et al. 2005).

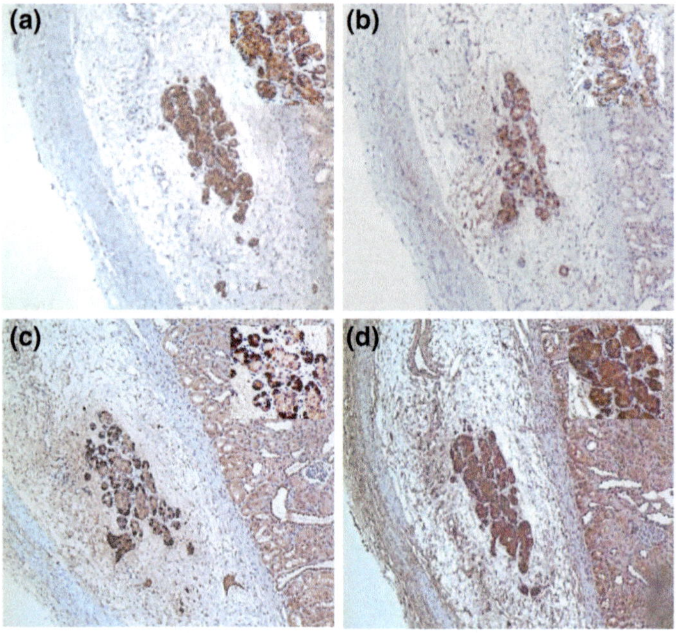

Fig. 12.6 Histopathology of the composite foetal pig pancreas fragment (FPPF). The FPPF/kidney graft cured the induced diabetic state and the composite kidney provided normal renal function of pancreatectomised (diabetic) nephrectomised Westran pigs. The FPPF demonstrated all the characteristics of islet cells with a strong positive staining for chromogranin A **(a)**, insulin **(b)**, glucagon **(c)**, and somatostatin **(d)**. **a–d** Magnification ×100, insert showing same stain at Magnification ×400

One of the drawbacks to using NICC is the time required to achieve normoglycaemia after transplantation. This makes monitoring after transplantation difficult, and reducing the time for in vivo differentiation and function would be an advantage (Korbutt et al. 1996; Yoon et al. 1999). NICC are generally used within the first week following isolation. However, there have been studies systematically evaluating the benefits or otherwise of extended culture using free NICC. The advantages of extended culture include the potential for pooling of multiple preparations for clinical transplantation and a reduction in the lag time to function due to maturation of the cells in vitro. A rather eloquent study was able to determine the optimal time point for NICC culture using functional and molecular profiling of NICC (Jimenez-Vera et al. 2015). They demonstrated optimised in vivo functional outcomes by allowing maturation of the β-cells in culture prior to transplantation with the basis of the work being to establish the optimal conditions for subsequent non-human primate studies.

Choice of the Pig as a Donor Source

The pig is also the only large animal species where oocyte transgenesis and targeted manipulation of genes has been achieved by a number of research groups (Nottle et al. 2007; Rood et al. 2005). This allows stable genetic modification and strategies to potentially overcome the various barriers of xenotransplantation as well as to be able to potentially reduce the need for heavy immunosuppressive drug regimens.

Nevertheless, a major hurdle to the use of pigs remains hyperacute rejection, which appears to have significant resolution by abrogating the 1,3-galactosyltransferase gene in pigs (Nottle et al. 2007). However, there are a number of other key issues that we are yet to overcome in order to move forward to the clinic. These include preformed and stimulated antibodies, stimulated innate immune cell reactivity, dysregulated coagulation, and a strong T-cell-mediated adaptive response (Cowan et al. 2014). Furthermore, the susceptibility of the xenograft to proinflammatory and procoagulant stimuli is probably increased by cross-species molecular defects in regulatory pathways. However, we have in our own armoury a number of unique countermeasures to combat these barriers. We have a great tactical advantage in genetic modification of the donor pigs at our disposal, a rather unique weapon to fight particular rejection mechanisms and incompatibilities.

Specifically, the potential variants of genetic modification have occurred at similar rates to which the new technologies have pushed forward. By using these cutting edge technologies, we can make significant inroads to produce a myriad of new transgenic pigs to address the many other targets to be overcome. These have been necessary to improve overall graft survival, overcoming the triggering of coagulation and complement reactivity when islets are introduced into the blood stream. This phenomenon is more commonly referred to as the instant blood-mediated inflammatory reaction (IBMIR) Bennet et al. (2000) demonstrated that contact of adult pig islets with human blood either in vivo in a loop model as demonstrated in Fig. 12.7 or in vitro or similarly with non-human primate blood in vivo triggered an immediate inflammatory response that included the immediate and irreversible activation of the complement and coagulation cascades and rapid clot formation with islet cell destruction. The reaction occurs similarly in the allo-setting with human islets mixed with fresh human blood, suggesting that non-specific factors such as exposure of pro-coagulants contribute to this 'instant blood-mediated inflammatory reaction' (IBMIR) (Moberg et al. 2002; Ozmen et al. 2002). The loop model (see Fig. 12.7) has been used extensively by a number of groups to evaluate and further define the process of triggering the IBMIR process and then developing strategies to provide potential interventions to either inhibit or prevent it from occurring (Akima et al. 2009; Akima et al. 2006).

Complement activation is an initiating factor in IBMIR, and complement inhibitors such as low molecular weight dextran sulphate are being trialled in islet

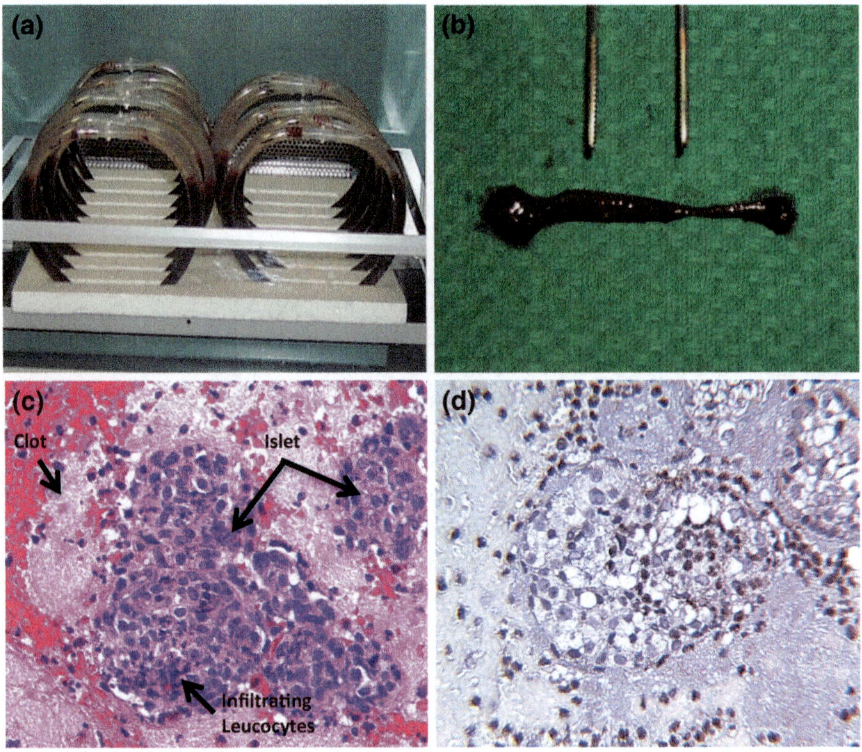

Fig. 12.7 Triggering of the instant blood-mediated inflammatory reaction (IBMIR) occurs immediately following the exposure of islet tissues with blood. It has been established in in vitro models such as the loop system where islets and freshly drawn blood are mixed in bypass tubing and placed in a shaking incubator **(a)**. The resulting clots occur within minutes and are removed for analysis **(b)**, sectioned and the histology as per the attached micrograph of a H&E-stained section **(c)** ×100 Magnification demonstrates adherence of platelets, entrapment of the islets in clots (*arrows*), and infiltration by neutrophils and monocytes into the islets (*arrows*). Additional histological analysis was performed to demonstrate staining for neutrophils and the presence of C4d as per micrograph **(d)**. ×400 Magnification

allotransplantation. However, it is seen in both the allo- and xeno-islet transplant settings with both having similar characteristics in the activation of complement and coagulation, upregulation of and adherence of platelets, entrapment of the islets in clots, and infiltration by neutrophils and monocytes into the islets from the clot (Akima et al. 2009; Bennet et al. 2000). It appears to be far worse in pig islet xenotransplantation, with exacerbation by binding of complement-fixing anti-αGal antibodies compounded by molecular incompatibilities affecting the regulation of coagulation (Akima et al. 2006), and the potential for complement activation via the alternative pathway is probably greater in xenotransplantation (Miyagawa et al. 2010).

There is also an intricate network of mechanisms responsible for the upregulation of various cell types to both infiltrate and destroy the graft. Yi et al. (2005) demonstrated that recruitment of antigen primed and CD4+ T-cell-activated macrophages were capable of both recognizing and rejecting porcine xenografts and that quite clearly that CCR5 was involved in both the activation and recruitment of macrophages to rejecting islet xenografts (Yi et al. 2005). It has been reported in a number of non-human primate models and appears most pronounced in non-transgenic porcine neonatal islet cell cluster transplants into non-immunosuppressed baboons (Hawthorne et al. 2014).

Genetic Manipulation of the Donor Source

There have been numerous approaches to tackle the problems associated with xeno-islet transplantation specifically the problems revolving around the hyperacute rejection and triggering of IBMIR when the β-cell replacement source is infused into the recipients blood stream. The basis of these approaches has been underpinned by the deletion of the αGal xenoantigen and transgenic expression of human complement regulatory proteins and anti-thrombotic/anti-inflammatory molecules.

The major step forward in this tactical approach was the elimination of αGal by the GalT gene knockout (GTKO), which provided a significant protective effect for neonatal islet xenografts. Following the development of the GTKO pigs, a number of landmark studies demonstrated surprisingly variable results when using the newly developed neonatal GTKO xenografts in non-human primate transplant experiments. One of these early studies showed the neonatal GTKO xenografts provided little or no benefit when compared to wild-type xenografts (van der Windt et al. 2009). However, subsequent studies showed significantly improved survival with little or no immediate intrahepatic inflammation or clotting by using the neonatal GTKO xenografts in non-human primate recipients when compared to wild-type xenografts (Thompson et al. 2011).

However, it appeared that despite the significant advantage and protection provided by the GTKO, there were still a number of significant hurdles to be overcome before there would be any significant advance in the prevention of these grafts from the ongoing attack of the other armamentarium of the immune system. It appeared that it would take a more directed and multipronged approach to also prevent the humoral and cellular events occurring in these grafts.

Supporting this approach, a number of studies have clearly demonstrated much greater protection provided by the addition of human complement regulators such as hCD46 on the pig islets. hCD46 alone did not substantially reduce the initial loss of islet mass, but was effective in limiting antibody-mediated rejection of the porcine islets in streptozotocin induced diabetic Cynomolgus monkeys. This was of great benefit as it allowed reduction in unsustainably high levels of immunosuppression to preserve the islets (van der Windt et al. 2009). Taking an even broader multitransgenic approach, investigators were able to achieve complete control of

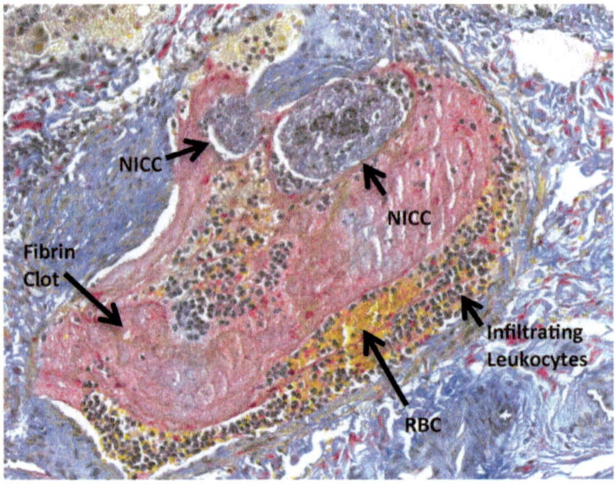

Fig. 12.8 Triggering of IBMIR in the xeno setting. Porcine neonatal islet cell clusters (NICCs) transplanted into the circulation of the liver of a baboon showing immediate triggering of IBMIR and HAR. Its typical features are shown here with the adherence of platelets, entrapment of the NICCs in well-defined clot, and infiltration by neutrophils and monocytes into the NICCs. Magnification ×200

IBMIR in neonatal porcine islet xenotransplantation in baboons (Hawthorne et al. 2014). A very difficult model to achieve graft survival, the baboon provides the basis by which we can provide good baseline data to be able to move forward to the clinic (Cooper et al. 2014).

As discussed earlier, the major problem in xenotransplantation is the immediate and unstoppable triggering of IBMIR and HAR when the major xenoantigens are not removed from the transplanted tissue source. The importance of this is that it is pivotal for the success of any programme to be able to reach the clinic with xenotransplantable tissues. This was very clearly demonstrated in the study by Hawthorne et al. where wild-type (WT) pig NICC xenografts were transplanted intraportally into the circulation of baboons. Transplantation of the wild-type porcine NICC triggered profound IBMIR and HAR in the liver of the transplanted baboons, with immediate intravascular clotting and widespread graft entrapment and destruction occurring within minutes and the resulting histopathology can be seen in Fig. 12.8. This was not prevented by any means of anticoagulation such as heparin or antithrombin treatment.

In contrast, IBMIR has been prevented when recipients were transplanted with NICC from αGal-deficient pigs transgenic for two human complement regulators CD55 and CD59 and immunosuppressed with a clinically relevant protocol. These genetically modified NICC clearly demonstrated they were not susceptible to immediate graft loss such as triggered by humoral injury despite the fact that the recipients developed some minor anti-pig antibody response. However, biopsies of the grafts 1 month after transplant demonstrated significant T-cell- and B-cell-mediated rejection, and insulin-, glucagon-, and somatostatin-positive cells along with strong expression of the two human complement regulators CD55 and CD59 on the NICC grafts are shown in Fig. 12.9.

This was nevertheless a significant step forward as these results indicated that both the immediate hyperacute rejection and IBMIR could be attenuated by using multitransgenic pigs with GalT gene knockout and expression of multiple

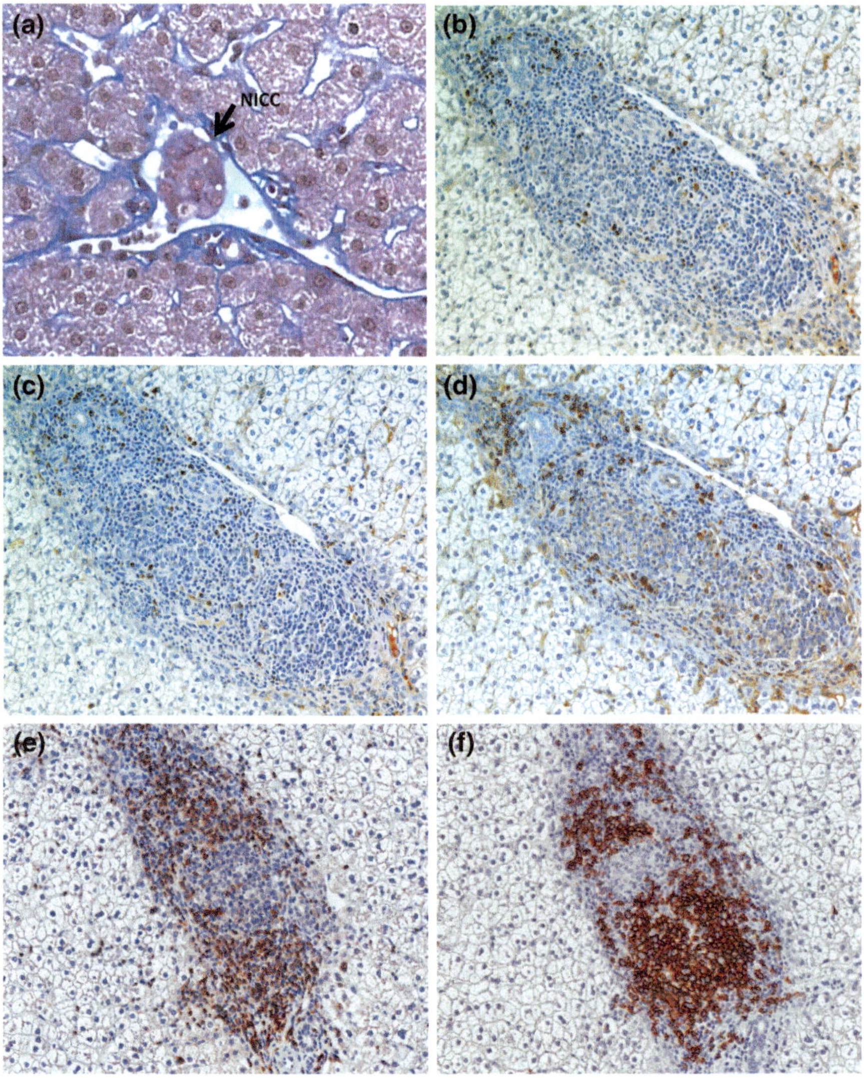

Fig. 12.9 Baboon liver biopsies demonstrating the TG NICC grafts from αGal-deficient CD55 and CD59 pigs at 1 h (**a**, magnification ×400) and 1 month post-transplant (**b**–**f**, magnification ×200). Insulin (**b**)-, glucagon (**c**)-, and somatostatin (**d**)-positive cells were observed as well as T and B lymphocyte cellular rejection (**e**–**f**, respectively)

human complement regulators. However, in the long term, graft survival may require more effective immunosuppression or further donor genetic modification. Clearly T cell responses continue to be a major cause of xenograft rejection, and current immunosuppression alone is unlikely to completely suppress these responses (Scalea et al. 2012). Additionally, both the direct and indirect pathways of antigen presentation appear to be involved in these anti-donor processes (Scalea et al. 2012). These issues have been discussed in previous sections of this chapter with the potential role of new immunosuppressive strategies being highlighted. Furthermore, proof of principle has been shown in our ability at least in the short term to be able to overcome this cellular rejection by utilising extremely heavy non-clinically applicable immunosuppression. This was shown by two groups who were able to achieve relative success in macaque models using novel anti T cell antibodies targeting the CD154/CD40 pathways, with both achieving insulin independence for up to 6 months post-transplant (Cardona et al. 2006; Hering et al. 2006). More recently, these studies have been followed up with the Emory group using a more specific chimeric anti-CD40 monoclonal antibody Chi220 and the IL-2 receptor-specific antibody basiliximab. They also used maintenance therapy of sirolimus and belatacept to achieve xeno-islet engraftment and survival with macaque recipients achieving insulin-independent normoglycaemia for up to 203 days post-transplant (Thompson et al. 2011, 2012).

Of late at the World Transplant Congress, our group took a significant step forward with results showing survival of genetically modified porcine neonatal islet xenografts in baboons. We demonstrated very positive results that transgenic expression of human complement regulatory proteins and deletion of GTKO can protect porcine neonatal islet cell cluster xenografts in baboons. They used GTKO piglets transgenic for human CD55, CD59, and H-transferase as donors. Recipient baboons received GTKO/CD55-CD59-HT NICC under standard (ATG, tacrolimus, mycophenolate mofetil; $n = 5$) or costimulation blockade-based immunosuppression (anti CD2, anti CD154, belatacept, tacrolimus; $n = 3$). The early inflammatory/thrombotic response was compared to that induced by wild-type (WT) NICC ($n = 4$). Graft survival was evaluated by immunohistochemical analysis up to 3 months post-transplant. GTKO/CD55-CD59-HT xenografts exhibited no signs of early thrombosis or infiltrate, and recipient platelet counts, fibrinogen, and D-dimer levels were unchanged from baseline. In contrast, WT xenografts triggered widespread thrombosis within 12 h, with substantial neutrophil and mononuclear cell infiltrate, accompanied by transient decreases in platelet count and fibrinogen and increased D-dimer levels. Analysis of liver biopsies from recipients under standard immunosuppression revealed loss of GTKO/CD55-CD59-HT NICC within 1 month, with heavy T cell and B cell infiltrates. However, the change to costimulation blockade-based immunosuppression reduced cellular infiltration, and cells staining positive for insulin, glucagon, and somatostatin were present in all GTKO/CD55-CD59-HT xenografts at more than three months post-transplant as can be seen in Fig. 12.10. Deletion of αGal and expression of human CD55 and CD59 can prevent early thrombotic destruction of porcine NICCs in the baboon model (Hawthorne 2014).

Quite clearly, we have demonstrated that we can avoid or prevent the very significant initial barriers to xenotransplantation by genetically modifying the donor source pigs and utilising cutting edge immunosuppression. At this point, the use of costimulation blockade-based immunosuppression appears to be more effective than standard immunosuppression in prolonging the survival of genetically

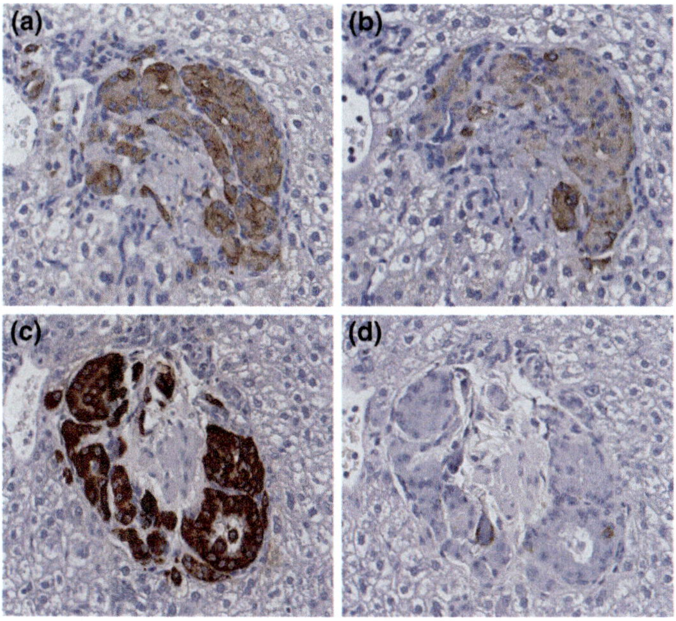

Fig. 12.10 Baboon liver biopsies 3 months post-transplant demonstrating functioning NICC grafts from αGal-deficient CD55 and CD59 transgenic pigs treated with costimulation blockade-based immunosuppression. Grafts stained positive for CD55 (**a**), CD59 (**b**), insulin (**c**), and glucagon (**d**). Magnification ×200

modified porcine NICC xenografts. This therefore provides the basis for us to be able to pursue preclinical studies to confirm their efficacy and safety before we can finally realise the dream of taking xenotransplantation to the clinic.

Discussion

The future holds many interesting potential new therapies that may or may not yield appropriate and safe methods for the treatment for type 1 diabetes. From what has been outlined in this chapter, we can see in the not-so-distant future a great potential for the prevention of type 1 diabetes. If, unfortunately, patients cannot be treated prior to the advent of their type 1 diabetes, then they may still be treated afterwards through one of the myriad of evolving therapies. These will target the disease at various stages and eventually be able to treat patients with severe type 1 diabetes and its secondary complications. Moving forward, researchers and clinicians have numerous fronts and multiple strategies arising at different stages of development in which to be able to offer patients treatments tailored for their stage of disease. Of the technologies discussed in this chapter, the most likely form of therapy to provide a broadly applicable treatment is the genetic manipulation of the patient's own cells to produce insulin or through techniques to substitute them with the latest types of β-islet cell replacement. By far the closest at this point-in-time are the xeno-islet cell transplants that are now currently undergoing preclinical trials in non-human primates. Ultimately, these technologies provide a view of the Holy Grail, in that we may be able to provide β-cells or surrogates as a cure for type 1 diabetes in the quite foreseeable future.

References

Akima S et al (2006) Differences in the responses to recombinant human activated protein C against allogeneic and xenogeneic IBMIR. Transplantation 82(1):992

Akima S et al (2009) Tirofiban and activated protein C synergistically inhibit the instant blood mediated inflammatory reaction (IBMIR) from allogeneic islet cells exposure to human blood. Am J Transplant 9(7):1533–1540

Allen RD et al (1997) Diabetic neuropathy after pancreas transplantation: determinants of recovery. Transplantation 63(6):830–838

Allen RD et al (2001) Pancreas and islet transplantation: an unfinished journey. Transplant Proc 33(7–8):3485–3488

Anazawa T et al (2010) Improved method of porcine pancreas procurement with arterial flush and ductal injection enhances islet isolation outcome. Transplant Proc 42(6):2032–2035

Ansari MJ et al (2003) The programmed death-1 (PD-1) pathway regulates autoimmune diabetes in nonobese diabetic (NOD) mice. J Exp Med 198(1):63–69

Babizhayev MA et al (2013) Diabetes mellitus: novel insights, analysis and interpretation of pathophysiology and complications management with imidazole-containing peptidomimetic antioxidants. Recent Pat Drug Deliv Formul 7(3):216–256

Balamurugan AN et al (2014) Islet product characteristics and factors related to successful human islet transplantation from the Collaborative Islet Transplant Registry (CITR) 1999–2010. Am J Transplant 14(11):2595–2606

Banting FG et al (1922) Pancreatic extracts in the treatment of diabetes mellitus. Can Med Assoc J 12(3):141–146

Barbu AR et al (2006) A perfusion protocol for highly efficient transduction of intact pancreatic islets of Langerhans. Diabetologia 49(10):2388–2391

Becker TC et al (1994) Overexpression of hexokinase I in isolated islets of Langerhans via recombinant adenovirus. Enhancement of glucose metabolism and insulin secretion at basal but not stimulatory glucose levels. J Biol Chem 269(33):21234–21238

Bennet W et al (2000a) Damage to porcine islets of Langerhans after exposure to human blood in vitro, or after intraportal transplantation to cynomologus monkeys: protective effects of sCR1 and heparin. Transplantation 69(5):711–719

Bennet W et al (2000b) Isolated human islets trigger an instant blood mediated inflammatory reaction: implications for intraportal islet transplantation as a treatment for patients with type 1 diabetes. Ups J Med Sci 105(2):125–133

Bergenstal RM et al (2010) Effectiveness of sensor-augmented insulin-pump therapy in type 1 diabetes. N Engl J Med 363(4):311–320

Bobisse S, Zanovello P, Rosato A (2007) T-cell receptor gene transfer by lentiviral vectors in adoptive cell therapy. Expert Opin Biol Ther 7(6):893–906

Boitard C (2012) Pancreatic islet autoimmunity. Presse Med 41(12 p 2):e636–e650

Brady JL et al (2013) Anti-CD2 producing pig xenografts effect localized depletion of human T cells in a huSCID model. Xenotransplantation 20(2):100–109

Bregenholt S et al (2003) The cholera toxin B subunit is a mucosal adjuvant for oral tolerance induction in type 1 diabetes. Scand J Immunol 57(5):432–438

Bruni A et al (2014) Islet cell transplantation for the treatment of type 1 diabetes: recent advances and future challenges. Diabetes Metab Syndr Obes 7:211–223

Callewaert H et al (2007) Cell loss during pseudoislet formation hampers profound improvements in islet lentiviral transduction efficacy for transplantation purposes. Cell Transplant 16(5):527–537

Cardona K et al (2006) Long-term survival of neonatal porcine islets in nonhuman primates by targeting costimulation pathways. Nat Med 12(3):304–306

Carrington EM et al (2010) BH3 mimetics antagonizing restricted prosurvival Bcl-2 proteins represent another class of selective immune modulatory drugs. Proc Natl Acad Sci USA 107(24):10967–10971

Chow VC et al (1999) Diabetic retinopathy after combined kidney-pancreas transplantation. Clin Transplant 13(4):356–362

Cole CN, Crawford LV, Berg P (1979) Simian virus 40 mutants with deletions at the 3' end of the early region are defective in adenovirus helper function. J Virol 30(3):683–691

Contreras JL et al (2002) Cytoprotection of pancreatic islets before and early after transplantation using gene therapy. Kidney Int 61(1 Suppl):S79–S84

Cooper DK et al (2014) Progress in pig-to-non-human primate transplantation models (1998-2013): a comprehensive review of the literature. Xenotransplantation 21(5):397–419

Couri CE et al (2009) C-peptide levels and insulin independence following autologous nonmyeloablative hematopoietic stem cell transplantation in newly diagnosed type 1 diabetes mellitus. JAMA 301(15):1573–1579

Cowan PJ, Cooper DK, d'Apice AJ (2014) Kidney xenotransplantation. Kidney Int 85(2):265–275

Coyle ME, Francis K, Chapman Y (2013) Self-management activities in diabetes care: a systematic review. Aust Health Rev 37(4):513–522

Creusot RJ et al (2008) Tissue-targeted therapy of autoimmune diabetes using dendritic cells transduced to express IL-4 in NOD mice. Clin Immunol 127(2):176–187

Csete ME et al (1995) Efficient gene transfer to pancreatic islets mediated by adenoviral vectors. Transplantation 59(2):263–268

Cullen KS et al (2014) Glucagon induces translocation of glucokinase from the cytoplasm to the nucleus of hepatocytes by transfer between 6-phosphofructo 2-kinase/fructose 2,6-bisphosphatase-2 and the glucokinase regulatory protein. Biochim Biophys Acta 1843(6):1123–1134

Denes B et al (2006) Suppression of hyperglycemia in NOD mice after inoculation with recombinant vaccinia viruses. Mol Biotechnol 34(3):317–327

Dirice E et al (2009) Adenovirus-mediated TRAIL gene (Ad5hTRAIL) delivery into pancreatic islets prolongs normoglycemia in streptozotocin-induced diabetic rats. Hum Gene Ther 20(10):1177–1189

du Toit DF et al (1998) Immunosuppression with cyclosporin A in combination with mycophenolate mofetil suppresses rejection of allogeneic fetal rat pancreatic allografts. Transplant Proc 30(8):4092–4093

Dupraz P et al (1999) Lentivirus-mediated Bcl-2 expression in betaTC-tet cells improves resistance to hypoxia and cytokine-induced apoptosis while preserving in vitro and in vivo control of insulin secretion. Gene Ther 6(6):1160–1169

Falorni A et al (1996) Culture maintenance of isolated adult porcine pancreatic islets in three-dimensional gel matrices: morphologic and functional results. Pancreas 12(3):221–229

Feng X et al (2000) Inducible nitric oxide synthetase is expressed in adult but not fetal pig pancreatic islets. Xenotransplantation 7(3):197–205

Fernandes JR et al (2004) Transplantation of islets transduced with CTLA4-Ig and TGFbeta using adenovirus and lentivirus vectors. Transpl Immunol 13(3):191–200

Fodor A et al (2007) Adult rat liver cells transdifferentiated with lentiviral IPF1 vectors reverse diabetes in mice: an ex vivo gene therapy approach. Diabetologia 50(1):121–130

Franceschi RT, Ge C (2008) Gene delivery by adenoviruses. Methods Mol Biol 455:137–147

Georgiou HM et al (1997) Genetic modification of an islet tumor cell line inhibits its rejection. Transplant Proc 29(1–2):1032–1033

Gerace D et al (2013) Pancreatic transdifferentiation in porcine liver following lentiviral delivery of human furin-cleavable insulin. Transplant Proc 45(5):1869–1874

Giannoukakis N et al (2000) Prevention of beta cell dysfunction and apoptosis activation in human islets by adenoviral gene transfer of the insulin-like growth factor I. Gene Ther 7(23):2015–2022

Ginn SL et al (2004) 405. Efficient lentivirus vector-mediated transduction of neonatal porcine islet cell clusters: a powerful tool for xenotransplantation research. Mol Ther 9(S1):S155–S156

Glaser B (2007) Type 2 diabetes: hypoinsulinism, hyperinsulinism, or both? PLoS Med 4(4):e148

Golden B et al (2005) Genetic analysis of families with autoimmune diabetes and thyroiditis: evidence for common and unique genes. J Clin Endocrinol Metab 90(8):4904–4911

Grey ST et al (1999) A20 inhibits cytokine-induced apoptosis and nuclear factor kappaB-dependent gene activation in islets. J Exp Med 190(8):1135–1146

Grinberg-Bleyer Y et al (2010) IL-2 reverses established type 1 diabetes in NOD mice by a local effect on pancreatic regulatory T cells. J Exp Med 207(9):1871–1878

Groth CG et al (1998) Pig-to-human islet transplantation. Transplant Proc 30(7):3809–3810

Gubitosi-Klug RA (2014) The diabetes control and complications trial/epidemiology of diabetes interventions and complications study at 30 years: summary and future directions. Diabetes Care 37(1):44–49

Halban PA et al (2001) Gene and cell-replacement therapy in the treatment of type 1 diabetes: how high must the standards be set? Diabetes 50(10):2181–2191

Hamaguchi K, Gaskins HR, Leiter EH (1991) NIT-1, a pancreatic beta-cell line established from a transgenic NOD/Lt mouse. Diabetes 40(7):842–849

Hawthorne WJ (2014) STATE OF THE ART 4: Janssen-Cilag symposium: islet xenotransplantation—update. Immunol Cell Biol 92:A1–A29

Hawthorne WJ et al (2000) A large-animal model to evaluate the clinical potential of fetal pig pancreas fragment transplantation. Cell Transplant 9(6):867–875

Hawthorne WJ et al (2011) Pre-clinical model of composite foetal pig pancreas fragment/renal xenotransplantation to treat renal failure and diabetes. Xenotransplantation 18(6):390–399

Hawthorne W et al (2014) Survival of genetically modified porcine neonatal islet xenografts in baboons. Transplantation 98:414

He Z et al (2006) Long-term gene expression and metabolic control exerted by lentivirus-transduced pancreatic islets. Xenotransplantation 13(3):195–203

Hering BJ et al (2006) Prolonged diabetes reversal after intraportal xenotransplantation of wild-type porcine islets in immunosuppressed nonhuman primates. Nat Med 12(3):301–303

Holm P et al (2004) Interaction and association analysis of a type 1 diabetes susceptibility locus on chromosome 5q11-q13 and the 7q32 chromosomal region in Scandinavian families. Diabetes 53(6):1584–1591

Hsu PY, Kotin RM, Yang YW (2008) Glucose- and metabolically regulated hepatic insulin gene therapy for diabetes. Pharm Res 25(6):1460–1468

Hu ZQ et al (1994) Mast cells display natural suppressor activity partially by releasing transforming growth factor-beta. Immunology 82(3):482–486

Jacobs-Tulleneers-Thevissen D et al (2013) Sustained function of alginate-encapsulated human islet cell implants in the peritoneal cavity of mice leading to a pilot study in a type 1 diabetic patient. Diabetologia 56(7):1605–1614

Jimenez-Vera E, Davies S, Phillips P, O'Connell PJ, Hawthorne WJ (2015) Long-term cultured neonatal islet cell clusters demonstrate better outcomes for reversal of diabetes: in-vivo and molecular profiles. Xenotransplantation 22:114–123

Johnson MC et al (2013) beta-cell-specific IL-2 therapy increases islet Foxp3 + Treg and suppresses type 1 diabetes in NOD mice. Diabetes 62(11):3775–3784

Karamitsos DT (2011) The story of insulin discovery. Diabetes Res Clin Pract 93(Suppl 1):S2–S8
Katagi M et al (2014) Hyperglycemia induces abnormal gene expression in hematopoietic stem cells and their progeny in diabetic neuropathy. FEBS Lett 588(6):1080–1086
Kawabata K, Tashiro K, Mizuguchi H (2010) Differentiation of functional cells from iPS cells by efficient gene transfer. Yakugaku Zasshi 130(11):1527–1534
Kawamoto K et al (2008) Prolonged survival of pig islets xenograft by adenovirus-mediated expression of either the membrane-bound human FasL or the human decoy Fas antigen gene. Xenotransplantation 15(5):333–343
Kin T et al (2005) Reversal of diabetes in pancreatectomized pigs after transplantation of neonatal porcine islets. Diabetes 54(4):1032–1039
Klein D et al (2000) Inhibition of Fas-mediated apoptosis in mouse insulinoma betaTC-3 cells via an anti-Fas ribozyme. Hum Gene Ther 11(7):1033–1045
Kodama K et al (2008) Tissue- and age-specific changes in gene expression during disease induction and progression in NOD mice. Clin Immunol 129(2):195–201
Koizumi M et al (2006) Forced expression of PDX-1 induces insulin production in intestinal epithelia. Surgery 140(2):273–280
Kojima S et al (2009) Central leptin gene therapy, a substitute for insulin therapy to ameliorate hyperglycemia and hyperphagia, and promote survival in insulin-deficient diabetic mice. Peptides 30(5):962–966
Korbutt GS et al (1996) Large scale isolation, growth, and function of porcine neonatal islet cells. J Clin Invest 97(9):2119–2129
Korsgren O et al (1991) Functional and morphological differentiation of fetal porcine islet-like cell clusters after transplantation into nude mice. Diabetologia 34(6):379–386
Lam VW et al (2010) Evolution of pancreas transplant surgery. ANZ J Surg 80(6):411–418
Lee KM et al (2014) TGF-beta-producing regulatory B cells induce regulatory T cells and promote transplantation tolerance. Eur J Immunol 44(6):1728–1736
Lei P et al (2007) Efficient production of bioactive insulin from human epidermal keratinocytes and tissue-engineered skin substitutes: implications for treatment of diabetes. Tissue Eng 13(8):2119–2131
Leibowitz G et al (1999) Gene transfer to human pancreatic endocrine cells using viral vectors. Diabetes 48(4):745–753
Levine F (1997) Gene therapy for diabetes: strategies for beta-cell modification and replacement. Diabetes Metab Rev 13(4):209–246
Liu YP, Berkhout B (2014) HIV-1-based lentiviral vectors. Methods Mol Biol 1087:273–284
Loganathan G et al (2014) Pretreatment of donor pigs with a diet rich in soybean oil increases the yield of isolated islets. Transplant Proc 46(6):1945–1949
Lu Y (2004) Recombinant adeno-associated virus as delivery vector for gene therapy–a review. Stem Cells Dev 13(1):133–145
MacKenzie DA, Hullett DA, Sollinger HW (2003) Xenogeneic transplantation of porcine islets: an overview. Transplantation 76(6):887–891
Mandel TE, Koulmanda M (1995) Xenotransplantation of fetal pig pancreas and reversal of diabetes in spontaneously diabetic NOD mice. Transplant Proc 27(3):2179–2180
Manolios N et al (1997) T-cell antigen receptor transmembrane peptides modulate T-cell function and T cell-mediated disease. Nat Med 3(1):84–88
McCulloch D (2007) Classification of diabetes mellitus and genetic diabetic syndromes. In: Rose B (ed) UpToDate. UpToDate: Waltham, MA
Meagher C et al (2007) CCL4 protects from type 1 diabetes by altering islet beta-cell-targeted inflammatory responses. Diabetes 56(3):809–817
Mia S et al (2014) An optimized protocol for human M2 macrophages using M-CSF and IL-4/IL-10/TGF-beta yields a dominant immunosuppressive phenotype. Scand J Immunol 79(5):305–314

Mitanchez D et al (1997) Glucose-stimulated genes and prospects of gene therapy for type I diabetes. Endocr Rev 18(4):520–540
Miyagawa S et al (2010) Complement regulation in the GalT KO era. Xenotransplantation 17(1):11–25
Miyazaki J et al (1990) Establishment of a pancreatic beta cell line that retains glucose-inducible insulin secretion: special reference to expression of glucose transporter isoforms. Endocrinology 127(1):126–132
Moberg L et al (2002) Production of tissue factor by pancreatic islet cells as a trigger of detrimental thrombotic reactions in clinical islet transplantation. Lancet 360(9350):2039–2045
Moore HP et al (1983) Expressing a human proinsulin cDNA in a mouse ACTH-secreting cell. Intracellular storage, proteolytic processing, and secretion on stimulation. Cell 35(2 Pt 1):531–538
Morrissey PE, Monaco AP (2014) Donation after circulatory death: current practices, ongoing challenges, and potential improvements. Transplantation 97(3):258–264
Naldini L (2011) Ex vivo gene transfer and correction for cell-based therapies. Nat Rev Genet 12(5):301–315
Nankivell BJ et al (1997) Recovery of diabetic neuropathy after pancreas transplantation. Transplant Proc 29(1–2):658–659
Nayerossadat N, Maedeh T, Ali PA (2012) Viral and nonviral delivery systems for gene delivery. Adv Biomed Res 1:27
Niu L et al (2008) Gene therapy for type 1 diabetes mellitus in rats by gastrointestinal administration of chitosan nanoparticles containing human insulin gene. World J Gastroenterol 14(26):4209–4215
Nottle MB et al (2007) Production of homozygous alpha-1,3-galactosyltransferase knockout pigs by breeding and somatic cell nuclear transfer. Xenotransplantation 14(4):339–344
O'Connell PJ et al (2006) Clinical islet transplantation in type 1 diabetes mellitus: results of Australia's first trial. Med J Aust 184(5):221–225
O'Connell PJ et al (2013) Transplantation of xenogeneic islets: are we there yet? Curr Diab Rep 13(5):687–694
Olson DE et al (2008) Hepatic insulin gene therapy normalizes diurnal fluctuation of oxidative metabolism in diabetic BB/Wor rats. Mol Ther 16(7):1235–1242
Ozmen L et al (2002) Inhibition of thrombin abrogates the instant blood-mediated inflammatory reaction triggered by isolated human islets: possible application of the thrombin inhibitor melagatran in clinical islet transplantation. Diabetes 51(6):1779–1784
Panakanti R, Mahato RI (2009) Bipartite adenoviral vector encoding hHGF and hIL-1Ra for improved human islet transplantation. Pharm Res 26(3):587–596
Plesner A et al (2010) XIAP inhibition of beta-cell apoptosis reduces the number of islets required to restore euglycemia in a syngeneic islet transplantation model. Islets 2(1):18–23
Prud'homme GJ, Draghia-Akli R, Wang Q (2007) Plasmid-based gene therapy of diabetes mellitus. Gene Ther 14(7):553–564
Rabinovitch A et al (1999) Transfection of human pancreatic islets with an anti-apoptotic gene (bcl-2) protects beta-cells from cytokine-induced destruction. Diabetes 48(6):1223–1229
Rafeullah N, Cackett N, Hussain K (2012) Fasting hypoglycaemia and postprandial hyperglycaemia as a prodrome of type 1 diabetes mellitus. Horm Res Paediatr 78(5–6):332–335
Rahmati S, Alijani N, Kadivar M (2013) In vitro generation of glucose-responsive insulin producing cells using lentiviral based pdx-1 gene transduction of mouse (C57BL/6) mesenchymal stem cells. Biochem Biophys Res Commun 437(3):413–419
Ravassard P et al (2011) A genetically engineered human pancreatic beta cell line exhibiting glucose-inducible insulin secretion. J Clin Invest 121(9):3589–3597
Ren B et al (2007) Long-term correction of diabetes in rats after lentiviral hepatic insulin gene therapy. Diabetologia 50(9):1910–1920
Ren B et al (2013) Long-term reversal of diabetes in non-obese diabetic mice by liver-directed gene therapy. J Gene Med 15(1):28–41

Rood PP et al (2005) Preformed antibodies to alpha1,3-galactosyltransferase gene-knockout (GT-KO) pig cells in humans, baboons, and monkeys: implications for xenotransplantation. Transplant Proc 37(8):3514–3515
Sachs DH, Galli C (2009) Genetic manipulation in pigs. Curr Opin Organ Transplant 14(2):148–153
Sadeghi H et al (2002) Genetic fusion of human insulin B-chain to the B-subunit of cholera toxin enhances in vitro antigen presentation and induction of bystander suppression in vivo. Immunology 106(2):237–245
Scalea J et al (2012) T-cell-mediated immunological barriers to xenotransplantation. Xenotransplantation 19(1):23–30
Schambach A et al (2013) Biosafety features of lentiviral vectors. Hum Gene Ther 24(2):132–142
Scharfmann R et al (2014) Development of a conditionally immortalized human pancreatic beta cell line. J Clin Invest 124(5):2087–2098
Selden RF et al (1987) Regulation of insulin-gene expression. Implications for gene therapy. N Engl J Med 317(17):1067–1076
Shahrestani S WA, Lam VW, Yuen L, Ryan B, Pleass HC, Hawthorne WJ (2016) Outcomes from pancreatic transplantation in donation after cardiac death: a systematic review and meta-analysis. Transplantation 2016, Mar 4. [Epub ahead of print] PMID: 26950713
Shapiro AM et al (2006) International trial of the Edmonton protocol for islet transplantation. N Engl J Med 355(13):1318–1330
Shternhall-Ron K et al (2007) Ectopic PDX-1 expression in liver ameliorates type 1 diabetes. J Autoimmun 28(2–3):134–142
Sigalla J et al (1997) Adenovirus-mediated gene transfer into isolated mouse adult pancreatic islets: normal beta-cell function despite induction of an anti-adenovirus immune response. Hum Gene Ther 8(13):1625–1634
Sobel DO, Henzke A, Abbassi V (2010) Cyclosporin and methotrexate therapy induces remission in type 1 diabetes mellitus. Acta Diabetol 47(3):243–250
Soltani N et al (2007) In vivo expression of GLP-1/IgG-Fc fusion protein enhances beta-cell mass and protects against streptozotocin-induced diabetes. Gene Ther 14(12):981–988
Sun JB, Holmgren J, Czerkinsky C (1994) Cholera toxin B subunit: an efficient transmucosal carrier-delivery system for induction of peripheral immunological tolerance. Proc Natl Acad Sci USA 91(23):10795–10799
Sutherland RM et al (2000) Protective effect of CTLA4Ig secreted by transgenic fetal pancreas allografts. Transplantation 69(9):1806–1812
Talebi S et al (2012) Derivation of islet-like cells from mesenchymal stem cells using PDX1-transducing lentiviruses. Biotechnol Appl Biochem 59(3):205–212
Tang DQ et al (2013) Genetically reprogrammed, liver-derived insulin-producing cells are glucose-responsive, but susceptible to autoimmune destruction in settings of murine model of type 1 diabetes. Am J Transl Res 5(2):184–199
Thompson DM et al (2011a) Reduced progression of diabetic microvascular complications with islet cell transplantation compared with intensive medical therapy. Transplantation 91(3):373–378
Thompson P et al (2011b) Islet xenotransplantation using gal-deficient neonatal donors improves engraftment and function. Am J Transplant 11(12):2593–2602
Thompson P et al (2012) Alternative immunomodulatory strategies for xenotransplantation: CD40/154 pathway-sparing regimens promote xenograft survival. Am J Transplant 12(7):1765–1775
Thwaites SE et al (2012) Excellent outcomes of simultaneous pancreas kidney transplantation in patients from rural and urban Australia: a national service experience. Transplantation 94(12):1230–1235
Tian C et al (2007) Induction of robust diabetes resistance and prevention of recurrent type 1 diabetes following islet transplantation by gene therapy. J Immunol 179(10):6762–6769

Tian J et al (2008) Regulated insulin delivery from human epidermal cells reverses hyperglycemia. Mol Ther 16(6):1146–1153

Tiwari JL et al (2012) Islet cell transplantation in type 1 diabetes: an analysis of efficacy outcomes and considerations for trial designs. Am J Transplant 12(7):1898–1907

Torchilin VP (2006) Recent approaches to intracellular delivery of drugs and DNA and organelle targeting. Annu Rev Biomed Eng 8:343–375

Toso C et al (2000) Isolation of adult porcine islets of Langerhans. Cell Transplant 9(3):297–305

Tuduri E, Bruin JE, Kieffer TJ (2012) Restoring insulin production for type 1 diabetes. J Diabetes 4(4):319–331

Vågesjö E, Christoffersson G, Waldén TB, Carlsson PO, Essand M, Korsgren O, Phillipson M (2015) Immunological shielding by induced recruitment of regulatory T-lymphocytes delays rejection of islets transplanted to muscle. Cell Transplant 24(2):263–276

van der Windt DJ et al (2009) Long-term controlled normoglycemia in diabetic non-human primates after transplantation with hCD46 transgenic porcine islets. Am J Transplant 9(12):2716–2726

Vizzardelli C et al (2002) Neonatal porcine pancreatic cell clusters as a potential source for transplantation in humans: characterization of proliferation, apoptosis, xenoantigen expression and gene delivery with recombinant AAV. Xenotransplantation 9(1):14–24

Voltarelli JC et al (2007) Autologous nonmyeloablative hematopoietic stem cell transplantation in newly diagnosed type 1 diabetes mellitus. JAMA 297(14):1568–1576

Waldron-Lynch F, Herold KC (2011) Immunomodulatory therapy to preserve pancreatic beta-cell function in type 1 diabetes. Nat Rev Drug Discov 10(6):439–452

Wirth T, Parker N, Yla-Herttuala S (2013) History of gene therapy. Gene 525(2):162–169

Xia J et al (2011) Evaluation of biolistic gene transfer methods in vivo using non-invasive bioluminescent imaging techniques. BMC Biotechnol 11:62

Xu J et al (2007) Reversal of diabetes in mice by intrahepatic injection of bone-derived GFP-murine mesenchymal stem cells infected with the recombinant retrovirus-carrying human insulin gene. World J Surg 31(9):1872–1882

Xu AJ et al (2010) Effects of adenovirus-mediated interleukin-10 gene transfer on apoptosis and insulin secretion function of beta cell. Zhonghua Yi Xue Za Zhi 90(24):1711–1715

Yi S et al (2003) T cell-activated macrophages are capable of both recognition and rejection of pancreatic islet xenografts. J Immunol 170(5):2750–2758

Yi S et al (2005) Involvement of CCR5 signaling in macrophage recruitment to porcine islet xenografts. Transplantation 80(10):1468–1475

Yi S et al (2012) Adoptive transfer with in vitro expanded human regulatory T cells protects against porcine islet xenograft rejection via interleukin-10 in humanized mice. Diabetes 61(5):1180–1191

Yoon KH et al (1999) Differentiation and expansion of beta cell mass in porcine neonatal pancreatic cell clusters transplanted into nude mice. Cell Transplant 8(6):673–689

Zhang Y et al (2008) Genetically engineered K cells provide sufficient insulin to correct hyperglycemia in a nude murine model. Acta Biochim Biophys Sin (Shanghai) 40(2):149–157

Index

A.A. Hardikar (ed.), *Pancreatic Islet Biology*, Stem Cell Biology and Regenerative Medicine, DOI 10.1007/978-3-319-45307-1

Printed by Printforce, the Netherlands